FLUID
MECHANICS

FLUID MECHANICS

Fourth Edition

Irving Granet, P.E.
Queensborough Community College
City University of New York

PRENTICE HALL
Englewood Cliffs, New Jersey   Columbus, Ohio

Library of Congress Cataloging-in-Publication Data

Granet, Irving.
    Fluid Mechanics / Irving Granet. — 4th ed.
        p.   cm.
    Includes bibliographical reference and index.
    ISBN 0-13-352170-2
    1. Fluid mechanics. I. Title.
  TA357.G7   1996
    620.1'06—dc20   95-6721
                     CIP

Cover photo: Degginger/H. Armstrong Roberts
Editor: Stephen Helba
Production Editor: Rex Davidson
Text Designer: Ruttle Graphics, Inc./Martin R. Murphy
Production Manager: Laura Messerly
Marketing Manager: Debbie Yarnell
Illustrations: Ruttle Graphics, Inc./Academy Artworks
Production Coordination: Ruttle Graphics, Inc.

This book was set in Garamond by Ruttle Graphics, Inc. and was printed and bound by Quebecor Printing/Book Press. The cover was printed by Phoenix Color Corp.

 ©1996 by Prentice Hall, Inc.
A Simon & Schuster Company
Englewood Cliffs, New Jersey 07632

Earlier editions, entitled *Fluid Mechanics for Engineering Technology,* © 1989, 1981, and 1971 by Prentice-Hall, Inc.

Printed in the United States of America.

10 9 8 7 6 5 4 3 2 1

ISBN: 0-13-352170-2

PRENTICE-HALL INTERNATIONAL (UK) LIMITED, *London*
PRENTICE-HALL OF AUSTRALIA PTY. LIMITED, *Sydney*
PRENTICE-HALL CANADA INC., *Toronto*
PRENTICE-HALL HISPANOAMERICANA, S.A., *Mexico*
PRENTICE-HALL OF INDIA PRIVATE LIMITED, *New Delhi*
PRENTICE-HALL OF JAPAN, INC., *Tokyo*
SIMON & SCHUSTER ASIA PTE. LTD., *Singapore*
EDITORA PRENTICE-HALL DO BRASIL, LTDA., *Rio de Janeiro*

This book is dedicated to my devoted wife, Arlene, whose continued love, patience, and forbearance made its completion possible.

CONTENTS

Contents

PREFACE

A textbook must serve the needs of both the student and the instructor to be an effective teaching tool. Since the publication of the last edition, I have obtained comments from both students and instructors to find ways to enhance its presentation. As a consequence, the overall arrangement and features have been retained while several major changes have been made. This effort has been directed to making the book more "user friendly." Some of the major items of change are as follows:

- The text is written and can be used without the use of calculus. At various places, however, calculus supplements for enrichment have been inserted. This has been done to show the student how calculus can be applied to fluid mechanics and to allow the instructor to enhance classroom presentation.

- The Learning Goals at the beginning of each chapter are now action oriented and indicate positive actions that the student should be able to accomplish after completing the chapter. In this way they provide a self evaluation for students to determine whether they have mastered the material.

- A unified approach to the solution of all the Illustrative Problems has been adopted. Using and repeating this methodology provides a rational approach to solving problems for the student.

- Spreadsheets are being widely used to solve engineering problems. Therefore, in addition to the use of BASIC programming, many Illustrative Problems are also solved using the Microsoft Excel® spreadsheet.

- The total number of problems has been increased by approximately 60%. In particular, the topics of conversions, buoyancy, viscosity, fluid dynamics, the energy equation, and flow in pipes have had substantial increases in the number of problems for student solution.

- The sections on buoyancy, viscosity, pipe flow, and flow losses have been clarified and expanded.

I am deeply grateful to those colleagues, instructors, and students who have made contributions during the preparation of this edition. It is impossible for me to acknowledge all of these persons individually, but each one should know that I am aware of and value his or her contribution.

To my wife, Arlene, I am especially grateful. Without her constant support and love I would not have been able to complete this work. My children and grandchildren exhibit the patience, love, and understanding that she has so unselfishly given to all of us.

Irving Granet
North Bellmore, N.Y.

SYSTEMS OF UNITS AND DIMENSIONAL CONSISTENCY

LEARNING GOALS

After reading and studying this chapter you should be able to:

1. Define the meaning of the term *fluid* as used in this text.
2. State the distinction between a *liquid* and a *gas*.
3. Indicate the role played by fluids in our lives.
4. Use the SI system of units, including its styling.
5. Use the conventional English system of units to solve problems in fluid mechanics.
6. Apply the concept of *dimensional consistency* to assure that the units in a problem are correct.
7. Use dimensional consistency in equations.
8. Convert from SI to English units and from English units to SI.

1.1   INTRODUCTION

Fluid mechanics is the study of the behavior of fluids whether they are at rest or in motion; the study of fluids at rest is best known as *fluid statics* and the study of fluids in motion is termed *fluid dynamics*. In this book we use the term *fluid* to refer to both gases and liquids. To distinguish between a liquid and a gas, we note that while both will occupy the container in which they are placed, a liquid presents a free surface if it does not completely fill the container, but a gas will always fill the volume of the container in which it is placed. For gases it is important to take into account the change in volume that occurs when either the pressure or the temperature is changed, whereas in most cases it is possible to neglect the change in volume of a liquid when there is a change in pressure.

It is apparent that almost every part of our lives and the technology of modern life involves some dependence on and knowledge of the science of fluid mechanics. Whether we consider the flow of blood in the minute blood vessels of the human body or the motion of an aircraft or missile at speeds exceeding the velocity of sound, we need to utilize some branch of fluid mechanics to describe the motion. The literature of this subject is so vast that a brief description cannot adequately reflect its scope. At one time the subject was treated from a purely mathematical approach by one group of investigators and from an entirely empirical experimental approach by another group of investigators. In this text we use the modern technique of coordinating both approaches by supplementing theory with experiment.

Since all measurements as well as theoretical developments must explicitly state the units being used, we start our study with a discussion of systems of units. As is obvious from Figure 1.1, we find that in the United States it is necessary at present to have a knowledge of both SI (metric) and English systems of units.

1.2   THE SI SYSTEM OF UNITS

At the time of the French Revolution, the systems of weights and measures used throughout the world were an incoherent and almost hopeless jumble. International trade and the interchange of scientific information both suffered greatly because of this condition. French scientists and scholars of this era developed a rational system of weights and measures called the *metric system,* which was adopted by most countries of the world. In 1960, the General Conference of Weights and Measures extensively revised and simplified the older metric system and gave it the French title, *Système International d'Unités* (International System of Units), commonly abbreviated SI. The latest revisions and additions were made in an international conference in 1971, and work still continues on these standards.

For the engineer, the greatest confusion has been the units for mass and weight. The literature abounds with units such as slugs, pounds mass, pound force, poundal, kilogram force, kilogram mass, dyne, and so on. In the SI system, the base unit for *mass* (not weight or force) is the kilogram, which is equal to the mass of the international standard kilogram

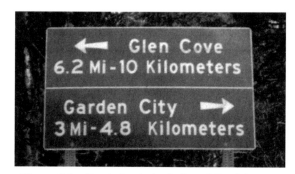

Figure 1.1   Road sign in both SI and English units.

located at the International Bureau of Weights and Measures. It is used to specify the quantity of matter in a body. The mass of a body never varies, and it is independent of gravitational force.

The SI *derived* unit for force is the newton (N). The unit of force is defined from *Newton's second law of motion*: force is equal to mass times acceleration *(F = ma)*. By this definition, 1 newton applied to a mass of 1 kilogram gives the mass an acceleration of 1 meter per second squared ($N = kg{\cdot}m/s^2$). The newton is used in all combinations of units that include force: pressure or stress (N/m^2), energy ($N{\cdot}m$), power ($N{\cdot}m/s$), and so on. By this procedure, the unit of force is not related to gravity as was the older kilogram force.

Weight is defined as a measure of gravitational force acting on a material object at a specified location. Thus, a constant mass has an approximately constant weight on the surface of the earth. The agreed standard value (standard acceleration) of gravity is 9.806 650 m/s^2. Figure 1.2 illustrates the difference between mass (kilogram) and force (newton).

The term "mass" or "unit mass" should be used only to indicate the quantity of matter in an object. The old practice of using weight in such cases should be avoided in engineering and scientific practice. The general relation that ties together mass *(m)* and weight *(w)* is found from Newton's second law of motion.

$$F = ma \tag{1.1}$$

If we now perform a simple experiment on the earth that consists of dropping a mass (in a vacuum to eliminate the effects of the air on the body) and measuring its acceleration, we would obtain from equation (1.1) that the unbalanced force acting on the body is its weight, *w*, and its acceleration is the acceleration of gravity, *g*. Thus we can write equation (1.1) as

$$w = mg \tag{1.2}$$

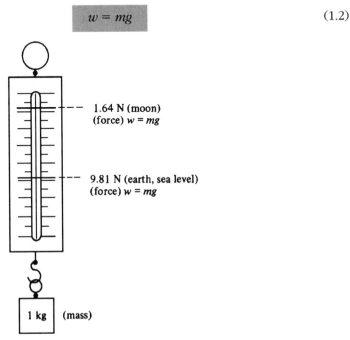

Figure 1.2   Mass and force.

where w is the weight of the body in newtons, m is its mass in kilograms, and g is the local acceleration of gravity in meters per second squared. For the surface of the earth we will use $g = 9.81$ m/s^2.

ILLUSTRATION 1.1   CONCEPTS OF MASS AND WEIGHT

What is the weight on the earth of a body that has a mass of 10 kg?

ILLUSTRATIVE PROBLEM 1.1

Given: A mass of 10 kg

Find: Its weight on the earth

Solution: The weight of the body will be its mass multiplied by the acceleration of gravity; that is,

$$w = mg$$

In terms of SI units,

$$w = 10 \text{ kg} \times 9.81 \frac{\text{m}}{\text{s}^2}$$

$$= 98.1 \frac{\text{kg·m}}{\text{s}^2}$$

and since

$$1 \text{ N} = 1 \frac{\text{kg·m}}{\text{s}^2}$$

then

$$w = 98.1 \text{ N}$$

The SI system consists of three classes of units:

1. Base units
2. Supplementary units
3. Derived units
   (a) With special names
   (b) Without special names

Table 1.1 gives the seven base units of the SI system. Several observations concerning this table should be noted. The unit of length is the meter, and the kilogram is a unit of mass, not weight. Also, symbols are never pluralized; never written with a period; and upper- and lowercase symbols *must* be used as shown *without exception.*

Table 1.2 gives the supplementary units of the SI system. These units can be regarded as either base units or as derived units.

Table 1.1   Base SI Units

Quantity	Base SI unit	Symbol
Length	meter	m
Mass	kilogram	kg
Time	second	s
Electric current	ampere	A
Thermodynamic temperature	kelvin	K
Amount of substance	mole	mol
Luminous intensity	candela	cd

Source: Reprinted with permission from *Strength of Materials for Engineering Technology* by Irving Granet, 2nd ed. Reston Publishing Co.

Table 1.2   Supplementary SI Units

Quantity	Supplementary SI unit	Symbol
Plane angle	radian	rad
Solid angle	steradian	sr

Source: Reprinted with permission from *Strength of Materials for Engineering Technology* by Irving Granet, 2nd ed. Reston Publishing Co.

Table 1.3 gives the derived units (with and without symbols) often used in fluid mechanics. These derived units are formed by the algebraic combination of base and supplementary units. It is noted that where the unit is named for a person, the first letter of the symbol appears as a capital (e.g., newton is N). Otherwise, the convention is to make the symbol lowercase.

In order for the SI system to be universally understood, it is important that the symbols for the SI units and the conventions governing their use be strictly adhered to. Care should be taken to use the correct case for symbols, units, and their multiples (e.g., K for kelvin, k for kilo; m for milli, M for mega). As noted earlier, unit *names* are never capitalized except at the beginning of a sentence. SI unit *symbols* derived from proper names are written with the first letter in uppercase; all other symbols are written in lowercase (for example, m for meter, s for second, K for kelvin, and Wb for weber). Also, unit names form their plurals in the usual manner. Unit symbols are always written in singular form: for example, 350 megapascals, or 350 MPa; 50 milligrams, or 50 mg. Since the unit symbols are standardized, the symbols should always be used in preference to the unit names. An exception is made when a number written out in words precedes the unit (e.g., seven meters, not seven m). Unit symbols are not followed by a period unless they occur at the end of a sentence, and the numerical value associated with a symbol should be separated from that symbol by a space (e.g., 1.81 mm, *not* 1.81mm). The period is only to be used as a decimal marker. Since the comma is used by some countries as a decimal marker, the SI system does not use the comma. A space is used to separate large numbers in groups of threes starting from the decimal in either direction. Thus, 3 807 747.0 and 0.030 704 254 indicate this type of grouping.

Table 1.3   Derived SI Units

Quantity	Derived SI unit	Symbol	Formula	Expressed in terms of base units
Acceleration	—	m/s^2	m/s^2	m/s^2
Area	square meter	m^2	m^2	m^2
Density	kilogram per cubic meter	—	kg/m^3	$kg{\cdot}m^{-3}$
Energy or work	joule	J	$N{\cdot}m$	$m^2{\cdot}kg{\cdot}s^{-2}$
Force	newton	N	$m{\cdot}kg{\cdot}s^{-2}$	$m{\cdot}kg{\cdot}s^{-2}$
Length	meter	m	m	m
Mass	kilogram	kg	kg	kg
Moment	newton meter	$N{\cdot}m$	$N{\cdot}m$	$m^2{\cdot}kg{\cdot}s^{-2}$
Moment of inertia of area	—	m^4	m^4	m^4
Plane angle	radian	rad	rad	rad
Power	watt	W	J/s	$m^2{\cdot}kg{\cdot}s^{-3}$
Pressure or stress	pascal	Pa	N/m^2	$N{\cdot}m^{-2}$
Rotational frequency	revolutions per second	rev/s	s^{-1}	s^{-1}
Temperature	degree celsius	°C	°C	$1\ °C = 1\ K$
Time	second	s	s	s
Torque (*see* Moment)	newton meter	$N{\cdot}m$	$N{\cdot}m$	$m^2{\cdot}kg{\cdot}s^{-2}$
Velocity (speed)	meter per second	m/s	m/s	$m{\cdot}s^{-1}$
Volume	cubic meter	m^3	m^3	m^3

Source: Reprinted with permission from *Strength of Materials for Engineering Technology* by Irving Granet, 2nd ed. Reston Publishing Co.

Notice that for numerical values of less than 1, the decimal point is preceded by a zero. For a number of four digits, the space can be omitted.

In addition, certain style rules should be adhered to:

1. When a product is to be indicated, use a space between unit names (e.g., newton meter).

2. When a quotient is indicated, use the work "per" (e.g., meter per second).

3. When a product is indicated, use the words "square," "cubic," and so on (e.g., square meter).

4. In designating the product of units, use a centered dot (e.g., N·s, kg·m).

5. For quotients, use a solidus (/) or a negative exponent (e.g., m/s or m·s⁻¹). The solidus (/) should not be repeated in the same expression unless ambiguity is

avoided by using parentheses. Thus, one should use m/s^2 or $m \cdot s^{-2}$ but *not* m/s/s; also, use $m \cdot kg/(s^3 \cdot A)$ or $m \cdot kg \cdot s^{-3} \cdot A^{-1}$ but *not* $m \cdot kg/s^3/A$.

One of the features of the older metric system and the current SI system that is most useful is the fact that multiples and submultiples of the units are in terms of factors of 10. Thus, the prefixes given in Table 1.4 are used in conjunction with SI units to form names and symbols of multiples of SI units. Certain general rules apply to the use of these prefixes:

1. The prefix becomes part of the name or symbol with no separation (e.g., kilometer, megagram, etc.)

2. Compound prefixes should not be used: use GPa, not kMPa.

3. In calculations, use powers of 10 in place of prefixes.

4. Try to select a prefix where the numerical value will fall between 0.1 and 1000. This rule may be disregarded when it is better to use the same multiple for all items. It is also recommended that prefixes representing 10 raised to a power that is a multiple of 3 be used (e.g., 100 mg, not 10 cg).

5. The prefix is combined with the unit to form a new unit, which can be provided with a positive or negative exponent. Therefore, mm^3 is $(10^{-3}\,m)^3$ or $10^{-9}\,m^3$.

6. Where possible, avoid the use of prefixes in the denominator of compound units. The exception to this rule is the prefix k in the base unit kg (kilogram).

Table 1.4   Factors of Ten for SI Units

Prefix	Symbol		Factor
tera	T	10^{12}	1 000 000 000 000
giga	G	10^{9}	1 000 000 000
mega	M	10^{6}	1 000 000
kilo	k	10^{3}	1 000
hecto	h	10^{2}	100
deka	da	10^{1}	10
deci	d	10^{-1}	0.1
centi	c	10^{-2}	0.01
milli	m	10^{-3}	0.001
micro	μ	10^{-6}	0.000 001
nano	n	10^{-9}	0.000 000 001
pico	p	10^{-12}	0.000 000 000 001
femto	f	10^{-15}	0.000 000 000 000 001
atto	a	10^{-18}	0.000 000 000 000 000 001

Source: Reprinted with permission from *Strength of Materials for Engineering Technology* by Irving Granet, 2nd ed. Reston Publishing Co.

ILLUSTRATION 1.2   CONCEPTS OF MASS AND WEIGHT

One kilogram of lead is taken to the moon, where the local acceleration of gravity is approximately one-sixth that of the earth's gravity. What is its mass and weight on the moon?

ILLUSTRATIVE PROBLEM 1.2

Given: One kilogram mass

Find: Its mass and weight on the moon if $g_{moon} = g_{earth}/6$

Solution: As shown in Figure 1.2, the body weighs 9.81 N on the earth. On the moon the mass will still be 1 kg, since the amount of matter in the body stays constant. However, since the local acceleration of gravity on the moon is one-sixth of the earth's gravity, it will weigh one-sixth of its earth's weight on the moon. Therefore,

$$w = mg$$

$$\text{weight (moon)} = 1 \text{ kg} \times \frac{9.81}{6} \text{ m/s}^2 = 1.635 \text{ N}$$

It is almost universally agreed that when a new language is to be learned, the student should be completely immersed and made to "think" in the new language. In dealing with the SI system, the student should not "think" in terms of customary units and then perform a mental conversion. It is better to learn to "think" in terms of the SI system, which will then become a second language. There will be times, however, when it may be necessary to convert from customary U.S. units to SI units. To facilitate such conversions, Appendix C gives some commonly used conversion factors.

ILLUSTRATION 1.3   CONVERSIONS

The volume rate of flow is often specified in liters per second and it is necessary to know the flow rate in cubic meters per second. Derive a conversion factor to convert from liters per second to cubic meters per second.

ILLUSTRATIVE PROBLEM 1.3

Given: Flow in liters per second (L/s)

Find: Flow in cubic meters per second (m³/s)

Solution: We will use an approach in the solution of this problem that will be used throughout this book, namely, dimensional consistency. Units must be consistent, and we can use the familiar rules of algebra to manipulate units. The definitions that we need are

$$1 \text{ liter} = 1000 \text{ cm}^3$$

$$100 \text{ cm} = 1 \text{ m}$$

Thus

$$\frac{1 \text{ liter}}{\text{second}} \times \frac{1000 \text{ cm}^3}{\text{liter}} \times \left(\frac{\text{meter}}{100 \text{ cm}}\right)^3 = \frac{\text{meter}^3}{\text{second}}$$

Giving us

$$\frac{L}{s} \times 10^{-3}\frac{m^3}{L} = \frac{m^3}{s}$$

ILLUSTRATION 1.4   CONVERSIONS

Derive the conversion factor to convert gallons per minute to cubic meters per second.

ILLUSTRATIVE PROBLEM 1.4

Given: Gallons per minute (gpm)

Find: Cubic meters per second (m³/s)

Solution: The needed definitions are,

$$1 \text{ gallon} = 231 \text{ in.}^3$$

$$1 \text{ in.} = 2.54 \text{ cm}$$

$$1 \text{ minute} = 60 \text{ seconds}$$

$$100 \text{ cm} = 1 \text{ meter}$$

Using the procedure of Illustrative Problem 1.3, we have

$$\frac{1 \text{ gallon}}{\text{minute}} \times \left(\frac{231 \text{ in.}^3}{\text{gallon}}\right) \times \left(\frac{2.54 \text{ cm}}{\text{in.}}\right)^3 \times \left(\frac{m}{100 \text{ cm}}\right)^3 \times \left(\frac{\min}{60 \text{ s}}\right) = \frac{m^3}{s}$$

This yields

$$\frac{1 \text{ gallon}}{\text{minute}} \times 6.31 \times 10^{-5}\frac{m^3 \cdot \text{minute}}{\text{gallon} \cdot s} = \frac{m^3}{s}$$

1.3   THE ENGLISH SYSTEM OF UNITS

After using both the English system of units and the SI system, the student will find that the English system of units is not as coherent or as easy to use as the SI system. The difficulty arises from the fact that mass and force can be defined independently of each other. This is due to the fact that two laws attributed to Newton—the law of universal gravitation and the second law of motion—each involve force and mass. We therefore find two basic systems of units to exist, consisting of various combinations of the *same* term to describe both force and weight. Table 1.5 is a simple listing of these combinations.

Table 1.5

Mass	pound mass	slug
Length	foot	foot
Force	pound force	pound force
Time	second	second

At present, in the United States, the accepted consistent set of units commonly used in fluid mechanics is the pound (force), the slug (mass), the foot (length), and the second (time). For this set of units the slug is the derived unit that equals the mass that a force of 1 pound will accelerate at the rate of 1 foot per second per second. From Newton's second law of motion,

$$F = ma \tag{1.1}$$

we find that

$$m = \frac{F}{a} = \frac{\text{lb}}{\text{ft/s}^2} = \frac{\text{lb} \cdot \text{s}^2}{\text{ft}} \tag{1.3}$$

Thus the units of the slug (mass) in the conventional English system are $\text{lb} \cdot \text{s}^2/\text{ft}$.

If we now again perform the same simple experiment on the earth that consists of dropping a mass (in a vacuum to eliminate the effects of the air on the body) and measuring its acceleration, we would obtain the acceleration of gravity, which is generally accepted as 32.174 ft/s². If we compare these results with those we obtained earlier, we can write equation (1.2) as a general result, independent of the system of units,

$$w = mg \tag{1.2}$$

where g = 9.806 m/s² or 32.174 ft/s² on the surface of the earth, depending on the system of units being used. For this book we usually round off these numbers to 9.81 and 32.17, respectively.

ILLUSTRATION 1.5   THE RELATION OF MASS AND WEIGHT— ENGLISH UNITS

What is the mass of a body that weighs 20 lb? Assume that the local acceleration of gravity is 32.17 ft/s².

ILLUSTRATIVE PROBLEM 1.5

Given: A body weighs 20 lb, where g = 32.17 ft/s².

Find: Its mass

Solution: Using equation (1.2), we obtain

$$m = \frac{w}{g} = \frac{20 \text{ lb}}{32.17 \text{ ft/s}^2} = 0.622 \frac{\text{lb} \cdot \text{s}^2}{\text{ft}} = 0.622 \text{ slugs}$$

ILLUSTRATION 1.6   THE RELATION OF MASS AND WEIGHT— ENGLISH UNITS

A mass of 10 slugs weighs 250 lb on a spring balance on an unknown planet. The spring balance was originally calibrated at a standard location on the earth. What is the value of the acceleration of gravity at this location on the unknown planet?

ILLUSTRATIVE PROBLEM 1.6

Given: A mass of 10 slugs weighs 250 lb on an unknown planet.

Find: The acceleration of gravity on this planet.

Solution: From equation (1.2),

$$g = \frac{w}{m} = \frac{250 \text{ lb}}{10 \text{ lb} \cdot \text{s}^2/\text{ft}} = 25.0 \frac{\text{ft}}{\text{s}^2}$$

1.4   DIMENSIONAL CONSISTENCY

The physical sciences are concerned with the discovery and formulation of exact relationships between the various quantities involved in recurring situations. Dimensional consistency, or dimensional analysis, is based upon the axiom that a general relationship may be established between two quantities only when these two quantities have the equivalent physical nature and are measured in the same units. Based on this axiom, an equation must at least be dimensionally homogeneous (have the same dimensions on both sides of the equality) if it is to be generally valid. Dimensional homogeneity, however, does not guarantee that an equation is valid, nor does an equation with different dimensions on both sides of the equality necessarily have to be invalid. Many empirical equations of this type (non-homogeneous) exist either for convenience or to express experimental data over a limited range of the variables involved.

We have already used the principle of dimensional consistency when we established the physical units of N and slugs. In each case we express the symbols in an equation in terms of mass, force, length, and time (at times, temperature too), and using the method of algebraic cancellation of units, establish the unit or the conversion factor needed to obtain the desired unit.

ILLUSTRATION 1.7   DIMENSIONAL CONSISTENCY

As we will see in Chapter 2, the pressure at the base of a uniform column of liquid having a specific weight of $\gamma\,(\text{N/m}^3)$ is

$$p = \gamma h$$

where h is the height in meters. What is the corresponding unit for p, pressure?

ILLUSTRATIVE PROBLEM 1.7

Given: γ in N/m^3 and h in meters

Find: Unit of pressure at base of a column

Solution: Applying the dimensions of each unit gives us

$$p = \gamma\!\left(\frac{\text{N}}{\text{m}^3}\right) \times h\,(\text{m})$$

$$p = \frac{\text{N}}{\text{m}^2}$$

Note that pressure is force divided by area (newtons per meter^2). The unit of pressure, N/m^2, is defined to be the pascal (Pa).

ILLUSTRATION 1.8   DIMENSIONAL CONSISTENCY

Solve Illustrative Problem 1.7 in the English system of units. Assume that the desired pressure unit is lb/in.^{2} (psi).

ILLUSTRATIVE PROBLEM 1.8

Given: γ in lb/ft^3 and h in feet

Find: Pressure at base of a column in lb/in.^{2} (psi)

Solution:

$$p = \gamma h$$

where γ is in lb/ft^3 and h is in feet.

$$p = \gamma\!\left(\frac{\text{lb}}{\text{ft}^3}\right) \times h\,(\text{ft})$$

$$= \frac{\text{lb}}{\text{ft}^2}\ (\text{psf})$$

The unit for pressure that we obtain is lb/ft² (psf). Since we want to have as our pressure unit lb/in.² (psi), we need to introduce the following conversion:

$$p\,(\text{psi}) = p\left(\frac{\text{lb}}{\text{ft}^2}\right) \times \frac{1}{(12\ \text{in./ft})^2} = p\left(\frac{\text{lb}}{\text{ft}^2}\right) \times \frac{1}{(12)^2\ \text{in.}^2/\text{ft}^2}$$

or

$$p\,(\text{psi}) = p\,(\text{psf}) \times \frac{1\,\text{ft}^2}{144\,\text{in}^2}$$

This conversion is a very common conversion in the English system and will appear often in later chapters.

ILLUSTRATION 1.9   DIMENSIONAL CONSISTENCY

What net force is required to accelerate a mass of 15 kg at the rate of 30 m/s²?

ILLUSTRATIVE PROBLEM 1.9

Given: Mass = 15 kg; acceleration = 30 m/s²

Find: acceleration

Solution: Using Newton's second law of motion,

$$F = ma$$

$$= 15\ \text{kg} \times 30\ \frac{\text{m}}{\text{s}^2} = 450\ \frac{\text{kg} \cdot \text{m}}{\text{s}^2}$$

But the unit of force in the SI system is the newton, which dimensionally has the units kg · m/s (see Illustrative Problem 1.1 and Table 1.3). Therefore,

$$F = 450\ \text{N}$$

In the study of fluids flowing in systems, we will encounter two energy terms, namely kinetic energy (K.E.) and pressure head. The equations for these terms are

$$\text{K.E.} = \frac{wV^2}{2g} = \frac{mV^2}{2} \tag{1.4}$$

and

$$\text{pressure head} = \frac{wp}{\gamma} \tag{1.5}$$

where V is velocity, g is the acceleration of gravity, p is pressure, and γ is specific weight (weight per unit volume). Kinetic energy has the units of N·m or ft·lb. Pressure head has the units of N·m or ft·lb. If a unit weight of 1 N or 1 lb is used, both terms will have units of N or ft. Since these terms are energy terms, they can also be written as $\dfrac{N \cdot m}{N}$ or $\dfrac{ft \cdot lb}{lb}$.

The following illustrative problems show the application of dimensional consistency to the solution of these equations.

ILLUSTRATION 1.10   DIMENSIONAL CONSISTENCY

A mass of 10 kg has a velocity of 12 m/s. Determine the kinetic energy of the mass in N·m.

ILLUSTRATIVE PROBLEM 1.10

Given: $m = 10$ kg, $V = 12$ m/s

Find: K.E.

Solution: From equation (1.4), K.E. $= mV^2/2$. For the conditions of this problem,

$$\text{K.E.} = \frac{(10 \text{ kg}) \times (12 \text{ m/s})^2}{2} = 720 \text{ kg·m}^2/\text{s}^2$$

But 1 N = 1 kg·m/s². Using this with our result yields

$$\text{K.E.} = 720 \frac{\text{kg·m}^2}{\text{s}^2} = 720 \text{ N·m}$$

ILLUSTRATION 1.11   DIMENSIONAL CONSISTENCY

A weight of 10 lb has a velocity of 6 ft/s. Determine its kinetic energy.

ILLUSTRATIVE PROBLEM 1.11

Given: $w = 10$ lb, $V = 6$ ft/s

Find: K.E.

Solution: Proceeding as we did in Illustrative Problem 1.10, K.E. $= wV^2/2g$, and for the data of this problem,

$$\text{K.E.} = \frac{(10 \text{ lb}) \times (6 \text{ ft/s})^2}{2 \times (32.17 \text{ ft/s}^2)}$$

giving us

$$\text{K.E.} = 5.6 \text{ ft·lb}$$

ILLUSTRATION 1.12 DIMENSIONAL CONSISTENCY

Water has a specific weight of 9810 N/m³. If it exerts a pressure of 500 kPa, determine its pressure head.

ILLUSTRATIVE PROBLEM 1.12

Given: $\gamma = 9810$ N/m³, $p = 500$ kPa

Find: Pressure head

Solution: From equation (1.5), pressure head = wp/γ. For a weight of 1 N,

$$\text{pressure head} = \frac{1 \text{ N} \times (500 \text{ kPa}) \times (1000 \text{ Pa/kPa}) \times [(1 \text{ N/m}^2)/\text{Pa}]}{9810 \text{ N/m}^3}$$

$$= 50.97 \text{ N·m per newton or } 50.97 \text{ N·m/N}$$

If we cancel the N terms, we obtain a pressure head of 50.97 m. Since this represents a height, the term "head" was given to it. We will discuss this in detail in a later chapter.

1.5 REVIEW

Both fluid statics and fluid dynamics are integral parts of our lives. In this introductory chapter, we began our study of fluid mechanics by looking at the systems of units that are currently used in the United States. We noted that even though most of the world uses the SI system of units, the use of the customary English system of units has persisted in this country. Thus we find that both the SI and English systems are used in fluid mechanics applications. This led us to devote much of our time to defining the elements of both of these systems of measurement. We further noted that all applications in the technologies require that dimensional consistency must exist in both equations and the conversion of units. Dimensional consistency is the key to obtaining meaningful results in calculations, and you are strongly urged to form the habit of putting dimension into equations to ensure that the desired result is obtained. The systematic use of this procedure when doing the problems at the end of a chapter will enable you to form a habit that will be of immense help to you professionally.

KEY TERMS

Terms of importance in this chapter:

Acceleration of gravity (g): the rate of acceleration of a body that is freely falling in a vacuum on the earth, taken as either 9.81 m/s² or 32.17 ft/s².

Centimeter (cm):   one hundredth of a meter; 2.54 cm = 1 in.

Dimensional consistency: a general relationship that may be established between two quantities only when these two quantities have the equivalent nature and are measured in the same units.

English system: the system of units that uses the units for mass, force, length, and time to be the slug, the pound, the foot, and the second, respectively.

Gallon: a unit of volume defined to be 231 in.[3].

Kilogram (kg):   the unit of mass in the SI system.

Kinetic energy (K.E.): $mV^2/2$ or $wV^2/2g$.

Mass (m):   the amount of matter contained in a body.

Newton (N):   the unit of force in the SI system of units $= mg$.

Newton's second law: the relation between force, mass, and acceleration, namely, $F = ma$.

Pound (lb):   the unit of force in the English system $= mg$.

Pressure (p):   force divided by area (F/A). In the SI system the unit of pressure is N/m^2 or pascal, Pa. In the English system the usual units of pressure are $lb/in.^2$ (psi) or lb/ft^2 (psf).

Pressure head: wp/γ, where w is the weight, p is the pressure, and γ is the specific weight of the substance.

SI system: the modern metric system that uses the units of mass, force, length, and time to be the kilogram, the newton, the meter, and the second, respectively.

Slug: the unit of mass in the English system, which equals weight in pounds divided by the acceleration of gravity.

Specific weight (γ):   the weight of a body divided by its volume—expressed as newtons per cubic meter or pounds per cubic foot.

Weight (w):   the force exerted on a body by the gravitational pull of the earth or the planet or satellite that the body is on.

KEY EQUATIONS

Newton's second law	$F = ma$	(1.1)
Mass–weight relation	$w = mg$	(1.2)
Pressure–height relation	$p = \gamma h$	Ill. Prob. 1.7
Kinetic energy	$\text{K.E.} = \dfrac{wV^2}{2g} = \dfrac{mV^2}{2}$	(1.4)
Pressure head	$\text{pressure head} = \dfrac{wp}{\gamma}$	(1.5)

QUESTIONS

1. State the two states of matter that the term *fluid* includes.

2. Would you expect that your weight would be greater, less, or equal to your earth weight on Jupiter? Why?

3. A mechanic states that the pressure in one of the tires on your car is 24 lb. Comment on this statement.

4. On which factor or factors does the pressure on the bottom of a column of fluid depend: (a) its height, (b) the acceleration of gravity, (c) the specific weight of the fluid, or (d) the cross-sectional area of the column?

5. If the velocity of an object is doubled, you would expect that its kinetic energy would (a) remain the same, (b) double, (c) be half of its original value, (d) be four times its original value?

6. Comment on what will occur to a person's mass and weight as the person climbs up a mountain. Does this seem reasonable to you?

7. Is it better to buy gasoline in the morning or evening, or do you think that it doesn't matter? Why?

8. What unit in the English system does the liter come closest to: (a) the pound, (b) the gallon, (c) the quart, or (d) the slug?

9. If the kilometer is approximately 5/8 of a mile, to what speed in kilometers per hour would 55 miles per hour correspond?

10. Comment on the general validity of the equations, (a) $V = at^2/2$ and (b) $d = at^2/2$, where V is velocity in m/s, a is acceleration in m/s², t is time in seconds (s), and d is displacement in m.

PROBLEMS

More difficult problems are denoted by an asterisk (*).

Conversion of Units

1.1. Convert a length of 10 ft to meters.

1.2. Convert a length of 135 cm to meters.

1.3. Convert a length of 1355 mm to meters.

1.4. Convert 3.85 in. to centimeters.

1.5. Convert 90.68 cm to inches.

1.6. Convert 5787 lb to tons.

1.7. Convert 6.3 tons to pounds

1.8. Convert 35.4 in. to feet.

1.9. Convert 3.87 ft to inches.

1.10. Convert 8.6 lb to grams.

1.11. Convert 462 g to pounds.

1.12. Convert 2 yd 2 ft 5 in. to centimeters.

1.13. Convert 550 mi/h to m/s.

1.14. Convert a flow rate of 2000 ft^3/min to L/s.

1.15. Convert 200 gal/h (gph) to m^3/s.

1.16. Convert 200 gph to L/s.

1.17. Convert 200 gph to L/min.

1.18. Convert 200 L/min to m^3/s.

1.19. Convert 200 gpm to L/min.

1.20. Convert 50.7 cm^3/s to m^3/s.

1.21. Convert 861 gpm to L/min.

1.22. Convert 1.71 m^3/s to L/min.

1.23. Convert 100 m^3/s to gpm.

1.24. Convert 3.52 ft^3/s to gpm.

1.25. Convert 3.52 ft^3/s to m^3/s.

1.26. Convert 150 gpm to m^3/s.

1.27. Convert 6.52 m^3/s to gpm.

1.28. Convert 3.65 m^3/s to L/s.

1.29. Convert 1.2 ft^3/s to gpm.

1.30. Convert 4.8 ft^3/s to L/s.

1.31. Convert 2000 kg to slugs.

1.32. Convert 2000 kg/m^3 to slugs/ft^3.

1.33. Convert 5 slugs to kg.

1.34. Convert 5 slugs/ft^3 to kg/m^3.

1.35. Convert 101 kPa (kN/m^2) to lb/in^2.

1.36. A rectangle has the dimensions of 12 in. $\times$ 10 in. Calculate its area in square meters.

1.37. A rectangle has the dimensions of 2 m $\times$ 3.2 m. Determine the area of the rectangle in square inches.

1.38. A rectangle has the dimensions of 15 cm $\times$ 18 cm. Determine the area of the rectangle in square meters.

1.39. A box has the dimensions of 12 in. $\times$ 12 in. $\times$ 6 in. Determine the number of cubic meters contained in the box.

1.40. A cylinder has a diameter of 10 cm and is 20 cm high. Determine its volume in cubic meters.

1.41. A cylinder has a diameter of 5 in. and is 10 in. high. Determine its volume in cubic meters.

1.42. A cube is 6 in. on a side. Calculate the volume of the cube in cubic meters.

1.43. A woman runs in a 1500-m race, which is sometimes called the "metric mile." If a mile is equal to 5280 ft, determine what part of a mile she ran.

1.44. A liter is a volume measurement that equals 1000 cm³ (10^{-3} m³). How many liters are there in a cubic foot?

1.45. A room is 15 ft wide and 20 ft long. Calculate its area in square meters.

1.46. A man weighs 175 lb and stands 6 ft 0 in. tall. What is his weight in newtons and his height in meters?

1.47. A driveway requires 38 yd³ of concrete. To how many cubic meters of concrete does this correspond?

1.48. An automobile engine has a displacement of 240 in³. Determine its displacement in liters.

1.49. A car weighs 2650 lb and is 18 ft long. What is its weight in newtons and its length in meters?

1.50. A block of metal weighing 1200 lb is hung from the end of a vertical rod and causes the rod to lengthen 0.001 inches. Express these data in terms of newtons and meters.

*1.51. If a gallon is a volume measurement of 231 in.³, determine the number of liters in a gallon. Note that the liter is sometimes called the metric quart. Does your answer support this name?

*1.52. A barrel of oil contains 55 gal of oil. If a gallon has 231 in.³, determine how many cubic meters there are in 10 barrels of oil.

1.53. Derive a conversion factor to convert cubic meters to cubic feet.

1.54. To what velocity in ft/s does 10 m/s correspond?

1.55. If a car is operated at a speed of 50 mi/h, determine its speed in feet per second.

1.56. A car travels at a speed of 55 mi/h. What is its speed in meters per second?

1.57. A car travels at a speed of 64 km/h. Determine its speed in miles per hour.

*1.58. A car is claimed to obtain a fuel usage in Europe of 12 km/L. Determine its equivalent fuel usage in miles per gallon.

1.59. One gigapascal (GPa) is (a) 10^6 Pa; (b) 10^8 Pa; (c) 10^3 Pa; (d) 10^9 Pa; (e) none of these.

1.60. One megapascal (MPa) is (a) 10^5 Pa; (b) 10^4 Pa; (c) 10^6 Pa; (d) 10^7 Pa; (e) none of these.

Mass and Weight

1.61. A mass of 5 kg is placed on a planet whose gravitational force is 10 times that of the earth. What is its weight on this planet?

1.62. A mass of 10 slugs weighs 320 lb on a spring balance. The acceleration of gravity at this location is (a) 32.7 ft/s²; (b) 0.03125 ft/s²; (c) 320 ft/s²; (d) 32.0 ft/s²; (e) none of these.

1.63. A mass of 10 kg weighs 90 N. The acceleration of gravity at this location is (a) 1/9 m/s^2; (b) 9.0 m/s^2; (c) 90 m/s^2; (d) 10 m/s^2; (e) none of these.

1.64. A weight of 160.85 lb corresponds to a mass of (a) 0.2 slug; (b) 0.5 slug; (c) 7.5 slugs; (d) 0.75 slug; (e) none of these.

1.65. What is the weight in newtons of a body whose mass is 12 kg?

1.66. What is the weight in newtons of a body whose mass is 8 slugs?

1.67. If a body has a mass of 200 g, what is its weight in newtons?

1.68. A body weighs 150 lb. What is its mass in slugs?

1.69. If a body has a mass of 25 slugs, it will weigh how many newtons?

1.70. At a location on the earth where $g = 32.2$ ft/s^2, a body weighs 161 lb. At another locations where $g = 32.0$ ft/s^2, the same body will weigh (a) 161 lb; (b) 162 lb; (c) 160 lb; (d) 163 lb; (e) none of these.

1.71. What is the weight of 10 kg at a location on the earth where $g = 9.7$ m/s^2?

1.72. If the body of Problem 1.71 is moved to another location on the earth where $g = 10$ m/s^2, what will its mass be?

*1.73. A mass of 100 kg is hung from a spring in a local gravitational field where $g = 9.806$ m/s^2, and the spring is found to deflect 25 mm. If the same mass is taken to a planet where $g = 5.412$ m/s^2, how much will the spring deflect if its deflection is directly proportional to the force applied?

*1.74. A balance scale is used to weigh a sample on the moon. If the value of g is one-sixth of the earth's gravity and the "standard" weights (earth weights) add up to 20 lb, what is the mass of the body?

*1.75. A balance scale is used to weigh a sample on the moon. If the value of g is one-sixth of the earth's gravity and the "standard" weights (earth weights) add up to 100 N, what is the mass of the body?

*1.76. Solve Problem 1.74 if a spring balance is used that was calibrated on the earth and reads 20 lb.

*1.77. Solve Problem 1.75 if a spring balance is used that was calibrated on the earth and reads 100 N.

Newton's Second Law, F = ma

1.78. A net force of 10 N acts on a body whose mass is 0.5 kg. What is the acceleration of the body?

1.79. A net force of 25 lb acts on a body whose mass is 5 slugs. What is the acceleration of the body?

1.80. If a body is accelerating at the rate of 5 m/s^2 when acted on by a net force of 10 N, determine its mass.

1.81. If a body whose mass is 10 slugs is accelerated at the rate of 5 ft/s^2, determine the unbalanced force acting on it.

1.82. A mass of 2 kg is accelerated at a rate of 1.2 m/s². What is the magnitude of the unbalanced force acting on it?

Kinetic Energy and Pressure Head

K.E. $= mV^2$ or $wV^2/2g$; pressure head $= wp/\gamma$, where m is mass, V is velocity, w is weight, g is the acceleration of gravity, p is pressure, and γ is specific weight.

1.83. A mass of 5 kg has a velocity of 2.1 m/s. Determine its kinetic energy in N·m.

1.84. A car having a mass of 1000 kg is operated at a speed of 90 km/h. Determine its kinetic energy in N·m.

1.85. A car weighs 2500 lb and is operated at 55 mi/h. Determine its kinetic energy in ft·lb.

1.86. A mass of 2 slugs has a speed of 20 ft/s. Determine the kinetic energy of this mass in ft·lb.

1.87. A bullet has a weight of 0.25 lb. If the bullet travels at a speed of 1000 ft/s, determine its kinetic energy in ft·lb.

1.88. What velocity is required to give a 5-kg body a kinetic energy of 10 N·m?

1.89. A body weighing 10 lb has a kinetic energy of 6.4 ft·lb. What is its velocity?

1.90. A body having a mass of 2 slugs has a kinetic energy of 25.2 ft·lb. Determine its velocity.

1.91. What velocity is required to give a 14-N body a kinetic energy of 76.2 N·m?

1.92. If a body is moving with a velocity of 21 ft/s and has a kinetic energy of 152 ft·lb, determine its mass.

1.93. A body has a kinetic energy of 91 ft·lb when it has a velocity of 12 ft/s. What is its weight?

1.94. Calculate the weight of a body that has a kinetic energy of 48 N·m when going at a speed of 40 km/h.

*1.95. Mercury weighs 13.6 times the weight of water. Determine the pressure head per newton of mercury if the pressure that it exerts is 1 MPa. The specific weight of water can be taken as 9810 N/m³.

*1.96. Water has a specific weight of 62.4 lb/ft³. Calculate the pressure head of 1 lb of water if it exerts a pressure of 10 psi.

1.97. If the pressure head exerted by water is 3 N·m/N, determine the pressure.

*1.98. The pressure head of a column of water is 10 ft·lb/lb. Determine the pressure in psi.

FLUID PROPERTIES

LEARNING GOALS

After reading and studying this chapter you should be able to:

1. Define the term *system*.
2. State the concept of a *property* and show how properties are used to define the state of a system.
3. Define the term *temperature* and indicate how this property is measured.
4. Convert from and to the Fahrenheit and Celsius temperature scales.
5. Relate the absolute temperatures of Rankine and Kelvin to the Fahrenheit and Celsius temperature scales, respectively.
6. Define the terms *density, specific weight, specific volume,* and *specific gravity.*
7. State the definition of *pressure* and show how it is measured.
8. Differentiate between gage and absolute pressures.
9. Derive the pressure–height relation.
10. Use the concept of surface tension to obtain the rise or depression of fluids in capillaries.
11. Define the term *compressibility.*
12. Calculate the shear stress in a fluid.
13. Define the terms *viscosity, dynamic viscosity,* and *kinematic viscosity.*

2.1   INTRODUCTION

In Section 1.1 we decided that the general term "fluid" would be used to refer to both liquids and gases. To distinguish between a liquid and a gas, we also noted that although both will occupy the container in which they are placed, a liquid presents a free surface if it does not completely fill the container. A gas will always fill the volume of the container in which it is placed. For gases it is important to take into account the change in volume that occurs when either the pressure or temperature is changed, whereas in most cases it is possible to neglect the change in volume of a liquid when there is a change in pressure.

As a general concept applicable to all situations, we can simply define a *system* as a grouping of matter taken in any convenient or arbitrary manner. For the present we note that systems can have energy stored or transferred to or from them, and although we are at liberty to choose a system in an arbitrary manner, all forces and energies must be accounted for when deriving the equations governing the system and its motion.

Let us consider a given system and then ask ourselves how we can distinguish changes that occur to it. To answer this question, it is necessary for us to identify those *external characteristics* that enable us both to distinguish and evaluate system changes. Conversely, if the external characteristics of a system do not change, we should be able to infer that the system has not undergone any change. Some of the external characteristics that can be used to describe a system are *temperature, pressure,* and *volume.* These observable external characteristics of a system are called *properties,* and we note that *when all the properties of a system are reproduced at different times and we are unable to distinguish any difference in the system, the system is in the same state at both times.* For the remaining portions of this chapter, we consider those fluid properties that play principal roles in our study of fluid mechanics.

2.2   TEMPERATURE

The *temperature* of a system is a measure of the random motion of the molecules of the system. If there are different temperatures within the body (or bodies composing the system), the question arises as to how the temperature at a given location is measured and how this measurement is interpreted. Let us examine this question, since similar questions will also have to be considered when other properties of a system are studied. In air at room pressure and temperature there are approximately 2.7×10^{19} molecules per cubic centimeter. If we divide the cube whose dimensions are 1 cm on a side into smaller cubes each of whose sides is one-thousandth of a centimeter, there will be about 2.7×10^{10} molecules in each of the smaller cubes, which is still an extraordinarily large number. Although we speak of temperature at a point, we really mean the average temperature of the molecules in the neighborhood of the point.

When measuring the temperature of a body with a thermometer, it should be noted that the thermometer measures only the temperature of its sensing head. For the thermometer to be able to measure the temperature of a system, it is necessary for both the thermometer and the system to be in thermal equilibrium. The measurement of temperature usually employs the measurement of some secondary characteristic (such as the height of a liquid in a

°F	°C	
212	100	Atmospheric boiling point
32	0	Ice-point
−460	−273	Absolute zero

Figure 2.1

bulb thermometer) when the system changes from one state to another. Other examples of physical effects that have been used to indicate temperature are the linear expansion of liquids and solids, the pressure change of a confined gas, the volume change of bodies, and the generation of voltages when dissimilar metals are placed in contact with each other. Although it is assumed that the student is familiar with the common temperature scales, a brief summary is given below.

The common scales of temperature are the Fahrenheit and the Celsius and are defined by using the ice point and the boiling point of water at atmospheric pressure. In the Celsius temperature scale, the interval between the ice point and the boiling point is divided into 100 equal parts. In addition, as shown in Figure 2.1, the Celsius ice point is zero and the Fahrenheit ice point is 32. The conversion from one scale to the other is directly derived from Figure 2.1 and results in the following relations:

$$°C = \tfrac{5}{9}(°F - 32) \tag{2.1}$$

$$°F = \tfrac{9}{5}(°C) + 32 \tag{2.2}$$

The ability to extrapolate to temperatures below the ice point and above the boiling point of water and to interpolate in these regions is provided by the *International Scale of Temperature*. This agreed-upon standard utilizes the boiling and melting points of different elements and establishes suitable interpolation formulas in the various temperature ranges between these elements. The data for these elements are given in Table 2.1.

ILLUSTRATION 2.1   TEMPERATURE

Determine the temperature at which the same value is indicated on both Fahrenheit and Celsius thermometers.

ILLUSTRATIVE PROBLEM 2.1

Given: °F = °C

Find: The temperature at which this occurs

Solution: Using equation (2.1) and letting °C = °F, we obtain

$$°F = \tfrac{5}{9}(°F - 32)$$

$$\tfrac{4}{9}\,°F = -\tfrac{160}{9}$$

$$°F = {}^-40$$

Therefore, both Celsius and Fahrenheit temperature scales indicate the same temperature at $^-40°$.

Table 2.1

Element	Melting or boiling point at 1 atm	Temperature °C	°F
Oxygen	Boiling	−182.97	−297.35
Sulfur	Boiling	444.60	832.28
Antimony	Melting	630.50	1166.90
Silver	Melting	960.8	1761.4
Gold	Melting	1063.0	1945.4
Water	Melting	0	32
	Boiling	100	212
Hydrogen	Boiling	−252.7	−422.86
Helium	Boiling	−268.9	−452.02
Nitrogen	Boiling	−195.8	−320.44

ILLUSTRATION 2.2   TEMPERATURE

Write a BASIC program to solve equations (2.1) and (2.2) for the temperature in one scale (°F or °C) when given a temperature in the other scale.

ILLUSTRATIVE PROBLEM 2.2

Given: Equations (2.1) and (2.2)

Find: Solutions using BASIC

Solution: BASIC is an acronym for *B*eginners *A*ll-purpose *S*ymbolic *I*nstruction *C*ode. The attached program is interactive and requests whether the input is Celsius or

Fahrenheit. Using this answer, the program goes to the appropriate line and solves the equation. It then prints out both the input and the solution. As a check on the program, 20°C and 68°F were used as test values.

```
10 INPUT "If Temperature is C, V = O, if F, V = 1"; H
20 IF H = 0 GOTO 40
30 IF H <>0 GOTO 70
40 INPUT "Temperature C = "; A
50 F = (9/5)*A + 32
60 GOTO 90
70 INPUT "Temperature F = "; B
80 C = (5/9)*(B − 32)
85 GOTO 110
90 LPRINT "Temperature C = "; A
100 LPRINT "Temperature F = "; F
105 END
110 LPRINT "Temperature F = "; B
120 LPRINT "Temperature C = "; C
130 END
```

```
Temperature C = 20
Temperature F = 68
```

```
Temperature F = 68
Temperature C = 20
```

ILLUSTRATION 2.3   TEMPERATURE

Write an Excel macro to solve equations (2.1) and (2.2).

ILLUSTRATIVE PROBLEM 2.3

Given: Equations (2.1) and (2.2)

Find: Solutions using Excel macro

Solution: The term *macro* is short for *macroinstruction,* a long series of commands that can be activated by one or two keystrokes on the computer. The program written is interactive, and when run requests input data from the user in the form of a dialog box. Note that the format used is similar to BASIC. The program has been written for the unknown temperature to be entered as -273 if it is Celsius or -460 if it is Fahrenheit.

```
 = INPUT("ENTER TEMPERATURE FAHRENHEIT",1)
 = INPUT("ENTER TEMPERATURE CELSIUS",1)
 = IF(A2>−460,GOTO(A10))
 = IF(A2 = −460,``GOTO(A6))
 = ((9*A3)/5) + 32
 = ALERT("TEMPERATURE FAHRENHEIT IS "&A6&".",2)
```

```
= ALERT("TEMPERATURE CELSIUS IS "&A3&".",2)
= RETURN( )
= IF(A3>−273,GOTO(A13))
= IF(A3 = −273,GOTO(A12))
= (A2 − 32)*5/9
= ALERT("TEMPERATURE FAHRENHEIT IS "&A2&".",2)
= ALERT("TEMPERATURE CELSIUS IS "&A12&".",2)
= RETURN( )
```

ILLUSTRATION 2.4   TEMPERATURE

Write an Excel worksheet to solve equations (2.1) and (2.2).

ILLUSTRATIVE PROBLEM 2.4

Given: Equations (2.1) and (2.2)

Find: Solutions using Excel worksheet

Solution: Both the BASIC program and the Excel macro are programming languages that require user knowledge and expertise. The use of a spreadsheet (worksheet) requires no special programming language and simplifies solutions. The program below starts with either °F placed in box A5 or °C in box A9. Boxes B5 and B9 are intermediate steps, and boxes C5 and C9 give the desired answers.

Conversion Program

	A	B	C	D
1				
2			Conversion Program	
3			Given °F	
4	°F			
5	68	36	20	
6				
7			Given °C	
8	°C			
9	20	36	68	
10				
	A5 is input	A9 is input		
	B5 = A5 − 32	B9 = 1.8 * A9		
	C5 = B5/1.8	C9 = B9 + 32		

The foregoing solutions have used Microsoft® Excel for the Macintosh© computer.

2.3   ABSOLUTE TEMPERATURE

Let us consider the case of a gas that is confined in a cylinder with a constant cross-sectional area by a piston that is free to move. If heat is now removed from the system, the piston will move down, but because of its weight, it will maintain a constant pressure on the gas. This procedure can be carried out for several gases, and if volume is plotted as a function of temperature, we shall obtain a family of straight lines that intersect at zero volume (Figure 2.2a). All of these lines intersect at one temperature. This unique temperature is known as the *absolute zero* temperature, and the accepted values on the Fahrenheit and Celsius temperature scales are $-459.69°$ and $-273.16°$, respectively, with the values $-460°$ and $-273°$ used for most engineering calculations.

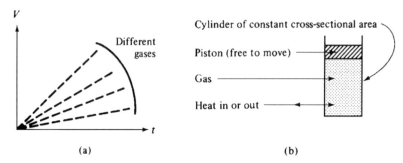

Figure 2.2   Gas thermometer.

Using these temperatures as the zeros of temperature, we arrive at two absolute temperature scales that are defined as

$$\text{degrees Rankine (°R)} = \text{°F} + 460 \qquad (2.3)$$

$$\text{degrees Kelvin (K)} = \text{°C} + 273 \qquad (2.4)$$

It is also possible to define the absolute temperature scale independent of the properties of any substance. The interested student is referred to texts on thermodynamics for this development.

ILLUSTRATION 2.5   TEMPERATURE

Convert 68°F to °R and K.

ILLUSTRATIVE PROBLEM 2.5

Given: °F

Find: °R and K

Solution: To use equation (2.4), we need to know the given temperature of 68°F in °C. We obtain this by using equation (2.1):

$$°C = \tfrac{5}{9}(°F - 32) = \tfrac{5}{9}(68 - 32) = 20$$

Therefore,

$$°R = 68 + 460 = 528$$

$$K = 20 + 273 = 293$$

As a check, note that $528/293 = 1.8$, which is $\frac{9}{5}$.

2.4   DENSITY, SPECIFIC WEIGHT, SPECIFIC VOLUME, AND SPECIFIC GRAVITY

The four properties of this section—density, specific weight, specific volume, and specific gravity—have been grouped together because they are in some way related to each other. The first of these items is *density*.

    Density is defined as the mass per unit volume of a substance. If we denote the mass of a body as m and its volume as V, the density ρ (rho) is

$$\rho = \frac{m}{V} \tag{2.5}$$

    In English units, where m is in slugs and V is in cubic feet, ρ has the units of slugs per cubic foot. In SI units, where m is in kg and V is in m³, ρ has the units kg/m³. The density of water at 4°C is 1000 kg/m³ or 1.94 slugs/ft³.

ILLUSTRATION 2.6   DENSITY

A cylinder has a diameter of $\frac{1}{2}$ m and is 1 m high. If it contains a fluid that has a mass of 200 kg, determine the density of the fluid.

ILLUSTRATIVE PROBLEM 2.6

Given: Cylinder with $d = \frac{1}{2}$ m, $h = 1$ m, mass = 200 kg

Find: Density of contents (ρ)

Solution: The volume of a cylinder is equal to the area of its base multiplied by its height. Therefore,

$$V = \frac{\pi d^2}{4} \times h$$

where d is the diameter and h is the height. For this problem,

$$V = \frac{\pi(\frac{1}{2}\text{m})^2}{4} \times 1 \text{ m} = 0.196 \text{ m}^3$$

From equation (2.5), the definition of density,

$$\rho = \frac{m}{V} = \frac{200 \text{ kg}}{0.196 \text{ m}^3} = 1020.4 \text{ kg/m}^3$$

ILLUSTRATION 2.7   DENSITY

Express your answer to Illustrative Problem 2.6 in English units if 1 kg = 2.205 lb.

ILLUSTRATIVE PROBLEM 2.7

Given: $\rho = 1020.4$ kg/m³ (Illustrative Problem 2.6)

Find: ρ in slugs/ft³

Solution: Using the procedure we established in Chapter 1, we will convert the mass and volume separately. Thus, the mass is

$$1 \text{ kg} \times 2.205 \, \frac{\text{lb}}{\text{kg}} \times \frac{\text{slug}}{32.17 \text{ lb}} = \text{slug}$$

$$\therefore \text{kg} \times 0.0685 = \text{slug}$$

and the volume is

$$1 \text{ m}^3 \times \left(100 \, \frac{\text{cm}}{\text{m}} \times \frac{\text{in.}}{2.54 \text{ cm}} \times \frac{\text{ft}}{12 \text{ in.}}\right)^3 = \text{ft}^3$$

$$\therefore 1 \text{ m}^3 \times 35.31 = \text{ft}^3$$

Therefore,

$$\rho = \frac{\text{kg}}{\text{m}^3} \times \frac{0.0685}{35.31} = \frac{\text{slug}}{\text{ft}^3} = 1.94 \times 10^{-3} \frac{\text{slug/ft}^3}{\text{kg/m}^3}$$

and using the results of Illustrative Problem 2.6,

$$\rho = 1020.4 \frac{\text{kg}}{\text{m}^3} \times 1.94 \times 10^{-3} \frac{\text{slug/ft}^3}{\text{kg/m}^3} = 1.98 \frac{\text{slugs}}{\text{ft}^3}$$

Also, knowing that the density of water at 4°C is 1000 kg/m³ or 1.94 slugs/ft³ gives us kg/m³ $\times$ 1.94 $\times$ 10⁻³ = slugs/ft³, which checks our calculations above.

Specific weight is defined as the weight per unit volume of a substance. Denoting the weight of the body as w and its volume as V, the specific weight γ (gamma) is

$$\gamma = \frac{w}{V} \tag{2.6}$$

If we compare equations (2.5) and (2.6), we find that

$$\frac{\gamma}{\rho} = \frac{w}{m} \tag{2.7}$$

But we have from equation (1.2) that $w/m = g$. Therefore,

$$\frac{\gamma}{\rho} = g \quad or \quad \boxed{\gamma = \rho g} \tag{2.8}$$

In English units w is in pounds, V is in cubic feet, and γ has the units of pounds force per cubic foot. In SI units, w is in newtons, V is in m³, and γ has the units N/m³.

The specific weights of mercury and water can be found in Table 8.3 as a function of temperature. Also see Tables 2.4 and 2.5 for the properties of water as a function of temperature.

ILLUSTRATION 2.8   DENSITY AND SPECIFIC WEIGHT

A cube has 6-in. sides. If the cube weighs 20 lb, what is its density and specific weight in both English and SI units?

ILLUSTRATIVE PROBLEM 2.8

Given: A cube 6 in. on a side weighing 20 lb

Find: γ in English and SI units

Solution: The volume of the cube is

$$6 \text{ in.} \times 6 \text{ in.} \times 6 \text{ in.} = 216 \text{ in.}^3 = \frac{216 \text{ in.}^3}{(12 \text{ in./ft})^3} = 0.125 \text{ ft}^3$$

The specific weight is

$$\gamma = \frac{w}{V} = \frac{20 \text{ lb}}{0.125 \text{ ft}^3} = 160 \frac{\text{lb}}{\text{ft}^3}$$

Since $\gamma = \rho g$,

$$\rho = \frac{\gamma}{g} = \frac{160 \text{ lb/ft}^3}{32.17 \text{ ft/s}^2} = 4.97 \text{ slugs/ft}^3$$

In SI units,

$$V = 216 \text{ in.}^3 \times \left(0.0254 \frac{\text{m}}{\text{in.}}\right)^3 = 3.54 \times 10^{-3} \text{ m}^3$$

Using 1 kg = 2.205 lb yields

$$\frac{20 \text{ lb}}{2.205 \text{ lb/kg}} = 9.070 \text{ kg}$$

Therefore,

$$\rho = \frac{m}{V} = \frac{9.070 \text{ kg}}{3.54 \times 10^{-3} \text{ m}^3} = 2562.2 \frac{\text{kg}}{\text{m}^3}$$

Using equation (2.8) yields

$$\gamma = \rho g = 2562.2 \, \frac{kg}{m^3} \times 9.81 \, \frac{N}{kg} = 25.14 \, \frac{kN}{m^3}$$

Specific volume is defined as a reciprocal of density or the volume per unit mass.

$$v = \frac{1}{\rho} \tag{2.9}$$

and the units of specific volume are ft^3/slug or m^3/kg.

The last of these four terms, *specific gravity,* is defined as the ratio of the specific weight or density of a substance to the specific weight or density of water at 4°C (39.2°F). From this definition we have

$$\text{specific gravity (sg)} = \frac{\gamma_{substance}}{\gamma_{water} \text{ at 4°C}} = \frac{\rho_{substance}}{\rho_{water} \text{ at 4°C}} \tag{2.10}$$

For water at 4°C,

$$\gamma_w = 62.4 \, \frac{lb}{ft^3} = 9.81 \text{ kN/m}^3$$

$$\rho_w = 1.94 \, \frac{slugs}{ft^3} = 1000 \text{ kg/m}^3$$

Therefore,

$$sg = \frac{\gamma}{62.4 \text{ lb/ft}^3} \quad \text{or} \quad \frac{\rho}{1.94 \text{ slugs/ft}^3} \tag{2.11}$$

and

$$sg = \frac{\gamma}{9.81 \text{ kN/m}^3} \quad \text{or} \quad \frac{\rho}{1000 \text{ kg/m}^3} \tag{2.12}$$

ILLUSTRATION 2.9   SPECIFIC GRAVITY

One gallon of an unknown substance weighs 7 lb. What is its specific gravity?

ILLUSTRATIVE PROBLEM 2.9

Given: $V = 1$ gallon, $w = 7$ lb

Find: sg

Solution: One gallon is 231 in.^{3} or

$$\frac{231 \text{ in.}^3}{1728 \text{ in.}^3/\text{ft}^3} = 0.1337 \text{ ft}^3$$

At 4°C, the weight of 1 gallon of water is

$$62.4 \frac{\text{lb}}{\text{ft}^3} \times 0.1337 \text{ ft}^3 = 8.343 \text{ lb}$$

Therefore, the specific gravity, by definition, is

$$sg = \frac{7}{8.343} = 0.839$$

ILLUSTRATION 2.10   SPECIFIC GRAVITY

If 10 kg of a substance occupies a volume of 2×10^{-2} m³, what is its specific gravity?

ILLUSTRATIVE PROBLEM 2.10

Given: $m = 10$ kg, $V = 2 \times 10^{-2}$ m³

Find: sg

Solution: The density of the substance is

$$\frac{10 \text{ kg}}{2 \times 10^{-2} \text{ m}^3} = 500 \text{ kg/m}^3$$

From equation (2.12),

$$sg = \frac{\rho}{1000 \text{ kg/m}^3} = \frac{500 \text{ kg/m}^3}{1000 \text{ kg/m}^3} = 0.5$$

Since we will use the properties of water extensively in the rest of this book, Table 2.2 lists the density, specific weight, specific volume, and specific gravity of water in both SI and English units.

Table 2.2 Properties of Water[†]

Property and symbol	Units		Value for water at 4°C	
	SI	English	SI	English
Density, ρ	kg/m³	slug/ft³	1000 kg/m³	1.94 slugs/ft³
Specific weight, γ	N/m³	lb/ft³	9810 N/m³	62.4 lb/ft³
Specific volume, v $v = 1/\rho$	m³/kg	ft³/slug	$\dfrac{\text{m}^3}{1000 \text{ kg}}$	$\dfrac{\text{ft}^3}{1.94 \text{ slugs}}$
Specific gravity, sg	—	—	1.0	1.0

†See also Tables 2.4 and 2.5 for the effect of temperature on density and specific weight.

2.5   PRESSURE

When a gas is confined in a container, molecules of the gas strike the sides of the container, and these collisions with the walls of the container cause the molecules of the gas to exert a force on the walls. Liquids behave in a similar fashion. Although there is a difference in the way a force is applied to a fluid and to a solid, we can define the term *pressure* to equal the force applied to the object divided by the area that is perpendicular to the force. Therefore,

$$p = \frac{F}{A} \tag{2.13}$$

It is sometimes necessary to define the pressure at a point. To do this mathematically, it is necessary to consider the area to be steadily shrinking. We now define pressure at a point to be the normal (perpendicular) force per unit area as the area shrinks to zero. Mathematically, this statement is given as

$$p = \left(\frac{F}{A}\right)_{\lim A \to 0} \tag{2.14}$$

In common English engineering, pressure is expressed in units of pounds force per square inch (psi) or pounds force per square foot (psf). In SI units, pressure is expressed as newtons per square meter (N/m^2) or pascals (Pa). One important observation must be made at this time: *When a fluid is at rest, the pressure at any boundary exerted by the fluid (and on the fluid) will be perpendicular to the boundary.*

Most mechanical pressure gages read zero when open to the atmosphere and, therefore, measure the pressure of fluids *above* local atmospheric pressure. This pressure is called *gage pressure* and is commonly designated in the English system as psig or psfg. *Absolute pressure* is the pressure in a system that would be measured from a base of a perfect vacuum at zero absolute pressure. Absolute pressure is designated as psia or psfa. Figure 2.3 shows the relationship between gage pressure and absolute pressure. The relation of gage pressure to absolute pressure is readily written as

$$\text{absolute pressure } (p_a) = \text{gage pressure } (p_g) + \text{atmospheric pressure } (p_{\text{atm}}) \tag{2.15}$$

ILLUSTRATION 2.11   PRESSURE

A pressure gage reads 20 psig when the local atmospheric pressure is 14.5 psia. What is the absolute pressure corresponding to this reading?

ILLUSTRATIVE PROBLEM 2.11

Given: p = 20 psig, p_{atm} = 14.5 psia

Find: p_a, psia

Solution: From Figure 2.3 or equation (2.15),

$$p_a = 20 + 14.5 = 34.5 \text{ psia}$$

In SI units the equivalent of 14.696 (14.7) psia is 101.325 kPa. All pressures in the SI system are absolute, and the terms "gage" and "vacuum" are not used. A unit that is sometimes used in conjunction with the SI system is the *bar.* The bar is equal to 10^5 Pa. Although use of the bar does reduce the powers of 10 in a calculation, it is not a recommended unit.

When pressure is measured below atmospheric pressure, it is commonly called *vacuum.* By referring to Figure 2.3, we obtain the relation between vacuum and absolute pressure as

$$\text{absolute pressure } (p_a) = \text{atmospheric pressure } (p_{atm}) - \text{vacuum } (p_g) \qquad (2.16)$$

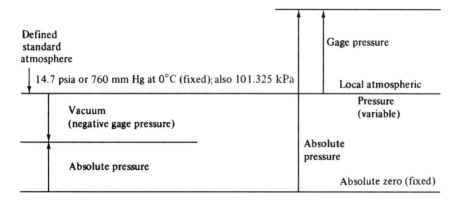

Figure 2.3

ILLUSTRATION 2.12   PRESSURE

If the atmospheric pressure is 14.7 psia and a vacuum gage reads a vacuum of 13.0 psig, what is the absolute pressure?

ILLUSTRATIVE PROBLEM 2.12

Given: $p_{atm} = 14.7$ psia, vacuum $= 13.0$ psig

Find: p_a, psia

Solution: From Figure 2.3 or equation (2.16), we obtain

$$p_a = 14.7 - 13.0 = 1.7 \text{ psia}$$

The *Bourdon gage* is the device most commonly used to measure pressure commercially. This type of gage is shown in Figure 2.4. It typically consists of a small-volume oval tube that is fixed at one end and is free at the other end to allow displacement under the deforming action of the pressure difference across the tube walls. In the most common model, shown in Figure 2.4, a tube with an oval cross section is bent in a circular arc. Under pressure the tube tends to become circular, with a subsequent increase in the radius of the arc. By an almost frictionless linkage, the free end of the tube rotates a pointer over a calibrated scale to give a mechanical indication of pressure.

The reference pressure in the case containing the Bourdon tube is usually atmospheric, so that the pointer indicates gage pressures. Absolute pressures can be measured directly without evacuating the complete gage casing by biasing a sensing Bourdon tube against a reference Bourdon tube that is evacuated and sealed as shown in Figure 2.5. Bourdon gages are available for a wide range of absolute, gage, and differential pressure measurements within a calibration uncertainty of 0.1% of the reading.

Very often pressure difference rather than pressure is desired. In this case it does not matter whether absolute or gage pressure is used, since *pressure difference is the same in either gage or absolute units.*

Columns of liquids are often used to measure pressure or pressure differences. So at this time let us address ourselves to the following problem. What will the pressure be at the base of a column of liquid due to the liquid if its specific weight is constant and equal to γ? Assume that the cross-sectional area is constant and equal to A and that the height of the fluid is h above the base of the column, as shown in Figure 2.6. From our definition of pressure as force divided by area, we must first find the force being exerted on the base of the column of fluid. This force will be equal to the weight of the fluid, which is given by equation (2.6) as $w = \gamma V$. The volume of fluid is $V = Ah$ and the weight is $w = \gamma Ah$. This is the force exerted on the base. If we divide this force by the area of the base, we obtain the pressure:

$$p = \frac{\gamma Ah}{A}$$

$$p = \gamma h \tag{2.17}$$

Equation (2.17) will be used extensively in Chapter 3 and should be thoroughly understood. Notice that as you go from the top of the column down, the pressure increases, or conversely, as you go up from the bottom of the column the pressure decreases. These changes in pressure are directly proportional to the changes in elevation. Also note that we have assumed that γ, the specific weight, is constant. If distance is measured up from the base, equation (2.17) should be written as

$$p = -\gamma h \tag{2.18}$$

Figure 2.7 shows a water tower where the water is pumped to the top and stored to maintain the pressure on the system and to have spare capacity.

The most common fluid other than water used to measure pressure differences, atmospheric pressure, and vacuum pressure is mercury. At approximately room temperature, the density of mercury is very nearly 13.6 g/cm^3. It will usually be assumed that the density of

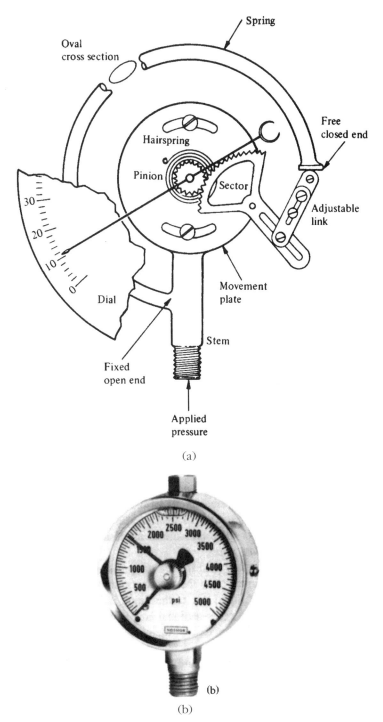

Figure 2.4   Bourdon gage. [(b) Courtesy of NOSHOK, Inc.]

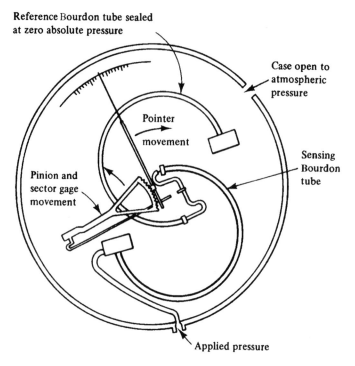

Figure 2.5 Bourdon gage for absolute pressure measurement.

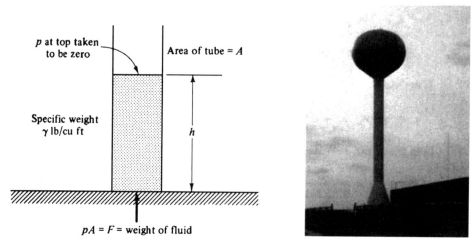

Figure 2.6 Pressure–height derivation.

Figure 2.7 Water tower.

mercury is 13.6 g/cm³ unless otherwise stated. The effect of temperature on the specific weight of mercury is given in Chapter 8.

ILLUSTRATION 2.13   FLUID PRESSURE

A glass tube open at the top contains a 1-in. high column of mercury. If the density of mercury is 13.6 g/cm³ = 13 600 kg/m³, determine the pressure at the base of the column in psi and Pa.

ILLUSTRATIVE PROBLEM 2.13

Given: ρ = 13.6 g/cm³, h = 1 in.

Find: p in psi and Pa

Solution: The specific weight is

$$\gamma = \rho g = 13\ 600\ \frac{\text{kg}}{\text{m}^3} \times 9.81\ \frac{\text{N}}{\text{kg}} = 133\ 416\ \text{N/m}^3$$

Therefore,

$$p = \gamma h = 133\ 416\ \frac{\text{N}}{\text{m}^3} \times 1\ \text{in.} \times 0.0254\ \frac{\text{m}}{\text{in.}} = 3388.8\ \frac{\text{N}}{\text{m}^2} = 3388.8\ \text{Pa}$$

In terms of psi,

$$3388.8\ \frac{\text{N}}{\text{m}^2} \times \frac{1}{9.81\ \text{N/kg}} \times 2.205\ \frac{\text{lb}}{\text{kg}} \times \left(\frac{\text{m}}{100\ \text{cm}} \times \frac{2.54\ \text{cm}}{\text{in.}}\right)^2 = 0.491\ \text{psi}$$

Thus, from Illustrative Problem 2.13 a column height of 1 in. of mercury is equivalent to a pressure of 0.491 psi. Notice that this enables us to express pressures in terms of equivalent heights of fluids. We will find this useful in Chapter 3.

In vacuum work it is common to express the absolute pressure in a vacuum chamber in terms of millimeters of mercury. Thus, a vacuum may be expressed as 10^{-5} mm Hg. If this is expressed in pounds force per square inch, it would be equivalent to 0.000000193 psi, with the assumption that the density of mercury (for vacuum work) is 13.6 g/cm³. The term *torr* is often found in the technical literature. A torr is defined as 1 mm Hg. Thus, 10^{-5} torr is the same as 10^{-5} mm Hg. Another unit of pressure used in vacuum work is the *micron*. A micron is defined as one-thousandth of 1 mm Hg, so that 10^{-3} mm Hg is equal to 1 micron, or 1 micron equals 10^{-6} m of mercury.

ILLUSTRATION 2.14   PRESSURE

A mercury manometer (vacuum gage) reads 26.5 in. of vacuum when the local barometer reads 30.0 in. Hg at standard temperature. Determine the absolute pressure in psia.

ILLUSTRATIVE PROBLEM 2.14

Given: Barometer = 30.0 in. Hg, vacuum = 26.5 in. of vacuum

Find: p_a, psia

Solution:

$$p = (30.0 - 26.5) \text{ in. Hg vacuum}$$

and from the preceding problem,

$$p = (30 - 26.5)0.491 = 1.72 \text{ psia}$$

If the solution is desired in psfa, it is necessary to convert from square inches to square feet. Since there are 12 in. in 1 ft, there will be 144 in.^{2} in 1 ft^2. Thus, the conversion to psfa requires that psia be multiplied by 144. This conversion is often necessary to keep the dimensions of equations consistent in English units.

2.6   SURFACE TENSION

When a liquid has a free surface, there is a discontinuity that takes place at the surface, since there is a liquid on one side of the surface and a different liquid or gas on the other side of the surface. For example, consider a drop of liquid in air. Within the drop, each molecule of liquid experiences forces due to the other liquid molecules. At the surface of the drop the liquid molecules experience forces due to both the air and the liquid molecules. If the drop is to maintain its shape, the attraction of the liquid molecules must exceed the attraction between the liquid molecules and air molecules. Effectively, this is the same as if the liquid were enclosed by a stretchable membrane under tension. Figure 2.8 shows a razor blade floating on water due to this effect. The force of this *surface tension* is always tangent to the interface. A liquid is said to wet a surface in contact with it if the attraction of the molecules for the surface exceeds the attraction of the molecules for each other; on the other hand, the liquid is classified as a nonwetting liquid if the attraction of other liquid molecules for each other is greater than their attraction to the surface. Figure 2.9 shows the action of wetting and nonwetting fluids.

Quantitatively, surface tension can be defined as the force of molecular attraction per unit length of free surface. It should be noted that surface tension is a function of both the liquid and the surface in contact with the liquid. Denoting surface tension by σ, we have, in general,

$$\sigma = \frac{F}{L} \tag{2.19}$$

Since the forces of attraction between molecules decrease with increasing temperature, it is found that surface tension also decreases with increasing temperature. Table 2.3 gives the surface tension of some fluids in contact with air at 68°F. Other data on surface tension will be found in Chapter 8 and Appendix B, Tables B.6 and B.7.

Figure 2.8   Effect of surface tension.

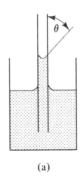 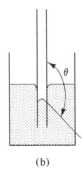

(a)   (b)

Figure 2.9   (a) Wetting and (b) nonwetting fluids.

Table 2.3 Surface Tension of Common Liquids in Contact with air at 68°F[†]

	Surface tension, σ	
	lb/ft	N/m
Alcohol, ethyl	0.00153	2.23×10^{-2}
Benzene	0.00198	2.89×10^{-2}
Carbon tetrachloride	0.00183	2.67×10^{-2}
Kerosene	0.0016−0.0022	$2.34 - 3.21 \times 10^{-2}$
Water	0.00498	7.27×10^{-2}
Mercury		
In air	0.0352	5.14×10^{-1}
In water	0.0269	3.93×10^{-1}
In vacuum	0.0333	4.86×10^{-1}
Oil		
Lubricating	0.0024−0.0026	$3.50 - 3.79 \times 10^{-2}$
Crude	0.0016−0.0026	$2.34 - 3.79 \times 10^{-2}$

Source: Reproduced with permission from *Fluid Mechanics* by V. L. Streeter, McGraw-Hill Book Company, New York, 1958, p. 14. SI units added.

[†]See Tables B.6 and B.7 for other data on surface tension.

Let us calculate the rise h shown in Figure 2.10a for the case of a wetting fluid in a circular tube. The basic consideration here is that the weight of the column of height h must be supported by an equal force in the vertical direction, where the source of this supporting force must be the vertical component of the surface tension. Therefore, the force downward (equal to the weight of the column of height h) is $\gamma h(\pi d^2/4)$, where γ is the specific weight of the fluid and d is the inside diameter of the tube. The total force due to the surface tension is the surface tension multiplied by the perimeter of the tube: $\sigma \pi d$. Since, however, we

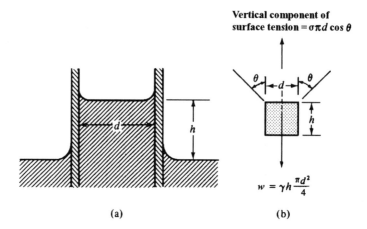

Figure 2.10   Surface tension for a wetting liquid.

only desire the vertical component of this force, it is necessary to multiply the total force by $\cos \theta$. Equating the two vertical forces, we obtain

$$\gamma h \frac{\pi d^2}{4} = \sigma \pi d \cos \theta$$

or

$$h = \frac{4\sigma \cos \theta}{\gamma d} = \frac{4\sigma \cos \theta}{\rho g d} \tag{2.20}$$

Water in contact with glass and air has a contact angle whose value is close to zero, while mercury in contact with glass and air has a contact angle of 129°, yielding a depression of the surface.

If we consider two infinite parallel plates instead of a circular tube, the equation of equilibrium becomes (for a unit depth)

$$\gamma h d = 2\sigma \cos \theta \tag{2.21}$$

or

$$h = \frac{2\sigma \cos \theta}{\gamma d} = \frac{2\sigma \cos \theta}{\rho g d} \tag{2.22}$$

Notice that the rise in the tube is twice the rise between the parallel plates.

ILLUSTRATION 2.15   SURFACE TENSION

A glass tube that has an inside diameter of 1.5×10^{-3} m is placed in a fluid whose contact angle is 10°. If the surface tension for this liquid in contact with glass is 7.3×10^{-2}

N/m, determine the rise of the liquid in the tube above the normal liquid level. Assume that the density of the fluid is 800 kg/m³.

ILLUSTRATIVE PROBLEM 2.15

Given: $d = 1.5 \times 10^{-3}$ m, $\theta = 10°$, $\sigma = 7.3 \times 10^{-2}$ N/m, $\rho = 800$ kg/m³

Find: h

Solution: The specific weight is

$$\gamma = \rho g = 800 \frac{kg}{m^3} \times 9.81 \frac{N}{kg}$$

and from equation (2.20),

$$h = \frac{4\sigma \cos \theta}{\gamma d}$$

$$= \frac{4 \times 7.3 \times 10^{-2} \text{ N/m} \times \cos 10°}{800 \text{ kg/m}^3 \times 9.81 \text{ N/kg} \times 1.5 \times 10^{-3} \text{ m}} = 0.0245 \text{ m} = 24.5 \text{ mm}$$

If the problem had specified two parallel plates rather than a tube, the rise would have been one-half of the value for the rise in a tube.

2.7   COMPRESSIBILITY

When a liquid is confined, it will offer resistance to any external agency that tends to change its volume. This resistance to change in volume is called the *compressibility* of the fluid. The change in volume that occurs is measured as a function of the original volume and the applied pressure. Mathematically, the *modulus of elasticity* or *bulk modulus* of the fluid at a point is defined to be the ratio of the stress (change in pressure) to the strain (change in volume divided by the original volume), or

$$\beta = -\frac{\Delta p}{\Delta V/V} \tag{2.23}$$

where V is the original volume, ΔV is the change in volume, and Δp is the change in pressure. The negative sign accounts for the fact that as the pressure increases, the volume decreases. It is usually assumed that the bulk modulus is insensitive to pressure but does vary with temperature. For water the bulk modulus is approximately 300,000 psi. Since steel has a modulus of elasticity of 30 million psi it is seen that water is relatively compressible. The physical properties of water are given in Tables 2.4 and 2.5 as a function of temperature. Figure 2.11 gives the variation of the bulk modulus of water with pressure and temperature.

Table 2.4 Physical Properties of Water in English Units

Temp. (°F)	Specific weight, γ (lb/ft^3)	Density, ρ (slugs/ft^3)	Viscosity, $10^5\mu$ (lb·s/ft^2)	Kine-matic viscosity $10^5\nu$ (ft^2/s)	Surface tension, 100σ (lb/ft)	Vapor pressure head, p_v/γ (ft)	Bulk modulus of elasticity, $10^{-3}\beta$ (lb/in.^{2})
32	62.42	1.940	3.746	1.931	0.518	0.20	293
40	62.43	1.940	3.229	1.664	0.514	0.28	294
50	62.41	1.940	2.735	1.410	0.509	0.41	305
60	62.37	1.938	2.359	1.217	0.504	0.59	311
70	62.30	1.936	2.050	1.059	0.500	0.84	320
80	62.22	1.934	1.799	0.930	0.492	1.17	322
90	62.11	1.931	1.595	0.826	0.486	1.61	323
100	62.00	1.927	1.424	0.739	0.480	2.19	327
110	61.86	1.923	1.284	0.667	0.473	2.95	331
120	61.71	1.918	1.168	0.609	0.465	3.91	333
130	61.55	1.913	1.069	0.558	0.460	5.13	334
140	61.38	1.908	0.981	0.514	0.454	6.67	330
150	61.20	1.902	0.905	0.476	0.447	8.58	328
160	61.00	1.896	0.838	0.442	0.441	10.95	326
170	60.80	1.890	0.780	0.413	0.433	13.83	322
180	60.58	1.883	0.726	0.385	0.426	17.33	313
190	60.36	1.876	0.678	0.362	0.419	21.55	313
200	60.12	1.868	0.637	0.341	0.412	26.59	308
212	59.83	1.860	0.593	0.319	0.404	33.90	300

Source: Reprinted from *Fluid Mechanics* by V. L. Streeter and E. B. Wylie, 7th ed., 1979, p. 534, by permission of McGraw-Hill Book Company, New York.

ILLUSTRATION 2.16   COMPRESSIBILITY

Determine the percentage change in the volume of water if its pressure is increased by 30,000 psi.

ILLUSTRATIVE PROBLEM 2.16

Given: Water, $\Delta p = 30{,}000$ psi

Find: $\Delta V/V$

Solution: From equation (2.23),

$$\beta = -\frac{\Delta p}{\Delta V/V}$$

or, since the bulk modulus of water is 300,000 psi and the change in volume is $\Delta V/V$, we can write

$$\frac{\Delta V}{V} = -\frac{\Delta p}{\beta} = \frac{-30,000}{300,000} = -0.1 \text{ or } -10\%$$

Table 2.5 Physical Properties of Water in SI Units

Temp. (°C)	Specific weight, γ (N/m³)	Density, ρ (kg/m³)	Viscosity, $10^3\mu$ (N · s/m²)	Kinematic viscosity $10^6\nu$ (m²/s)	Surface tension, 100σ (N/m)	Vapor pressure head, p_v/γ (m)	Bulk modulus of elasticity, $10^{-7}\beta$ (N/m²)
0	9805	999.9	1.792	1.792	7.62	0.06	204
5	9806	1000.0	1.519	1.519	7.54	0.09	206
10	9803	999.7	1.308	1.308	7.48	0.12	211
15	9798	999.1	1.140	1.141	7.41	0.17	214
20	9789	998.2	1.005	1.007	7.36	0.25	220
25	9779	997.1	0.894	0.897	7.26	0.33	222
30	9767	995.7	0.801	0.804	7.18	0.44	223
35	9752	994.1	0.723	0.727	7.10	0.58	224
40	9737	992.2	0.656	0.661	7.01	0.76	227
45	9720	990.2	0.599	0.605	6.92	0.98	229
50	9697	988.1	0.549	0.556	6.82	1.26	230
55	9679	985.7	0.506	0.513	6.74	1.61	231
60	9658	983.2	0.469	0.477	6.68	2.03	228
65	9635	980.6	0.436	0.444	6.58	2.56	226
70	9600	977.8	0.406	0.415	6.50	3.20	225
75	9589	974.9	0.380	0.390	6.40	3.96	223
80	9557	971.8	0.357	0.367	6.30	4.86	221
85	9529	968.6	0.336	0.347	6.20	5.93	217
90	9499	965.3	0.317	0.328	6.12	7.18	216
95	9469	961.9	0.299	0.311	6.02	8.62	211
100	9438	958.4	0.284	0.296	5.94	10.33	207

Source: Reprinted from *Fluid Mechanics* by V. L. Streeter and E. B. Wylie, 7th ed., 1979, p. 535, by permission of McGraw-Hill Book Company, New York.

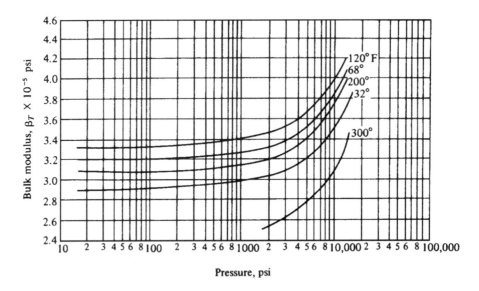

Figure 2.11   Variation of the bulk modulus of water with pressure and temperature. (Reproduced with permission from *Fluid Mechanics* by Arthur G. Hansen, John Wiley & Sons, Inc., New York, 1967, p. 487.)

2.8   VISCOSITY

Because of the forces that exist between the molecules of a fluid, fluids show the ability to resist forces that tend to change the shape of the fluid. If a shearing force is applied to a fluid that is at rest, the fluid will deform as one layer is caused to slide over an adjacent layer with a different velocity. All fluids, be they liquids or gases, exhibit this property of resisting motion by the shearing of adjacent layers. The property of a fluid that expresses its ability to resist shear is called its *viscosity.*

Consider the situation shown in Figure 2.12. The upper plate is moving to the right with a velocity V relative to the lower plate. If the viscosity is constant, the variation of velocity between the plates will be linear. The particle in contact with the bottom plate will not be moving, while the particle touching the upper plate will be moving with the velocity

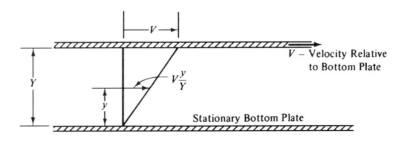

Figure 2.12   Viscosity.

of the upper plate, V. As shown, the velocity anywhere between the plates will be simply given by Vy/Y.

To determine the relationship between the variables involved, let us conduct three "thought" experiments:

1. If the distance between the plates and the reactive velocity of the upper plate to the lower plate are both kept constant, the force required will be directly proportional to the area of the moving plate, or

$$F \propto A \qquad (2.24)$$

2. If the distance between the plates is kept constant and the area of the moving plate is kept constant, the force required will be directly proportional to the relative velocity of the plates, that is,

$$F \propto V \qquad (2.25)$$

3. If the relative velocity and the area of the moving plate are kept constant, but the distance between the plates is varied, the force required will be found to be inversely proportional to the distance between the plates. This arises from the fact that as the distance between the plates decreases, the shear stress will increase. We write this as

$$F \propto \frac{1}{Y} \qquad (2.26)$$

Combining equations (2.24) through (2.26) yields

$$F \propto \frac{AV}{Y} \qquad (2.27)$$

Introducing the proportionality constant μ gives us

$$F = \mu \frac{AV}{Y} \qquad (2.28)$$

We can also note that in general force/area (F/A) is a stress and that in this case F/A is a shear stress. If we now denote F/A, the shear stress, in equation (2.28) as τ

$$\tau = \mu \left(\frac{V}{Y} \right) \qquad (2.29)$$

The proportionality constant μ in equations (2.28) and (2.29) is called the *coefficient of viscosity, dynamic viscosity, absolute viscosity,* or simply the *viscosity of the fluid*.

Note that the term $\frac{V}{Y}$ in equation (2.29) represents the assumed linear velocity distribution in the fluid. It is also the rate of the shearing strain, since during each unit of time there is an angular change equal to $\frac{V}{Y}$. In words, we can use this to define μ from equation (2.29) as

$$\mu = \frac{\text{shearing stress}}{\text{rate of shearing strain}} \qquad (2.29a)$$

Fluids that have the linear characteristic that viscosity is proportional to the ratio of the shearing stress to the shearing strain area are called newtonian fluids. Most gases and simple fluids are newtonian fluids and behave according to equation (2.29a). Pastes, greases and slurries are examples of fluids that do not behave as newtonian fluids.

If we refer to either equation (2.28) or (2.29), we see that viscosity must have the dimensions of force per unit area multiplied by time. The basic units of viscosity in the SI system are $N \cdot s/m^2$, $Pa \cdot s$, or $kg/m \cdot s$. In the English system the customary units are $lb \cdot s/ft^2$ or $slug/ft \cdot s$. Since much of the data on viscosity in reference tables is based on the older, now obsolete centimeter, gram, second (cgs) system, we need to note that the *poise,* the unit of viscosity in this system, has the units $dyne \cdot s/cm^2$ or $g/cm \cdot s$. It is found that at 68.4°F, the viscosity of water is 1 *centipoise* (1 cP), where the centipoise is one-hundredth of a poise. The unit of viscosity in the SI system is not given a name; however, 1 P equals $\frac{1}{10}$ of the SI viscosity unit. Table 2.6a gives some convenient conversion factors for viscosity.

Viscosity in fluids is due to either the attraction of the molecules of the fluid or the transfer of momentum between the molecules. If the molecules of a fluid attract each other, they tend to resist the flow of one layer of fluid over an adjacent layer. This gives rise to the shear forces that are observed in liquids. As the temperature of a liquid is increased, molecular attraction decreases. Thus we expect that *the viscosity of all liquids decreases as the temperature is increased.* Figure 2.13 shows that the viscosity of all liquids does indeed decrease as the temperature is increased. In the case of gases, the interaction between fast- and slow-moving molecules causes adjacent layers to accelerate and decelerate with respect to each other. This causes the observed viscosity effects in gases. As the temperature of a gas is increased, the molecular motion is increased, which in turn causes an increase in the viscosity. Thus we conclude that the *viscosity of all gases increases as the temperature increases.* This effect is also shown in Figure 2.13. The effect of pressure on the viscosity of fluids is small and is usually neglected.

Table 2.6a Viscosity Conversion Factors

1 P = 0.1 Pa·s

1 cP = 0.001 Pa·s

1 lb·s/ft² = 478.8 P

1 lb·s/ft² = 47.88 Pa·s

1 P = 0.002089 lb·s/ft²

1 Pa·s = 2.089 × 10⁻² lb·s/ft²

1 cP = 2.09 × 10⁻⁵ lb·s/ft²

Table 2.6b   Summary of Viscosity Systems

System	Viscosity Units
SI	N·s/m², Pa·s, kg/m·s
U.S. Customary System (USCS)	lb·s/ft², slug/ft·s
cgs	poise = dyne·s/cm², g/cm·s

ILLUSTRATION 2.17   VISCOSITY

The viscosity of mercury at 68°F is 1.58×10^{-2} P, that is, 1.58×10^{-3} N·s/m². Determine the force necessary to maintain a relative velocity of 2 m/s between two plates that are separated by 0.1 m and whose area is 0.1 m². Consider only viscous effects. Refer to Figure 2.12. *(continues on page 50)*

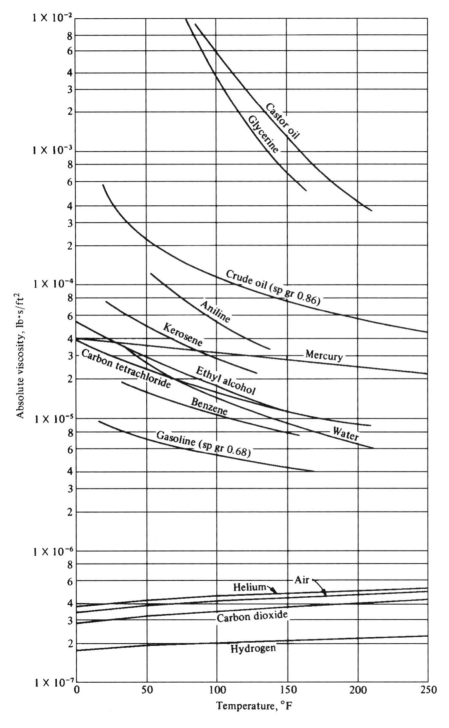

Figure 2.13 Absolute viscosities of certain gases and liquids. (Reproduced with permission from *Fluid Mechanics* by V.L. Streeter, McGraw-Hill Book Company, New York, 1962, p. 534.)

ILLUSTRATIVE PROBLEM 2.17

Given: $\mu = 1.58 \times 10^{-3}$ Pa·s, $Y = 0.1$ m, $A = 0.1$ m², $V = 2$ m/s

Find: F

Solution: From equation (2.28),

$$F = \mu \frac{AV}{Y}$$

Therefore,

$$F = \mu \frac{AV}{Y} = \frac{1.58 \times 10^{-3} \text{ N·s/m}^2 \times 0.1 \text{ m}^2 \times 2 \text{ m/s}}{0.1 \text{ m}} = 0.00316 \text{ N}$$

ILLUSTRATION 2.18   VISCOSITY

A solid cylinder having a mass of 1 kg slides down a pipe as indicated in Figure 2.14. An oil having the viscosity of 8×10^{-2} N·s/m² keeps the cylinder concentric in the pipe. Determine the steady-state speed of the cylinder, considering only the effects of viscosity.

ILLUSTRATIVE PROBLEM 2.18

Given: $d_o = 0.062$ m, $d_i = 0.060$ m, $\mu = 8 \times 10^{-2}$ N·s/m², m = 1 kg

Find: V

Solution: As we see from Figures 2.14b and 2.14c, the free-body diagrams, the cylinder will reach equilibrium when the weight of the cylinder just equals the summation of

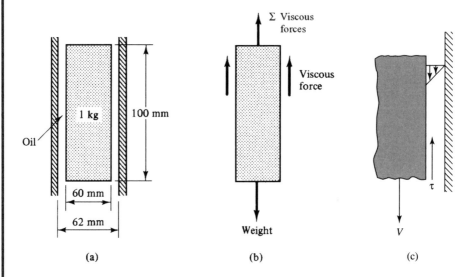

Figure 2.14   (a) Illustrative Problem 2.18. (b) Free-body diagram. (c) Velocity profile and shear stress.

the viscous forces that resist its motion. Assuming that the velocity is linear in the annulus between the cylinder and the pipe, we have

$$\text{weight} = 1 \text{ kg} \times 9.81 \frac{N}{kg} = 9.81 \text{ N}$$

$$\text{area} = \text{surface area of the cylinder}$$

$$= \pi dL = \pi \times 0.06 \text{ m} \times 0.1 \text{ m} = 0.0188 \text{ m}^2$$

$$\text{spacing: } Y = \frac{d_o - d_i}{2}$$

$$= \frac{0.062 \text{ m} - 0.060 \text{ m}}{2} = 0.001 \text{ m}$$

Since the sum of the forces in the vertical (y) direction must equal zero,

$$\text{viscous force} = \text{weight of cylinder}$$

$$\mu \frac{AV}{Y} = \text{weight of cylinder}$$

$$8 \times 10^{-2} \frac{N \cdot s}{m^2} \times 0.0188 \text{ m}^2 \times \frac{V}{0.001 \text{ m}} = 9.81 \text{ N}$$

Solving gives us

$$V = 6.52 \text{ m/s}$$

It is sometimes convenient to combine viscosity and density; viscosity divided by density is known as *kinematic viscosity*. It is usually denoted as ν. In the metric system the name *stoke* (or one-hundredth of its value, the *centistoke*) is used to designate kinematic viscosity; a centistoke is 1 cP divided by a density of 1 g/cm³. Figure 2.15 shows the kinematic viscosity of certain gases and liquids, where the gases are at atmospheric pressure. By definition,

$$\nu = \frac{\mu}{\rho} \tag{2.30}$$

In the SI system,

$$\nu = \frac{N \cdot s/m^2}{N \cdot s^2/m^4} = \frac{m^2}{s}$$

and in the English system,

$$\nu = \frac{lb \cdot s}{ft^2} \Big/ \frac{lb \cdot s^2}{ft^4} = \frac{ft^2}{s}$$

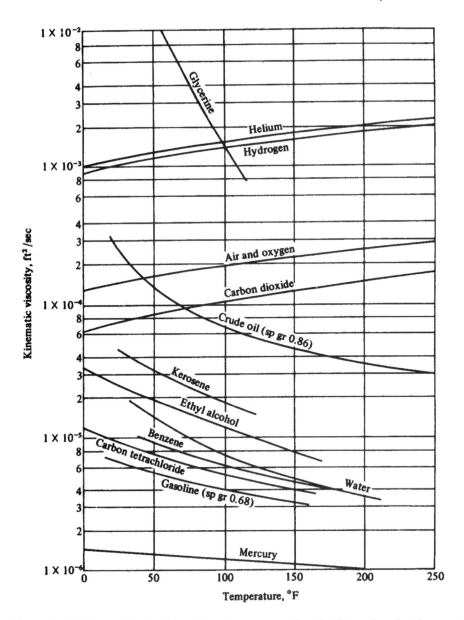

Figure 2.15   Kinematic viscosities of certain gases and liquids. (Reproduced with permission from *Fluid Mechanics* by V.L. Streeter, McGraw-Hill Book Company, New York, 1962, p. 535.)

The centistoke can be converted to SI units by noting that ν in centistokes $\times\ 10^{-6} = \nu$, SI. Data for the absolute and kinematic viscosity of water are given in Appendix B.5. Table 2.7 gives some useful kinematic viscosity conversions.

The stoke has the units of cm^2/s, with 1 stoke = 100 centistokes. The conversion from centistokes to the SI unit of m^2/s is found by the conversion $(cm^2/100\ s) \times (m/100\ cm)^2 = m^2/s$, or ν centistokes $\times\ 10^{-6} = \nu$, SI. Alternately, ν SI $\times\ 10^6 = \nu$ centistokes. The conversion from ν ft^2/s to centistokes is given as ν ft^2/s = 92,903 centistokes.

Table 2.7 Kinematic Viscosity Conversions

$ft^2/s \times 9.290 \times 10^{-2} = m^2/s$

$ft^2/s \times 929.0 = stoke$

$m^2/s \times 10.764 = ft^2/s$

$m^2/s \times 10^4 = stoke$

$stoke \times 1.076 \times 10^{-3} = ft^2/s$

$stoke \times 10^{-4} = m^2/s$

$centistoke \times 1.076 \times 10^{-5} = ft^2/s$

$centistoke \times 10^{-6} = m^2/s$

$centistoke \times 10^{-2} = stoke$

ILLUSTRATION 2.19   VISCOSITY

For Figure 2.16, calculate
a) the absolute viscosity, lb·s/ft²
b) the kinematic viscosity, m²/s

ILLUSTRATIVE PROBLEM 2.19

Given: μ = 0.6 cP, sg = 0.85

Find: μ (lb·s/ft²) and v (m²/s)
       Note: V and Y are not needed

Solution: a) μ = 0.6 cP $\times$ 2.09 $\times$ 10⁻⁵ (lb·s/ft²)/cP = 1.254 $\times$ 10⁻⁵ lb·s/ft²
b) Since $v = \mu/\rho$, $v = \mu/(sg \times \rho_w)$
Therefore $v = \mu$ (lb·s/ft²)/(0.85 $\times$ 1.94 slug/ft³)
However, slug = lb/(ft/s²) = lb·s²/ft, which leads to

$$1 = lb·s^2/slug·ft$$

The desired conversion proceeds in the following manner:

$$\mu = \left(\frac{lb·s/ft^2}{0.85 \times 1.94 \ slug/ft^3}\right) \times \frac{slug \cdot ft}{lb \cdot s^2} \times \left(\frac{12 \ in.}{ft}\right)^2 \times \left(\frac{2.54 \ cm}{in.}\right)^2\left(\frac{m}{100 \ cm}\right)^2$$

$$= 7.06 \times 10^{-7} \frac{m^2}{s}$$

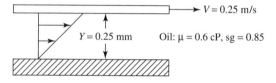

$V = 0.25$ m/s

$Y = 0.25$ mm   Oil: μ = 0.6 cP, sg = 0.85

Figure 2.16   Illustrative Problem 2.19

There are many methods of measuring viscosity or related quantities that can be compared to viscosity. In general, the devices used to make viscosity measurements (known as *viscometers* or *viscosimeters*) fall into three categories: *rotational, falling sphere,* or *flow-type* (capillary) units. Because of the effects of walls, concentricity, finite size, and so on, it is almost impossible to obtain analytically the constants of a given device. In almost every instance, the units are calibrated against liquids of known viscosity.

The Saybolt viscometer (shown in Figure 2.17) is the most commonly used device for the determination of the absolute viscosity of liquids. It consists basically of a temperature-controlled bath into which is placed a sample of the fluid to be tested. At the base of the device there is a standard orifice. The time required for 60 cm^3 of liquid to flow through the orifice and be collected is recorded. This time, known as Saybolt Seconds Universal, SSU, is a measure of the viscosity of the liquid. The relationship between SSU and kinematic viscosity is given by

    a) for $t < 100$ s, SSU = $0.226t - 195/t$ centistokes
    b) for $t > 100$ s, SSU = $0.220t - 135/t$ centistokes

The Society of Automotive Engineers has a system of rating lubricants that indicates the viscosity at a given temperature, and is related to viscosities in SSU. SAE viscosity numbers having the suffix "W" are based on oil viscosities at 0°F (−18°C). SAE ratings without the "W" are based on viscosities at 210°F (99°C). An oil having more than one viscosity number, such as 10W–40, meets the specifications at both temperatures.

For lubricating oils and for hydraulic fluids that operate through wide temperature ranges it is very important to know how the viscosity of the fluid changes with temperature. The viscosity index (VI) is an empirical criterion that is used to measure the viscosity change of a fluid with temperature. Viscosity index is found by measuring the viscosity of a sample of the fluid at 40°C and 100°C, and comparing these values with those of reference fluids. Oils having a low viscosity index show a large change of viscosity with temperature change, while oils having a high viscosity index show a small change of viscosity with change of temperature.

2.9   REVIEW

In this chapter we have studied and in some cases reviewed certain properties of fluids that we will need in our subsequent studies. These include temperature, density, specific weight, specific volume, specific gravity, pressure, surface tension, compressibility, and viscosity. These properties help us to establish the state of a given system and to understand the behavior of fluid systems. In many cases it is necessary to know the definition of the property

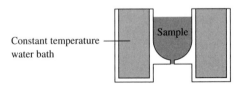

Figure 2.17   Saybolt Viscometer

or the specific behavior of the fluid that gives rise to the property. For example, the common scales of temperature are defined by use of the freezing point and boiling point of water at atmospheric pressure; surface tension causes the surface of a fluid to behave as if it is a stretched membrane; compressibility measures the resistance of a fluid to a change in its volume; and viscosity is a measure of the ability of a fluid to resist shear forces. Although the first two chapters in this book are primarily introductory in nature, the student should master this material in order to complete the material successfully in later chapters.

KEY TERMS

Terms of importance in this chapter:

Absolute pressure: pressure measured above a base of absolute zero of pressure.

Absolute temperature: temperature measured above a base of absolute zero, corresponding to $-459.69°F$ and $-273.16°C$.

Atmospheric pressure: the pressure exerted by the atmosphere—usually measured using a barometer.

Bourdon gage: a mechanical pressure gage that uses a bent, shaped element and mechanical linkages to give an analog output of pressure.

Bulk modulus (β):   the ratio of the stress (change in pressure) to the strain (change in volume divided by the original volume) of a fluid.

Celsius temperature scale: a temperature scale based on dividing the interval between the ice point and the boiling point of water at atmospheric pressure into 100 equal divisions.

Centipoise: one-hundredth of a poise—a cgs measure of viscosity.

Centistoke: one-hundredth of a stoke—a cgs measure of kinematic viscosity.

Compressibility: the resistance of a fluid to a change in its volume.

Density (ρ):   the mass per unit volume of a substance.

Fahrenheit temperature scale: a temperature scale based on dividing the interval between the ice point (set at 32) and the boiling point of water at atmospheric pressure into 180 equal divisions.

Gage Pressure: pressure measured above atmospheric pressure.

Kelvin temperature scale: the absolute temperature scale corresponding to the Celsius temperature scale: $K = °C + 273$.

Kinematic viscosity (v):   the ratio of the absolute viscosity divided by the density of a fluid: $\nu = \mu/\rho$

Nonwetting liquid: a liquid whose molecular attraction is greater than the molecules' attraction to the surface with which they are in contact.

Pascal (Pa):   the unit of pressure in the SI system that equals force in newtons divided by area in square meters: $Pa = N/m^2$.

Poise: the cgs unit of viscosity that equals dyne $\cdot s/cm^2$.

Pressure-height relation: the mathematical relation that expresses the pressure exerted on the base of a column of fluid as a function of the specific weight of the fluid and the height of the column: $p = -\gamma h$.

Rankine temperature scale: the absolute temperature scale corresponding to the Fahrenheit temperature scale: $°R = °F + 460$.

Specific gravity (sg):   the ratio of the specific weight or density of a substance to the specific weight or density of water at 4°C.

Specific volume (v):   the volume per unit mass—also, the reciprocal of density.

Specific weight (γ):   the weight of a unit volume—also, g times density.

Stoke: the cgs unit of kinematic viscosity.

Surface tension (σ):   the membrane-like behavior of the interface between a liquid and a gas.

Temperature: a measure of the random molecular motion of a system.

Torr: A unit of pressure equal to that of a column of mercury that is 1 mm high.

Vacuum: negative gage pressure.

Viscometer: a device to measure the viscosity of a fluid.

Viscosity (μ):   the resistance that a fluid has to shearing forces.

Wetting liquid: a liquid whose attraction for a surface exceeds the attraction of its molecules for each other.

KEY EQUATIONS

Temperature conversion	$°C = \frac{5}{9}(°F - 32)$	(2.1)
Temperature conversion	$°F = \frac{9}{5}(°C) + 32$	(2.2)
Temperature conversion	degrees Rankine $(°R) = °F + 460$	(2.3)
Temperature conversion	degrees Kelvin $(K) = °C + 273$	(2.4)
Density	$\rho = \dfrac{m}{V}$	(2.5)
Specific weight	$\gamma = \dfrac{w}{V}$	(2.6)
Specific weight—density	$\gamma = \rho g$	(2.8)
Specific volume	$\nu = \dfrac{1}{\rho}$	(2.9)
Specific gravity	$sg = \dfrac{\gamma_{substance}}{\gamma_{water} \text{ at } 4°C}$	(2.10)

$$sg = \frac{\rho_{substance}}{\rho_{water} \text{ at } 4°C} \tag{2.10}$$

Pressure $\qquad\qquad p_a = p_g + p_{atm} \tag{2.15}$

Pressure $\qquad\qquad p_a = p_{atm} - p_g \tag{2.16}$

Pressure–height relation $\qquad p = -\gamma h \tag{2.18}$

Surface tension $\qquad\qquad \sigma = \frac{F}{L} \tag{2.19}$

Surface tension rise in a tube $\qquad h = \frac{4\sigma \cos \theta}{\gamma d} \tag{2.20}$

Surface tension rise—parallel plates $\quad h = \frac{2 \sigma \cos \theta}{\gamma d} \tag{2.22}$

Compressibility $\qquad\qquad \beta = -\frac{\Delta p}{\Delta V/V} \tag{2.23}$

Viscosity $\qquad\qquad F = -\frac{\mu AV}{Y} \tag{2.28}$

Viscosity $\qquad\qquad \tau = \mu\left(\frac{V}{Y}\right) \tag{2.29}$

Kinematic viscosity $\qquad\qquad \nu = \frac{\mu}{\rho} \tag{2.30}$

QUESTIONS

1. State why you think that velocity is or is not a property of a system.
2. If a book is placed on a shelf that is 3 m above the floor, is its position a property?
3. A student wishes to take her temperature. After placing a thermometer in her mouth for 1 minute, will she obtain a correct reading of her temperature? Explain.
4. A temperature scale is proposed that will use the ice point of water as $10°$ and the boiling point as $90°$. How practical do you think that this scale is?
5. How would you propose to measure the temperature of a ladle of molten steel?
6. A person lifts two blocks of differing materials. If the blocks each have the same volume, comment on his statement that one block is denser than the other.
7. A spring balance when properly calibrated measures weight. How would you measure mass?
8. A gas station attendant states that you need a heavier-weight oil in your car. What is meant by this statement?

9. Why does your doctor state your blood pressure as 120 over 70?

10. A student reads a pressure gage and records a reading of 50 kg/cm². Did the student read the gage properly?

11. Two pressure gages are used to measure the pressure at different locations in a pipeline. Is the pressure difference between these readings gage or absolute pressure?

12. As the pressure is increased, the bulk modulus of water (a) increases, (b) decreases, or (c) remains the same.

13. State the condition that is required for a liquid to show surface tension.

14. Surface tension causes a liquid to (a) rise, (b) fall, (c) remain the same height, or (d) either rise or fall in a capillary tube.

15. Explain how a steel razor blade can float on water. Use a free-body diagram to illustrate your explanation.

16. *True or false:* Viscosity is a measure of the thickness of a fluid.

17. *True or false:* A clear liquid always is less viscous than a dark liquid.

18. The viscosity of liquids (a) increases, (b) remains the same, or (c) decreases, as temperature is increased.

19. The viscosity of gases (a) increases, (b) remains the same, or (c) decreases, as temperature is increased.

20. Give several reasons why the oil in a car should be changed at either a fixed mileage or in a fixed period of time, whichever comes first.

21. Comment on the statement that it takes a certain brand of ketchup longer to flow because it is thicker and richer.

22. The units of viscosity in SI are (a) m/s, (b) m/s², (c) N/m²·s, (d) N·m/s², (e) N·m²/s, (f) none of these.

23. A gas has (a) some viscosity, (b) zero viscosity, (c) viscosity at 20°C only, (d) none of these.

PROBLEMS

Temperature

2.1 Convert 20, 40, and 60°C to equivalent degrees Fahrenheit.

2.2 Change 0, 10, and 50°F to degrees Celsius.

2.3 Convert 500°R, 500 K, 600 K, and 650°R to degrees Celsius.

*2.4 A Fahrenheit and a Celsius thermometer are used to measure the temperature of a fluid. If the Fahrenheit reading is 1.5 times that of the Celsius reading, what are both readings?

*2.5 Derive a relation between degrees Rankine and degrees Kelvin, and, based on this result, show that (°C + 273)1.8 = °F + 460.

*2.6 An arbitrary temperature scale is proposed in which 20° is assigned to the ice point and 75° is assigned to the boiling point. Derive an equation relating this scale to the Celsius scale.

*2.7 For the temperature scale proposed in Problem 2.6, what temperature corresponds to absolute zero?

*2.8 A new thermometer scale on which the ice point of water at atmospheric pressure would correspond to a marking of 200 and the boiling point of water at atmospheric pressure would correspond to a marking of ⁻400 is proposed. What would the reading of this new thermometer be if a Fahrenheit thermometer placed in the same environment read 80°F?

Density, Specific Weight, Specific Volume, and Specific Gravity

2.9 Determine the density and specific volume of the contents of a 10-ft³ tank if the contents weigh 250 lb.

2.10 A tank contains 500 kg of a fluid. If the volume of the tank is 0.5 m³, what is the density of the fluid and what is the specific volume?

2.11 Calculate the specific weight of a fluid if a gallon of it weighs 20 lb. Also calculate its density.

2.12 The specific gravity of a fluid is 1.2. Calculate its density and its specific weight in SI units.

2.13 A fluid has a density of 0.90 g/cm³. What is its specific weight in both the English and SI systems?

2.14 It is proposed by gasoline dealers to sell gasoline by the liter. If gasoline has a density of 1.3 slugs/ft³, what is the weight of 60 liters of gasoline?

2.15 A household oil tank can hold 275 gallons of oil. If oil has a specific weight of 8800 N/m³, how many pounds of oil will there be in a full tank?

2.16 A car has a fuel tank that holds 14 gallons of gasoline. If the gasoline has a specific gravity of 0.70, calculate the weight of gasoline in units of pounds.

2.17 An oil has a specific gravity of 0.8 at 20°C. Determine its specific weight, density, and specific volume in SI units.

2.18 An oil has a specific weight of 0.025 lb/in.³. Calculate its specific gravity.

2.19 A can weighs 12 N when empty, 212 N when filled with water at 4°C, and 264 N when filled with an oil. Calculate the specific gravity of the oil.

2.20 The density of a liquid is 1.51 slugs/ft³. Calculate the specific weight and the specific gravity of the liquid.

2.21 A shipping company requires that packages have a base perimeter plus the height not to exceed 108 in. By calculation, a student determines that the optimum package size should be an 18-in.-square base and 36-in. height. If the company further specifies that the weight must not exceed 70 lb, determine the specific weight of the contents of the package.

2.22 A body has a specific gravity of 1.48 and a volume of 7.24×10^{-4} m^3. What is the weight of the body?

2.23 Oil has a specific gravity of 0.83. What is the weight of a liter of this oil?

2.24 What is the weight of a gallon of oil that has a specific gravity of 0.86?

2.25 A cylinder 6 in. in diameter and 10 in. high contains oil that has a density of 850 kg/m^3. Determine the weight of the oil in English units.

2.26 A liquid has a density of 1100 kg/m^3. Determine its specific gravity and its specific weight.

2.27 The fuel tank of a car holds 60 liters of gasoline. Assuming that the gasoline has a specific gravity of 0.74, determine the weight of the gasoline in the tank.

2.28 If the density of a liquid is 780 kg/m^3, calculate its specific gravity and its specific weight.

2.29 A liquid has a specific weight of 200 lb/ft^3. Calculate the volume needed to have a weight of 390 lb.

*2.30 In the older cgs system the density in g/cm^3 is numerically equal to the specific gravity of the fluid. Show that this is indeed true.

Pressure

2.31 A skin diver descends to a depth of 60 ft in fresh water. What is the pressure on his body? The specific weight of fresh water can be taken as 62.4 lb/ft^3.

2.32 A skin diver descends to a depth of 25 m in a salt lake, where the density is 1026 kg/m^3. What is the pressure on this person's body at this depth?

2.33 If a Bourdon gage reads 25 psi, what is the absolute pressure in Pa?

2.34 A column of fluid is 1 m high. The fluid has a density of 2500 kg/m^3. What is the pressure at the base of the column?

2.35 A column of fluid is 25 in. high. If the specific weight of the fluid is 60.0 lb/ft^3, what is the pressure in psi at the base of the column?

2.36 Convert 14.696 psia to 101 325 kPa.

2.37 A pressure gage indicates 25 psi when the barometer is at a pressure equivalent to 14.5 psia. Compute the absolute pressure in psi and feet of mercury when the specific gravity of mercury is 13.0.

2.38 Repeat Problem 2.37 if the barometer stands at 750 mm Hg and the specific gravity is 13.6.

2.39 A vacuum gage reads 8 in. Hg when the atmospheric pressure is 29.0 in. Hg. If the specific gravity of mercury is 13.6, compute the absolute pressure in psi.

2.40 A vacuum gage reads 10 in. Hg when the atmospheric pressure is 30 in. Hg. Assuming the density of mercury to be 13 595 kg/m^3, determine the pressure in Pa.

*2.41 A water main is 18 in. in diameter and carries water at 120 psi. If the pipe's top is buried 10 ft below the surface of the street, how high above the street will a plume of water reach if the pipe fails locally?

2.42 A tank is filled with fresh water until there is a depth of 35 ft of water. Determine the pressure in psig at the bottom of the tank.

2.43 The piston of a pump is 7.5 cm in diameter. If a force of 2000 N is applied to the piston, what is the pressure that is being developed at the face of the piston?

2.44 The door of a jet liner has the dimensions of 2.3 m × 1.2 m. If the inside pressure of the cabin is 98 kPa absolute, what force is exerted on the door when the plane flies at an altitude where the outside pressure is 15 kPa absolute?

*2.45 Water and oil are placed in a tank. The oil has a specific weight of 50 lb/ft³. If the height of the oil and water are each 5 ft, determine the pressure in psig at the bottom of the tank.

Surface Tension

2.46 A tube whose diameter is 0.001 ft is placed in a mercury pool. If the surface tension of mercury is 35×10^{-3} lb/ft, determine the depression of the mercury in the tube below the surface of the mercury pool. Assume that $\theta = 129°$ and that the specific weight of mercury is 13.6 times that of water.

2.47 If instead of the tube in Problem 2.46, two infinite parallel plates are used, determine the depression of the mercury between the plates below the surface of the pool.

2.48 Solve Problem 2.46 for water having a surface tension of 5×10^{-3} lb/ft.

2.49 Solve Problem 2.47 for water having a surface tension of 5×10^{-3} lb/ft.

2.50 If a fluid has a surface tension of 2.9×10^{-3} N/m and a specific gravity of 0.85, how high will it rise in a capillary tube having a diameter of 2×10^{-4} m if the wetting angle is 25°?

2.51 Solve Problem 2.50 for two parallel plates.

2.52 It is desired to limit the rise of water in a capillary tube to 0.5 mm. Calculate the diameter of the glass tube required if $\sigma = 7 \times 10^{-2}$ N/m.

*2.53 A thin platinum wire is built into the shape of a circle. If the diameter of the circle is 100 cm, determine the vertical force required to lift the ring from the surface of a pool of water. Neglect the weight of the wire and use 7.3×10^{-2} N/m for the surface tension of the water. (Note that if the wire is made of a material such as platinum, the water will contact the wire on both its inside and outside surfaces.)

*2.54 The pressure inside a spherical droplet of liquid exceeds the pressure outside the droplet due to surface tension. Derive the following expression for the excess pressure, Δp, if d is the diameter of the droplet and σ is the surface tension:

$$\Delta p = \frac{4\sigma}{d}$$

2.55 Determine the excess pressure in a water droplet if its diameter is 1 mm and the surface tension is 7×10^{-2} N/m. Use the results of Problem 2.54.

Compressibility

2.56 Determine the bulk modulus of a fluid if it undergoes a 1% change in volume when subjected to a pressure of 10,000 psi.

2.57 The bulk modulus of water is approximately 2.1×10^9 N/m^2 near room conditions. What will the percent change in volume of water be if the water is subjected to a pressure of 200 MPa?

2.58 Determine the reduction in volume of 5 m^3 of water if the pressure is increased by 5 MPa. Use 2.2×10^9 N/m^2 for the bulk modulus of the water.

2.59 A fluid has a volume of 2.000 ft^3 when subjected to a pressure of 1000 psi and a volume of 1.975 ft^3 when subjected to a pressure of 7000 psi. Determine the bulk modulus of the fluid.

Viscosity

For the following problems use Figures 2.13 and 2.15 as necessary.

2.60 Determine the viscosity of gasoline at 100°F.

2.61 Determine the kinematic viscosity of gasoline at 100°F.

2.62 Determine the viscosity of crude oil at 100°F.

2.63 Determine the kinematic viscosity of crude oil at 100°F.

2.64 Determine the viscosity of air at 100°F.

2.65 Determine the kinematic viscosity of air at 100°F.

2.66 The viscosity of an oil is measured as 35 SSU. Determine its kinematic viscosity in centistokes, m^2/s, and ft^2/s.

2.67 The viscosity of an oil is measured as 135 SSU. Calculate its kinematic viscosity in centistokes, m^2/s, and ft^2/s.

2.68 The viscosity of an oil is measured as 35 centipoise. Convert this to Pa·s (N·s/m^2) and lb·s/ft^2.

2.69 A fluid has a viscosity of 8.5×10^{-3} N·s/m^2. Convert this to units of lb·s/ft^2.

2.70 Convert 0.5 stoke to m^2/s and ft^2/s.

2.71 An oil has a dynamic viscosity of 7×10^{-4} lb·s/ft^2. If its density is 1.9 slugs/ft^3, determine the kinematic viscosity of the oil.

2.72 An oil having a dynamic viscosity of 4×10^{-5} ft^2/s has a specific gravity of 0.85. Determine the dynamic viscosity of the oil.

2.73 An oil with a specific weight of 60.0 lb/ft^3 has a dynamic viscosity of 1×10^{-3} lb·s/ft^2. Determine the kinematic viscosity of the oil.

2.74 Using Table 2.6a, convert 125 cP to Pa·s and lb·s/ft^2.

2.75 Two plates separated by 3 in. are placed in oil with a viscosity of 133×10^{-6} lb·s/ft^2. If the plates are parallel and have an area of 5 ft^2, determine the relative velocity (considering only viscous effects) if a force of 0.1 lb is applied to one of the plates parallel to the orientation of the plates.

For Problems 2.76, 2.77, and 2.78, refer to Figure 2.14.

2.76 A piston having a diameter of 6.000 in. slides inside a vertical pipe having an inside diameter of 6.008 in. The piston is 6 in. long and weighs 2.2 lb. Determine the terminal speed of the piston if it is lubricated with an oil having a viscosity of 0.020 lb·s/ft².

2.77 If the piston has a diameter of 100 mm and the cylinder has a diameter of 101 mm and weighs 8 N, determine the viscosity of the lubricating oil. Assume that the piston falls with a constant speed of 1 m/s and that its length is 90 mm.

2.78 If the system of Figure 2.14 is placed in a horizontal aspect and the piston is made to move horizontally, determine the force required. Neglect the weight of the piston. Compare your answer with Illustrative Problem 2.18 and comment. Assume that the piston remains concentric in the cylinder.

2.79 The 1-in. space between two parallel plates is filled with an oil that has a viscosity of 3×10^{-2} lb·s/ft². A thin plate having a surface area of 1.5 ft² is placed between the original plates and is pulled horizontally with a constant speed of 0.25 ft/s. If the thin plate is placed midway between the two fixed plates, calculate the force required.

2.80 If the thin plate in Problem 2.79 is placed $\frac{1}{4}$ in. away from one of the fixed plates, determine the force required.

2.81 A cube has a weight of 10 N and is 0.1 m on a side. It slides down an inclined plane that is sloped 30° from the horizontal. If there is a thin oil film of 0.01 mm thickness between the block and the plane and the oil has a viscosity of 5×10^{-5} lb·s/ft², calculate the terminal speed of the block.

2.82 If the cube of Problem 2.81 is pulled up the plane at a velocity of 1 m/s, determine the force required.

FLUID STATICS

LEARNING GOALS

After reading and studying this chapter you should be able to:

1. Define the branch of fluid mechanics that is known as *fluid statics.*

2. Calculate the pressure at the base of a column of fluid.

3. State the *equal level/equal pressure* concept.

4. Calculate the effect of elevation on the reading of the barometer for an atmosphere that is assumed to be at constant temperature.

5. Apply the equal level/equal pressure principle to manometers.

6. Show that all bodies experience a buoyant force when they are submerged in a fluid.

7. Calculate the buoyant force on a body using Archimedes' principle.

8. Determine the forces on plane surfaces that are submerged in fluids.

9. Locate the center of gravity for various simple shapes.

10. Show that the center of pressure is *always* below the center of gravity.

11. Locate the center of pressure for various shapes.

12. Calculate the stresses in thin cylinders and thin spheres due to internal pressure.

3.1   INTRODUCTION

The study of fluids at rest or having no velocity with respect to an observer in a gravitational field is known as *fluid statics*. Since the fluid will be at rest with respect to an observer, there will be no relative motion between adjacent fluid layers and viscosity will not enter into problems in fluid statics. These considerations enable us to treat fluids at rest mathematically, and the results of these mathematical studies will be very accurate for engineering purposes. To solve problems in fluid statics it is only necessary to use the principle of equilibrium of bodies that is developed in mechanics. Thus, for a body (or element of mass) to be in equilibrium, it is necessary that the sum of the external forces and moments acting on the body be zero. This principle, combined with a knowledge of the fluid density, enables one to analyze problems in fluid statics readily.

3.2   PRESSURE RELATIONSHIPS

In Chapter 2 the pressure at the base of a uniform column of liquid was derived as a function of the specific weight of the liquid and the height of the liquid column. It is useful at this time to consider this same problem again from a slightly different viewpoint. Consider a cylinder of fluid having a height h and specific weight γ. The height of the column will be measured as positive in the vertical direction, as shown in Figure 3.1. We now apply the condition of equilibrium that the sum of the forces in the vertical direction at the base of the column must be zero. Since we have the weight acting down as

$$w = \gamma h A \tag{3.1}$$

then

$$pA = \gamma h A \tag{3.2}$$

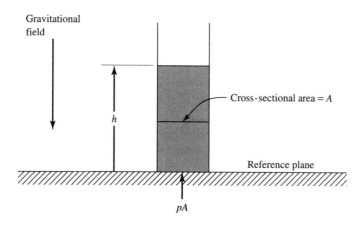

Figure 3.1

or

$$p = \gamma h \qquad (3.2a)$$

Since we are measuring height, h, to be positive in the "up" direction, it is necessary to introduce a minus sign into equation (3.2a) to account for this, since weight increases as we proceed "down" the column. Equation (3.2a) is therefore rewritten as

$$p = -\gamma h \qquad (3.2b)$$

which corresponds to equation (2.18). Note that equation (3.2b) is for the pressure due to the liquid column only; the negative sign is to be interpreted to mean that pressure decreases as we go up the column of liquid, and γ is taken to be constant.

Equations (3.2) can also be interpreted by referring to Figure 3.2, which is a plot of pressure against height. As can be seen, $-\gamma$ is simply the constant slope, and from Figure 3.2 we can write directly

$$p = p_a + \gamma h - \gamma x = p_a + \gamma(h - x) \qquad (3.2c)$$

where p_a is the pressure at the free surface on top of the column. Equations (3.2) represent the fundamental relations among pressure, specific weight, and column height. As stated earlier, the negative sign indicates that the pressure decreases as one goes up the column. Note that the pressure in a column can be specified by stating feet of a fluid of a given density. In the nomenclature of hydraulics, this was known as a "head," and this nomenclature is still used today. Thus, 3 ft of water represents the head of water or a gage pressure of 3 ft $\times$ 62.4 lb/ft^3 = 187.2 psfg or 1.3 psig.

If the liquid does not have a constant density (say due to temperature gradients or to pressure effects), it is necessary to evaluate equations (3.2) taking into account the fact that

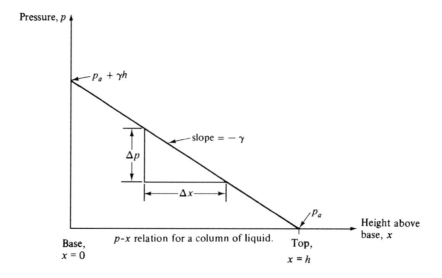

Figure 3.2 $p - x$ relation for a column of liquid.

γ is a variable. This can be done either by using the methods of calculus or by graphically solving the equation.

CALCULUS ENRICHMENT

The relation between the pressure and height of a fluid can be derived considering Figure A:

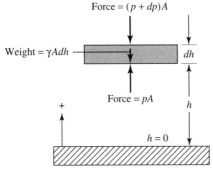

$$\text{Force} = (p + dp)A$$

$$\text{Weight} = \gamma A dh$$

$$dh$$

$$\text{Force} = pA$$

$$h$$

$$h = 0$$

Figure A The pressure–height relation

Summing the forces in the vertical direction and noting that up is positive yields

$$dp = -\gamma dh \tag{a}$$

where γ is also ρg. Thus,

$$dp = -\rho g dh \tag{b}$$

In order to solve Equations (a) or (b) it is necessary to know ρ as a function of h. In general, we can integrate these equations,

$$p(h) - p(0) = -\int_0^h \rho g dh \tag{c}$$

For the special case of a liquid having a constant density, Equation (c) can be integrated, starting at the surface, where $p = 0$, and noting that h is measured positive upward. This gives us the result that

$$p = -\rho g h \tag{d}$$

or

$$p = -\gamma h \tag{e}$$

Let us now consider the case of a fluid whose density decreases linearly as a function of height. For this case we can write

$$\rho = \rho_0(1 - \alpha h) \tag{f}$$

where ρ_0 corresponds to h_0 and α is the rate at which density decreases with elevation. Placing (f) into (c) gives us

$$p(h) - p(0) = -\int_0^h \rho_0(1 - \alpha h)dh \tag{g}$$

Integrating gives

$$p(h) - p(0) = -\left[\rho_0 h - \alpha\rho_0 \frac{h^2}{2}\right]_0^h \tag{h}$$

Simplifying gives

$$p(h) - p(0) = -\rho_0 h\left[1 - \frac{\alpha h}{2}\right] \tag{i}$$

In meteorology, the decrease of temperature with height is called the *lapse rate*. The lapse rate is related to Equations (c) and (i).

ILLUSTRATION 3.1   PRESSURE EQUIVALENTS

A weather broadcaster states that the barometer reads 29.92 inches of mercury and is steady. Calculate the atmospheric pressure in psi if the specific gravity of mercury is 13.6.

ILLUSTRATIVE PROBLEM 3.1

Given: p = 29.92 in. Hg, sg = 13.6

Find: p in psi

Solution: This problem represents the application of equation (3.2a) with appropriate attention to units. Thus,

$$p = \gamma h$$

$$p = \frac{29.92 \text{ in.}}{12 \text{ in./ft}} \times \frac{13.6 \times 62.4 \text{ lb/ft}^3}{144 \text{ in.}^2/\text{ft}^2}$$

$$p = 14.69 \text{ psi}$$

Obviously, the statements 29.92 inches of mercury and 14.69 psi are both equivalent statements of pressure.

At this time let us examine Figure 3.3, where four different containers hold fluids of the same specific weight, and are open to the atmosphere at the top. At the top, level A, the pressure is atmospheric. At some other level, level B, which is a distance x below level A, the pressure in *each* of the containers will be the *same*, since $p_B = p_A + \gamma x$. The variation in pressure is a function only of the specific weight of the fluid and the vertical depth. It is independent of the size, shape, or orientation of the container. This is also shown in Figure 3.4, where we see a tank having several different sections. Notice that $p_1 = p_2 = p_3$; that is, a line of constant height is also a line of constant pressure. This fact has been termed the

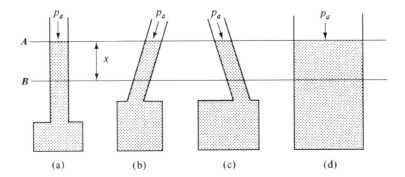

Figure 3.3   Equal level/equal pressure principle.

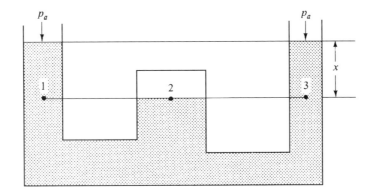

Figure 3.4   Equal level/equal pressure principle.

equal level/equal pressure principle and forms the basis for the measurement of pressure using a barometer or manometer. Notice that this principle holds for a given fluid. It does not hold if different fluids are involved.

ILLUSTRATION 3.2   EQUAL LEVEL/EQUAL PRESSURE PRINCIPLE

A tube is filled with mercury (specific gravity = 13.6) and is inverted into a well of mercury as shown in Figure 3.5. Neglecting the pressure of the vapor in the tube

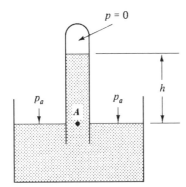

Figure 3.5   Illustrative Problem 3.2.

above the mercury, how high will the column of mercury be in the tube above the level in the well?

ILLUSTRATIVE PROBLEM 3.2

Given: sg = 13.6

Find: h

Solution: At level A in both the tube and the well, the pressure must be the same, and for this problem will be p_a, atmospheric pressure. If we proceed from the top of the tube down to level A, we have $p_A = 0 + \gamma h$. But this must equal p_a. Therefore,

$$p_a = 0 + \gamma h = \gamma h$$

and

$$h = \frac{p_a}{\gamma}$$

If atmospheric pressure is 101 325 Pa,

$$h = \frac{101\ 325\ \text{N/m}^2}{1000\ \text{kg/m}^3 \times 9.806\ \text{N/kg} \times 13.6} = 0.760\ \text{m} = 760\ \text{mm}$$

We can generalize this result in the following manner:

$$1 \text{ standard atmosphere} = \begin{cases} 14.69 \text{ psia} \\ 2116 \text{ psfa} \\ 29.92 \text{ in. Hg} \\ 760 \text{ mm Hg} \\ 101\ 325 \text{ Pa} \\ 10.34 \text{ m of water} \\ 33.91 \text{ ft of water} \\ 1.01 \text{ bars} \end{cases}$$

There is one case of variable density that is of special use. Consider a condition where the local atmospheric pressure at sea level is known. If an airplane is 1000 ft above sea level, what would the pressure be? Conversely, if the reading of a barometer is known at sea level and some unknown elevation, what is the elevation? To answer this problem we shall make the assumption that the air temperature remains constant and apply equation (3.2b), written to account for the variation in γ, as

$$\frac{\Delta p}{\Delta x} = -\gamma \qquad\qquad (3.3)$$

The relation that governs the pressure, temperature, and density for an *ideal gas* (air can be so considered for our purposes) is

$$pv = RT \qquad or \qquad \frac{p}{\gamma} = RT \qquad\qquad (3.4)$$

where p is the pressure in psfa, R is a constant for a gas (53.3 for air in these engineering units), and T is the absolute temperature in degrees Rankine. In SI units, p is in Pa, T is in Kelvin, γ is in N/m³, and R for air is 29.25. Inserting γ from equation (3.4) into equation (3.3) yields

$$\frac{\Delta p}{\Delta x} = \frac{-p}{RT} \qquad\qquad (3.5)$$

or

$$\frac{\Delta p}{p} = \frac{-\Delta x}{RT} \qquad\qquad (3.6)$$

To obtain a solution to equation (3.6) it is necessary to sum this equation between the required pressure limits. Graphically, this process of summation can be illustrated by referring to Figure 3.6. Plotting $1/p$ against p gives us the coordinates of this graph. By selecting a value of Δp as shown, the shaded area gives us $\Delta p/p$. Therefore, the sum of the $\Delta p/p$ values is the area under the curve between the limits of p_1 and p_2. By the methods of calculus, it can be shown that this area is ln (p_2/p_1).

   Returning to equation (3.6) and noting that R and T are constants for this problem, we can easily evaluate the right-hand side graphically by plotting x against $1/RT$, as shown in

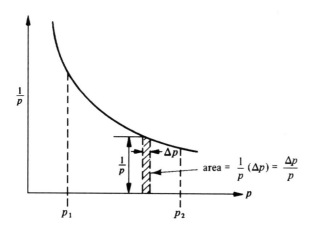

Figure 3.6   Evaluation of $\Delta p/p$.

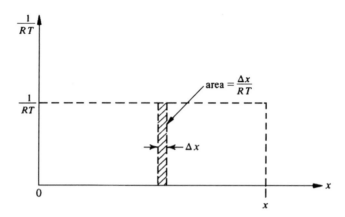

Figure 3.7   Evaluation of $\Delta x / RT$.

Figure 3.7. For all values of Δx, $1/RT$ is a constant, and the curve is simply a rectangle whose area is x/RT. Thus, from integrating equation (3.6) equation (3.4) becomes

$$\ln \frac{p_2}{p_1} = -\frac{x}{RT}$$

or

$$\ln \frac{p_1}{p_2} = \frac{x}{RT} \qquad (3.7)$$

The readings of a barometer are directly proportional to the local atmospheric pressure, and therefore the ratio p_1/p_2 can be replaced by the barometer readings b_1 and b_2 as the ratio b_1/b_2. Equation (3.7) becomes

$$\ln \frac{b_1}{b_2} = \frac{x}{RT} \qquad (3.8)$$

where x is the difference in elevation and b_1 and b_2 are the barometer pressures at sea level and elevation, respectively.

ILLUSTRATION 3.3   VARIABLE ATMOSPHERIC PRESSURE

A barometer reads 760 mm Hg at sea level and 750 mm Hg at some other elevation. If the air temperature can be taken to be constant and equal to 20°C (293 K), determine the unknown elevation.

ILLUSTRATIVE PROBLEM 3.3

Given: $b_1 = 760$ mm Hg, $b_2 = 750$ mm Hg, $T = 293$ K

Find: x

Solution: Using equation (3.8), we obtain

$$\ln \frac{760}{750} = \frac{x}{29.25 \times 293}$$

Therefore,

$$x = 29.25 \times 293 \ln \frac{760}{750}$$

$$= 113.51 \text{ m } (372 \text{ ft})$$

The student should note that for small elevation differences the assumption of constant temperature is quite good. For large elevation differences there is a large change in temperature and equation (3.7) is no longer applicable.

For convenience and for standardization of performance data, a *standard atmosphere* has been defined; it is given in Table 3.1. It will be seen from this table that the temperature

Table 3.1   ICAO Standard Atmosphere

Altitude (ft)	Temperature (°F)	Pressure (psia)	Specific weight (lb/ft³)	Density (slugs/ft³)	Viscosity $\times 10^7$ (lb·s/ft²)
0	59.00	14.696	0.07648	0.002377	3.737
5,000	41.17	12.243	0.06587	0.002048	3.637
10,000	23.36	10.108	0.05643	0.001756	3.534
15,000	5.55	8.297	0.04807	0.001496	3.430
20,000	−12.26	6.759	0.04070	0.001267	3.325
25,000	−30.05	5.461	0.03422	0.001066	3.217
30,000	−47.83	4.373	0.02858	0.000891	3.107
35,000	−65.61	3.468	0.02367	0.000738	2.995
40,000	−69.70	2.730	0.01882	0.000587	2.969
45,000	−69.70	2.149	0.01481	0.000462	2.969
50,000	−69.70	1.690	0.01165	0.000364	2.969
55,000	−69.70	1.331	0.00917	0.000287	2.969
60,000	−69.70	1.049	0.00722	0.000226	2.969
65,000	−69.70	0.826	0.00568	0.000178	2.969
70,000	−69.70	0.650	0.00447	0.000140	2.969
75,000	−69.70	0.512	0.00352	0.000110	2.969
80,000	−69.70	0.404	0.00277	0.000087	2.969
85,000	−65.37	0.318	0.00216	0.000068	2.997
90,000	−57.20	0.252	0.00168	0.000053	3.048
95,000	−49.05	0.200	0.00131	0.000041	3.099
100,000	−40.89	0.160	0.00102	0.000032	3.150

Source: Reproduced with permission from *Elementary Fluid Mechanics* by J. K. Vennard, McGraw-Hill Book Company, New York, 1961, p. 548.

is not constant and that the equations previously derived are valid only for small changes in elevation.

If h is the height above sea level in feet and the temperature is assumed to decrease linearly, it can be approximated by

$$T = (519 - 0.00357h) \tag{3.9}$$

where $T = °$R. At 35,000 ft, the temperature becomes $-67°$F, and is assumed to be constant above this level.

ILLUSTRATION 3.4   VARIABLE ATMOSPHERIC PRESSURE

A barometer reads 760 mm Hg at sea level and 20°C. It is taken up a hill to a location where the elevation is 300 m above sea level. If the temperature stays constant, what is the expected reading of the barometer?

ILLUSTRATIVE PROBLEM 3.4

Given: $b_1 = 760$ mm Hg, $T = 293$ K, $x = 300$ m

Find: b_2

Solution: Applying equation (3.8), we obtain

$$\ln \frac{b_1}{b_2} = \frac{x}{RT}$$

$$\ln \frac{760}{b_2} = \frac{300}{29.24 \times (20 + 273)}$$

Since $\ln (760/b_2) = \ln 760 - \ln b_2$,

$$\ln 760 - \ln b_2 = \frac{300}{29.24 \times (20 + 273)}$$

$$6.63332 - \ln b_2 = 0.03502$$

$$\ln b_2 = 6.59830$$

$$b_2 = 733.8 \text{ mm}$$

3.3   PRESSURE MEASUREMENT—MANOMETRY

Since pressure is a property of a system, its measurement is both desirable and necessary. In addition, as will be seen in later chapters, a knowledge of the pressure in various portions of a system is necessary to the determination of the flow and the state of a system. The simplest and most widely used mechanical pressure gage is the Bourdon gage. As is repeated in Figure 3.8, this gage consists simply of a bent tube closed on one end with a mechanism to indicate the movement of the closed end. The tube is usually noncircular in cross section, and pressure is applied at the open end. When a pressure in excess of atmospheric pressure is applied to the open end, the closed end will move as the tube tends to straighten out (similar to a New Year's Eve blowout noisemaker), with the reading of the gage proportional to the displacement of the tube. This type of gage can readily be calibrated and is rugged and easy to use. It is equally capable of measuring pressures above and below local atmospheric pressure. The student should note that the device illustrated measures gage pressure only. It is possible to evacuate the case of the gage and thereby create a gage capable of measuring absolute pressures.

If the gage is connected at an elevation different from the elevation where the pressure is to be measured, it is necessary to make a correction to the reading of the gage. Referring to Figure 3.8, it will be seen that if the gage is mounted above a pipe, the gage will read *low* by an amount equal to γh. The specific weight γ should be obtained at the temperature of the fluid in the connecting line to the gage, since this may be different from the temperature in the pipe. For commercial usage, this correction is usually negligible.

We have already shown that the pressure at the base of a column of liquid is simply a function of the height of the column and the specific weight of the liquid. Therefore, the

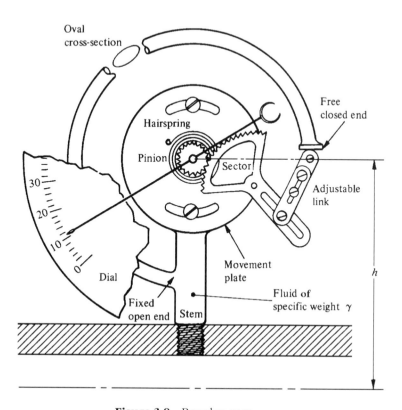

Figure 3.8   Bourdon gage.

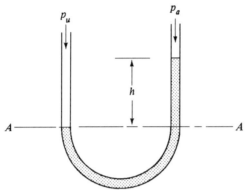

Figure 3.9 U-tube manometer.

height of a column of liquid of known specific weight can be and is used to measure pressure and pressure differences. Instruments that utilize this principle are known as manometers, and the study of these pressure-measuring devices is known as manometry. By properly arranging a manometer and selecting the fluid judiciously, it is possible to measure extremely small pressures, very large pressures, and pressure differences. A simple manometer is shown in Figure 3.9, where the right arm is exposed to the atmosphere while the left arm is connected to the unknown pressure. As shown, the fluid is depressed in the left arm and raised in the right arm until no unbalanced pressure forces remain. It has already been demonstrated that the pressure at a given level in either arm must be the same so that we can select any level as a reference and write a relation for the pressure. Actually, it is much easier and more convenient to select the interface between the manometer fluid and the unknown fluid as a common reference level. In Figure 3.9 the pressure at elevation AA is the same in both arms of the manometer based upon the equal level/equal pressure principle. Starting with the connected manometer arm (left), we have the unknown pressure p_u acting on the fluid. As one proceeds down the arm we arrive at level AA. We now use the equal level/equal pressure principle to go horizontally to the right arm. Going up the right arm, the pressure will decrease by an amount equal to γh until we finally reach p_a, atmospheric pressure. Writing this process from the left arm (p_u) to the right arm, we have

$$p_u - \gamma h = p_a \tag{3.10a}$$

or

$$p_u - p_a = \gamma h \tag{3.10b}$$

ILLUSTRATION 3.5   MANOMETER

The tank shown in Figure 3.10 is filled with water ($\gamma = 9810$ N/m^3). The manometer on the side of the tank has air trapped in it as the tank is filled. After filling the tank, the level of the water in the manometer is found to be 1 m below the level in the tank. Determine the pressure of the trapped air and the pressures at levels B and C

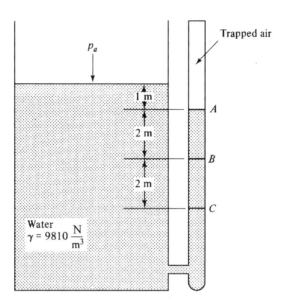

Figure 3.10 Illustrative Problem 3.5.

ILLUSTRATIVE PROBLEM 3.5

Given: Water, $\gamma = 9810 \dfrac{N}{m^3}$

Find: p of trapped air and at levels B and C

Solution: Starting at the level of the open part of the tank, the pressure is p_a. If we now proceed down the tank to level A, the pressure will be $p_a + \gamma h = p_a + 9810(1)$ N. This must also be the pressure of the trapped air, since level A is also the interface between the air and the water in the manometer. If we now proceed to level B, which is 3 m below the exposed surface of the tank, $p_B = p_a + \gamma h = p_a + 9810(3) = p_a + 29\ 430$ N. At level C the pressure is $p_a + 9810(5) = p_a + 49\ 050$ N. Notice that at each common level, the pressure in the tank and manometer must be the same.

ILLUSTRATION 3.6   MANOMETER

A manometer is connected to a tank as shown in Figure 3.11. Determine the pressure in the tank if atmospheric pressure is 100 kPa.

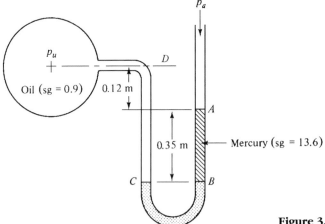

Figure 3.11   Illustrative Problem 3.6.

ILLUSTRATIVE PROBLEM 3.6

Given: Oil, sg = 0.9; mercury, sg = 13.6

Find: p_u

Solution: Starting with the top of the manometer at the leg open to the atmosphere, we have atmospheric pressure until elevation A. If we proceed down the right side of the manometer, we reach level B. The pressure at this level is $p_B = p_a + 0.35\gamma_{Hg}$. Since B and C are at the same level, the equal level/equal pressure principle yields $p_C = p_B$. We are now at the left leg of the manometer. If we now proceed up the column to level D, the pressure must have *decreased* by an amount of $\gamma_{oil}(0.12 + 0.35)$, and this must be p_u, the unknown pressure in the tank. Putting this information together,

$$p_a + \gamma_{Hg}(0.35) - \gamma_{oil}(0.12 + 0.35) = p_u$$

Using $\gamma_{water} = 9.81$ kN/m³ and the data of the problem,

$$100 \text{ kPa} + 13.6 \times 9.81 \text{ kN/m}^3 \times 0.35 \text{ m} - 0.9 \times 9.81 \text{ kN/m}^3 \times 0.47 \text{ m} = p_u$$

$$142.55 \text{ kPa} = p_u$$

This is obviously an absolute pressure. For gage pressure, which is not defined in SI units, we will simply use $p_u - p_a$. Therefore

$$p_u - p_a = 142.55 \text{ kPa} - 100 \text{ kPa} = 42.55 \text{ kPa}$$

The basic utility of the manometer lies in the fact that a linear distance readily permits us to obtain the pressure difference between two points. Therefore, a manometer can and is used to determine pressure differences between the pressures in two containers, or be-

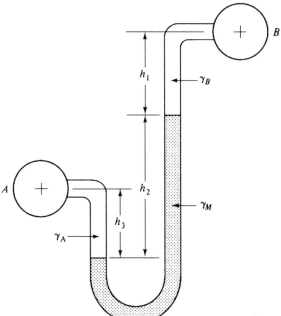

Figure 3.12   Differential manometer.

tween two points in the same pipe through which a fluid is flowing. Figure 3.12 shows one possible manometer arrangement for measuring the pressure difference between the pressures in two pipes through which different fluids are flowing. This type of manometer is called a *differential manometer.* We can use the principles already developed to obtain the pressure difference between the two pipes. The best way is to start systematically at one end of the system and to apply the equal level/equal pressure principle as required until the other end of the system is reached. Illustrative Problem 3.7 illustrates this procedure.

ILLUSTRATION 3.7   DIFFERENTIAL MANOMETER

If γ_A = 8830 N/m³, γ_B = 9806 N/m³, γ_M = 13 330 N/m³, h_1 = 1 m, h_2 = 2 m, and h_3 = 1.5 m, determine $p_A - p_B$ for the configuration shown in Figure 3.12.

ILLUSTRATIVE PROBLEM 3.7

Given: γ_A = 8830 N/m³, γ_B = 9806 N/m³, γ_M = 13 330 N/m³, h_1 = 1 m, h_2 = 2 m, h_3 = 1.5 m

Find: $p_A - p_B$

Solution: Starting at the left, we have p_A. Proceeding down a distance h_3 yields an *increase* in pressure of $\gamma_A h_3$. The same pressure must exist *at the same level* in the right arm. We now proceed up the right arm a distance h_2, giving us a pressure *decrease* of $\gamma_M h_2$. Proceeding further up the right arm yields another pressure *decrease* of $\gamma_B h_1$, and now we are at pipe B, whose pressure is p_B. If we now write these relationships in sequence, we have

$$p_A + \gamma_A h_3 - \gamma_M h_2 - \gamma_B h_1 = p_B$$

or

$$p_A - p_B = \gamma_M h_2 + \gamma_B h_1 - \gamma_A h_3$$

Substituting the given data yields

$$p_A - p_B = 13\ 330\ \frac{N}{m^3} \times 2\ m + 9806\ \frac{N}{m^3} \times 1\ m - 8830\ \frac{N}{m^3} \times 1.5\ m$$

and

$$p_A - p_B = 23\ 221\ \frac{N}{m^2} = 23.221\ kPa$$

In the simple U-tube manometer it was assumed that the pressure was acting on the manometer fluid without considering how this came about. Let us consider the case of water (or other fluid) flowing in a pipe and a different fluid (say, mercury or oil) being used in the manometer. This situation is shown in Figure 3.13. The pressure at level A is found from the equilibrium expression as follows:

$$p_a + \gamma h + \gamma_1 h_1 = p_A \tag{3.11}$$

If gage pressure is desired,

$$p_A = \gamma h + \gamma_1 h_1 \tag{3.12}$$

In equations (3.11) and (3.12) it is assumed that γ_1 is constant and the same in the manometer leg and the pipe. If this is not the case (due, say, to heat transfer), then a correction must be made.

ILLUSTRATION 3.8   DIFFERENTIAL MANOMETER

Water ($\gamma_1 = 62.4\ lb/ft^3$) flows in a pipe. If a manometer connected as in Figure 3.13 is used with mercury as the fluid ($\gamma = 850\ lb/ft^3$) and h is 5 in. when the level B is 15 in. above the centerline of the pipe, what is the pressure in the pipe?

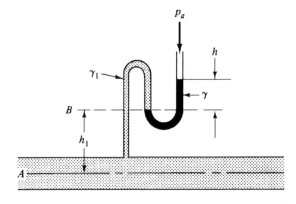

Figure 3.13   Manometer.

ILLUSTRATIVE PROBLEM 3.8

Given: $\gamma_1 = 62.4 \text{ lb/ft}^3$, $\gamma_{Hg} = 850 \text{ lb/ft}^2$, $h = 5 \text{ in.}$, $h_1 = 15 \text{ in.}$

Find: p_A

Solution: Referring to Figure 3.13 and using equation (3.12), we obtain

$$p_A = \gamma h + \gamma_1 h_1$$

$$= 850 \, \frac{\text{lb}}{\text{ft}^3} \times \frac{5 \text{ in.}}{12 \text{ in./ft}} + \frac{15 \text{ in.}}{12 \text{ in./ft}} \times 62.4 \, \frac{\text{lb}}{\text{ft}^3} = 432.2 \text{ psfg}$$

or

$$p_A = 432.2 \, \frac{\text{lb}}{\text{ft}^2} \times \frac{1}{(12 \text{ in./ft})^2} = 3 \text{ psig}$$

or

$$p_A = 14.7 + 3 = 17.7 \text{ psia}$$

To achieve greater accuracy and sensitivity in manometers, several different arrangements have been used. Perhaps the simplest of these is the inclined manometer. Consider a relatively large reservoir of liquid connected to a small-bore tube that makes an angle θ with the horizontal. The pressure or pressure differential to be measured is connected to the large reservoir, while the inclined tube is open-ended. This is shown schematically in Figure 3.14. The unknown pressure p_u is given by

$$p_A + \gamma h = p_u \tag{3.13a}$$

or

$$p_u - p_A = \gamma h \tag{3.13b}$$

But h is $h' \sin \theta$. Thus,

$$p_u - p_A = \gamma h' \sin \theta \tag{3.14}$$

Since θ is fixed, a scale placed along the tube can be calibrated to read directly in units of h of a fluid. Usually, this is done by reading directly inches of water or any other desired unit, such as centimeters of water, for $p_u - p_A$.

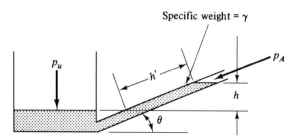

Specific weight = γ

Figure 3.14   Inclined manometer.

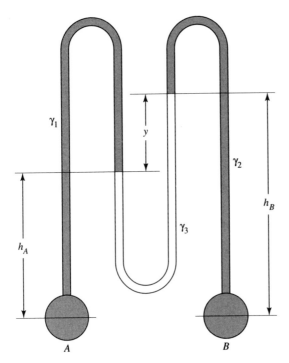

Figure 3.15   Three-fluid manometer.

Another method of achieving greater sensitivity and accuracy is to use a manometer with more than one fluid. The arrangement shown in Figure 3.15 can be used in this manner. Starting at level A yields

$$p_A - h_A\gamma_1 - y\gamma_3 + h_B\gamma_2 = p_B \tag{3.15a}$$

or

$$p_A - p_B = (-h_B\gamma_2 + h_A\gamma_1) + y\gamma_3 \tag{3.15b}$$

In the usual case the manometer is connected to different positions on the same pipe and γ_1 can be taken to be equal to γ_2. Also, $h_B - h_A = y$. Thus,

$$\boxed{p_A - p_B = y(\gamma_3 - \gamma_1) \quad \text{or} \quad y(\gamma_3 - \gamma_2)} \tag{3.16}$$

For small differences in $p_A - p_B$ it is apparent that the manometer fluid (γ_3) should have a specific weight very nearly equal to the specific weight of the fluid in the pipes. For large pressure differences one can use a heavy fluid such as mercury to increase $\gamma_3 - \gamma_2$ and reduce the manometer reading.

ILLUSTRATION 3.9   DIFFERENTIAL MANOMETER

A three-fluid manometer such as the one shown in Figure 3.15 is to be used to measure large pressure differences. Assuming that the same liquid is in the legs with $\gamma = 62.4$ lb/ft³ (water) and that the manometer fluid is mercury ($\gamma_3 = 850$ lb/ft³), determine the pressure difference between A and B if y is equal to 1 ft and A and B are at the same level.

ILLUSTRATIVE PROBLEM 3.9

Given: $\gamma_1 = \gamma_2 = 62.4 \text{ lb/ft}^3$, $\gamma_3 = 850 \text{ lb/ft}^3$, $y = 1$ ft

Find: $p_A - p_B$

From equation (3.16),

$$p_A - p_B = y(\gamma_3 - \gamma_1) = 1 \text{ ft} \times (850 - 62.4) \frac{\text{lb}}{\text{ft}^3} = 787.6 \text{ psf}$$

or

$$p_A - p_B = \frac{787.6 \text{ lb/ft}^2}{144 \text{ in.}^2/\text{ft}^2} = 5.47 \text{ psi}$$

We can also solve this problem using the equal level/equal pressure principle. Starting at B,

$$p_B - \gamma_2 h_B + \gamma_3 y + \gamma_1 h_A = p_A$$

Solving,

$$p_A - p_B = \gamma_3 y + \gamma_1 h_A - \gamma_2 h_B$$

Since $\gamma_1 = \gamma_2$ and $h_B - h_A = y$, $p_A - p_B = y(\gamma_3 - \gamma_1)$, which is the same as above.

ILLUSTRATION 3.10   DIFFERENTIAL MANOMETER

If in Illustration 3.9, A is 6 in. higher than B, determine $p_A - p_B$.

ILLUSTRATIVE PROBLEM 3.10

Given: Same as Illustrative Problem 3.9, with A 6 in. higher than B

Find: $p_A - p_B$

Solution: Refer to Figure 3.15. Starting from A, we have

$$p_A - h_A \gamma_1 - y \gamma_3 + h_B \gamma_2 = p_B$$

Assuming that $\gamma_1 = \gamma_2 = \gamma$, we obtain

$$p_A - p_B = \gamma(h_A - h_B) + y \gamma_3$$

From the statement of the problem,

$$h_B - (\tfrac{1}{2} + h_A) = y$$

or

$$h_B - h_A = y + \tfrac{1}{2}$$

and thus

$$p_A - p_B = \gamma(-y - \tfrac{1}{2}) + y(\gamma_3 - \gamma) - \tfrac{1}{2}\gamma$$

This result could almost have been anticipated by observation; that is, elevating one leg by $\tfrac{1}{2}$ ft is equivalent to decreasing the pressure differences by $\tfrac{1}{2}$ ft of fluid in the leg. Therefore,

$$p_A - p_B = 1(850 - 62.4) - \tfrac{1}{2}(62.4) = 756.4 \text{ psf}$$

or

$$p_A - p_B = \frac{756.4}{144} = 5.25 \text{ psi}$$

3.4   BUOYANCY OF SUBMERGED BODIES

If a body is weighed in a vacuum and then is weighed in a fluid, it will be found to have a different weight in each case. This is due to the buoyant action of the fluid displaced. To evaluate this effect, let us consider the body shown in Figure 3.16. This body will be considered to be floating in a submerged position, and therefore it will be in equilibrium in this position. Obviously, this is a problem in hydrostatics that can be resolved by applying the principles already developed in this chapter—by considering the small volume shown in Figure 3.16 and summing all such volumes.

From the free-body diagram shown, we need consider only the vertical forces, since all horizontal forces on the body are equal and opposite and give no net horizontal force. The

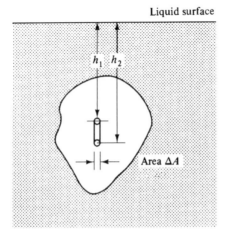

Figure 3.16   Submerged body.

vertical forces are the weight of the body and the forces on each end of the body, with the resultant of these forces equal to the sum of these three vectors. Thus,

$$\text{apparent weight (resultant)} = w + p_1\Delta A - p_2\Delta A \tag{3.17a}$$

$$\text{apparent weight (resultant)} = \gamma_s\Delta V + \gamma h_1\Delta A - \gamma h_2\Delta A \tag{3.17b}$$

where γ_s is the specific weight of the solid and γ is the specific weight of the liquid. The volume of the element is given as ΔV and equals $(h_2 - h_1)\Delta A$. Therefore,

$$\text{apparent weight} = \gamma_s\Delta V - \gamma\Delta V = \Delta V(\gamma_s - \gamma) \tag{3.18}$$

Notice that the apparent weight equals the weight of the body in a vacuum less the weight of the fluid displaced. If all the small volumes in the body are summed, the total volume of the body will be obtained. Thus

$$\text{apparent weight of the submerged body} = V(\gamma_s - \gamma) \tag{3.19}$$

Stated in words, *a submerged body is buoyed up by a force equal to the weight of the displaced fluid,* and the buoyant force must also act through the center of gravity of the displaced fluid for the body to be in equilibrium. This is known as *Archimedes' principle.*

ILLUSTRATION 3.11   BUOYANCY

A large stone weighs 100 lb in a vacuum, and when the stone is immersed in water it weighs 60 lb. Calculate the volume of the stone in cubic feet and its specific weight. Assume that the specific weight of water is 62.4 lb/ft^3.

ILLUSTRATIVE PROBLEM 3.11

Given: w_{stone} = 100 lb in vacuum, 60 lb in water; γ_{water} = 62.4 lb/ft^3

Find: V_{stone}, γ_{stone}

Solution: Refer to Figure 3.16, where the buoyant effect in water is the difference between the weight in a vacuum and the weight in water, or $100 - 60 = 40$ lb. This buoyant effect equals the weight of water displaced or the volume of displaced water multiplied by the specific weight of water:

$$\gamma V = 40 \text{ lb}$$

$$V = \frac{40 \text{ lb}}{62.4 \text{ lb/ft}^3} = 0.641 \text{ ft}^3$$

The specific weight is simply weight divided by volume:

$$\gamma_s = \frac{w}{V} = \frac{100 \text{ lb}}{0.641 \text{ ft}^3} = 156.01 \text{ lb/ft}^3$$

ILLUSTRATION 3.12   BUOYANCY

Solve Illustration 3.11 if the initial weighing of 100 lb is made in air having a specific weight (γ) of 0.075 lb/ft³.

ILLUSTRATIVE PROBLEM 3.12

Given: γ_{air} = 0.075 lb/ft³; other quantities the same as in Illustrative Problem 3.11

Find: V_{stone}, γ_{stone}

Solution: If we include the buoyant effect of the air, we need to write a separate equation for each weighing. Thus,

$$100 = \gamma_s V - \gamma_a V$$

where γ_a is the specific weight of the air and γ_s is the specific weight of the stone. Also,

$$60 = \gamma_s V - \gamma_w V$$

where γ_w is the specific weight of water. Subtracting the second equation from the first yields

$$40 = \gamma_w V - \gamma_a V = V(\gamma_w - \gamma_a)$$

and

$$V = \frac{40}{62.4 - 0.075} = 0.642 \text{ ft}^3$$

From our first equation,

$$100 = \gamma_s \times 0.642 - 0.075 \times 0.642$$

yielding

$$\gamma_s = 155.84 \text{ lb/ft}^3$$

   It is obvious that for this large object the buoyant effect of the air is essentially negligible. The error in weighing would probably exceed the buoyant effect of the air. For most applications, weighing in air can be taken to be the same as weighing in a vacuum. Only for the most precise analytical applications will it be necessary to include this effect.

ILLUSTRATION 3.13   BUOYANCY

A balloon uses hot air to provide its lifting capability. Prior to release it is tethered to the ground by a steel cable. If the gondola, weights, passengers, and material weigh 2000 lb, determine the tension in the cable when the balloon is inflated as a sphere having a diameter of 60 ft. Assume that the hot air has a specific weight of 0.040 lb/ft³ and the ambient air has a specific weight of 0.070 lb/ft³.

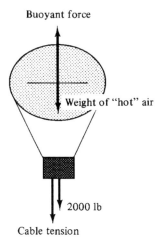

Buoyant force

Weight of "hot" air

2000 lb

Cable tension   **Figure 3.17**   Illustrative Problem 3.13.

ILLUSTRATIVE PROBLEM 3.13

Given: $d = 60$ ft, $\gamma_{\text{hot air}} = 0.040$ lb/ft³, $\gamma_{\text{air}} = 0.070$ lb/ft³, weight $= 2000$ lb

Find: Cable tension

Solution: From the free-body diagram of Figure 3.17, we see that the tension in the cable must equal the buoyant force on the balloon minus the weight of the hot air added to the weight of the loads. To calculate the buoyant force, we have

$$\text{buoyant force} = \gamma_{\text{air}} \times V_{\text{balloon}} = \gamma_{\text{air}} \times \frac{4}{3}\pi R^3$$

$$= 0.070 \; \frac{\text{lb}}{\text{ft}^3} \times \frac{4}{3}\pi(30)^3 \; \text{ft}^3 \quad = 7917 \; \text{lb}$$

$$\text{weight of "hot" air} = 0.040 \; \frac{\text{lb}}{\text{ft}^3} \times \frac{4}{3}\pi(30)^3 \; \text{ft}^3 \quad = 4524 \; \text{lb}$$

$$\text{other weights} \qquad\qquad\qquad\qquad\qquad = 2000 \; \text{lb}$$

Therefore,

$$\text{tension} = 7917 \; \text{lb} - (4524 \; \text{lb} + 2000 \; \text{lb}) = 1393 \; \text{lb}$$

ILLUSTRATION 3.14   BUOYANCY

A block of material having a specific weight of 35 kN/m³ is suspended in water by a cable. If the tension in the cable is 5.0 kN, determine the volume of the block.

ILLUSTRATIVE PROBLEM 3.14

Given: $T = 5.0$ kN, $\gamma_{water} = 9.81\dfrac{kN}{m^3}$, $\gamma = 35\dfrac{kN}{m^3}$

Find: Volume of block

Solution: From the free-body diagram of Figure 3.18,

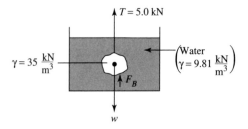

Figure 3.18   Illustrative Problem 3.14

$$T + F_B - w = 0$$

$$T + F_B = w = \gamma V = 35V$$

But the buoyant force $F_B = \gamma_{water}(V)$

$$T + 9.81V = 35V$$

$$5 = 25.19V$$

$$V = 0.198 \text{ m}^3$$

3.5   FORCES ON PLANE-SUBMERGED SURFACES

It was seen in Chapter 2 that the shear stress in a fluid is proportional to the velocity difference between the shearing surfaces and inversely proportional to the distance separating the surfaces. If a fluid is at rest, the shear stress must be zero, which means that no tangential forces can exist in the fluid. The only forces that can exist are forces that are normal to surfaces in contact with the fluid. If a plane is completely immersed in a fluid it will experience a normal force on both sides of the plane, which will tend to compress the plane. However, when a plane is placed on a fluid so that only one side is subjected to the fluid pressure, it will experience a net unbalanced force. Since pressure varies with depth, the pressure on a nonhorizontal plane will in general be variable.

First let us consider the special case of a vertical plane and subdivide this topic into (1) a vertical plane just touching a free surface, (2) a completely submerged vertical plane, and (3) a vertical plane with different liquid levels on each side.

For the vertical plane just touching a free surface, case 1, the pressure on one side will vary linearly from zero to γh due to the column of fluid on this side. This is shown schematically in Figure 3.19a, where the pressure is indicated to vary linearly, giving rise to a triangular pressure distribution. The force exerted on the area ΔA is the average pressure on this area multiplied by the area. The average pressure is

$$p_{avg} = \gamma x + \frac{\gamma \Delta x}{2} \tag{3.20}$$

The force on the area ΔA is the product of the pressure and area,

$$F_x = \gamma x (\Delta A) + \frac{\gamma (\Delta x)(\Delta A)}{2} \tag{3.21}$$

Since Δx is small, the term $(\Delta x)(\Delta A)$ is indeed very small and may be neglected when compared to terms containing only Δx. Therefore, the second term in equation (3.21) can be neglected when compared with the first term. Thus

$$F_x = \gamma x \Delta A \tag{3.22}$$

It is necessary to sum each of the F_x terms over all values of x in order to evaluate the total force on the plane. For the special case where the width of the plane is constant, it will be immediately seen that A, the sum of all the ΔA's, is not a function of x, and one need only use the average value of x to obtain the total force on the plane. Thus since x varies from zero to h,

$$F_x = \frac{\gamma h A}{2} \tag{3.23}$$

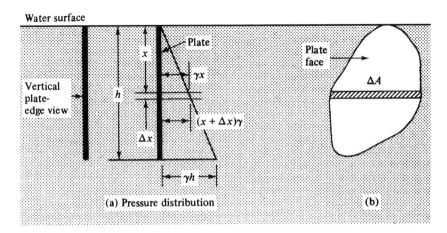

(a) Pressure distribution   **(b)**

Figure 3.19   Pressure on a vertical submerged plane with one edge touching a free surface.

Note that $h/2$ represents the *center of gravity* (or *centroid*) of a rectangle and that the average pressure on the area is just $\gamma h/2$ or $\gamma \bar{h}$, where $\bar{h}$ is the location of the center of gravity from the liquid surface.

Let us now consider a plane whose width is not constant but varies in some manner with liquid depth. This situation is shown in Figure 3.19b. The pressure on the small area ΔA can be written as

$$p = \frac{F_x}{A} = \frac{\gamma x \Delta A}{A} \tag{3.24}$$

and the pressure over the entire plate as

$$p = \frac{F_x}{A} = \sum \frac{\gamma x \Delta A}{A} \tag{3.25}$$

or simply,

$$p = \sum \frac{\gamma x \Delta A}{A} \tag{3.26}$$

where $\sum$ denotes the sum of all such quantities.

The student will note from elementary physics that $\sum(x \Delta A / A)$ is the definition of the *location of the center of gravity* or centroid of any area. Thus

$$p = \gamma \bar{h} \tag{3.27}$$

where $\bar{h}$ is the location of the center of gravity of the area from the surface. Thus, from equation (3.27) we see for a vertical submerged plane touching the surface of a fluid that the average pressure on the plane is simply the product of the specific weight of the fluid multiplied by the depth of the center of gravity from the surface of the fluid. The total force is the total area multiplied by this average pressure. Therefore,

$$F_{\text{total}} = \gamma \bar{h} A \tag{3.28}$$

The second case of a vertical plane, case 2, is one that is completely submerged. Figure 3.20 shows a vertical plane of any shape whose top edge is located some distance below the free surface of the liquid. Again, the pressure on one side will vary linearly with the depth of fluid, and the plot of pressure on one side of the plane against the distance below the top edge of the plate is a trapezoid whose value of pressure on the top edge is γh_1 and on the bottom edge is γh_2. The force on the area ΔA is the average pressure multiplied by the area. Therefore,

$$F_x = pA \tag{3.29}$$

or

$$p = \frac{F_x}{A} \tag{3.30}$$

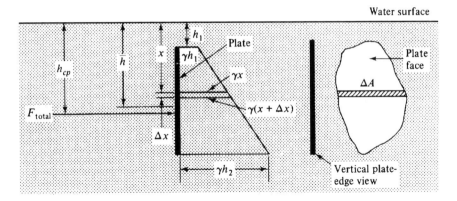

Figure 3.20 Submerged vertical surface.

and

$$p = \frac{F_x}{\Delta A} = \gamma x + \frac{\gamma \Delta x}{2} \qquad (3.31)$$

The force on ΔA is therefore

$$F_x = \left(\gamma x + \frac{\gamma \Delta x}{2}\right)\Delta A \qquad (3.32)$$

Carrying out the multiplication indicated by equation (3.32) and neglecting products of small numbers and adding all such items yields

$$F_x = \sum \gamma x \Delta A \qquad (3.33)$$

Since pressure is force divided by area, the average pressure on the plate is

$$p = \sum \frac{\gamma x \Delta A}{A} \qquad (3.34)$$

where A is the total plate area and $\sum$ is the sum of all such quantities over the plate. Again $\sum(x\Delta A/A)$ is simply the location of the center of gravity or centroid of the plate. Thus, we have the result that for a vertical submerged plate the average pressure on a surface is simply the product of the specific weight of the fluid multiplied by the depth of the center of gravity below the surface of the fluid. Also, as before, the total force is the total area multiplied by the average pressure.

This leads to the result that equation (3.18) is general for any vertical surface, namely $F_{total} = \gamma \bar{h} A$, *where $\bar{h}$ is the location of the center of gravity of the plane below the surface of the fluid.*

ILLUSTRATION 3.15   FORCE ON A VERTICAL SUBMERGED SURFACE

As shown in Figure 3.21, a vertical plate is used to dam a channel 2 ft wide. If the plate is $2\frac{1}{2}$ ft high and the water on one side of the plate is at a level $\frac{1}{2}$ ft below the top, what is the total force on the plate?

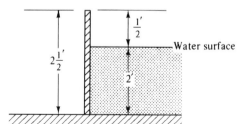

Figure 3.21   Illustrative Problem 3.15.

ILLUSTRATIVE PROBLEM 3.15

Given: Channel 2 ft wide, water = 2 ft high, plate = $2\frac{1}{2}$ ft high

Find: F_{total}

Solution: For this problem, the $2\frac{1}{2}$-ft dimension is not important. Only the portion of the plate that is submerged is important. Its center of gravity is 1 ft below the free liquid surface and the area in contact with the liquid is 2 × 2 ft, or 4 ft². The total force on the plate is therefore

$$F_{total} = \gamma \bar{h} A = 62.4 \, \frac{\text{lb}}{\text{ft}^3} \times 4 \text{ ft}^2 \times 1 \text{ ft} = 249.6 \text{ lb}$$

As the last case of a vertical surface subjected to hydrostatic forces, case 3, consider the plane subjected to different liquid levels (but the same liquid) on both sides. It is possible to obtain the net force acting on this plane by considering each side separately and vectorially summing the forces. By considering Figure 3.22 a simple expression for the net force for this case can readily be developed. The pressure distribution for the left side is a triangle extending from zero at the upper free surface to γh_1 at the lower edge of the plate. On the right-hand side the pressure distribution is also a triangle extending from zero at its free surface to γh_3 at the base of the plate. By superposing these two diagrams as shown by the dashed line on the figure, we reach the conclusion that from h_2 down the pressure is constant and equal to γh_2 on the plate. The total force on the plate is therefore given by

$$F_{total} = \gamma h_2 A_1 + \gamma \frac{h_2}{2} A_2 \tag{3.35}$$

where A_1 is the plate area below the right liquid surface and A_2 the area above the right liquid surface.

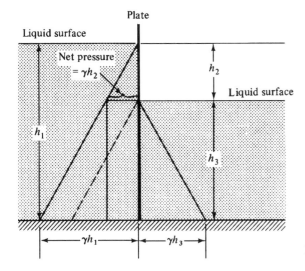

Figure 3.22   General submerged vertical surface.

ILLUSTRATION 3.16   FORCE ON VERTICAL SUBMERGED SURFACES

As shown in Figure 3.23, a vertical plate extends from the surface of water on one side to a depth of 10 m. On the other side there is water to within 2 m of the top. If the plate is 5 m wide, what is the net force on the plate?

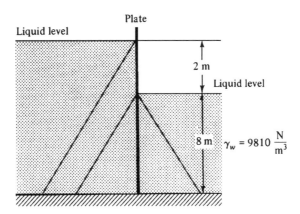

Figure 3.23   Illustrative Problem 3.16

ILLUSTRATIVE PROBLEM 3.16

Given: Vertical plate with $h = 10$ m on one side, 8 m on the other side; width = 5 m, $\gamma = 9810$ N/m^3.

Find: F_{total} on plate

Solution: This problem can be solved by considering each side separately. For the left side,

$$F_{total} = \gamma \bar{h} A = 9810 \, \frac{N}{m^3} \times 5 \text{ m} \times 10 \text{ m} \times 5 \text{ m} = 2 \, 452 \, 500 \text{ N}$$

On the right side,

$$F_{total} = \gamma \bar{h} A = 9810 \, \frac{N}{m^3} \times 4 \text{ m} \times 8 \text{ m} \times 5 \text{ m} = 1 \, 569 \, 600 \text{ N}$$

Thus,

$$\text{net force} = 2 \, 452 \, 500 \text{ N} - 1 \, 569 \, 600 \text{ N} = 882 \, 900 \text{ N} = 882.9 \text{ kN}$$

An alternative solution is obtained by considering the superposed pressure distribution as shown in Figure 3.23. For the upper 2 m, we have

$$F_{total} = 1 \text{ m} \times 2 \text{ m} \times 5 \text{ m} \times 9810 \, \frac{N}{m^3} = 98 \, 100 \text{ N}$$

The lower 8 m can be taken to have a uniform pressure distribution equal to 10 m minus 8 m, or 2 m over the entire surface. The force is therefore

$$9810 \, \frac{N}{m^3} \times 2 \text{ m} \times 8 \text{ m} \times 5 \text{ m} = 784 \, 800 \text{ N}$$

The total force is the sum of these two forces, or

$$F_{total} = 98 \, 100 \text{ N} + 784 \, 800 \text{ N} = 882 \, 900 \text{ N} = 882.9 \text{ kN}$$

Both methods yield the same result, and it is left to the student to use the one of his or her choice, or, better yet, to do this type of problem by both methods to obtain an independent check on the solution.

Vertical and horizontal surfaces represent special cases of submerged planes, and, in general, the submerged plane can make any angle with the free surface of the liquid. Consider the plane shown in Figure 3.24, which makes an angle θ with the liquid surface. Also recall that it was proved earlier in this chapter that the pressure intensity on any horizontal plane must be the same anywhere on the plane. Thus, the pressure at the level x is the same regardless of the orientation of the surface. However, this pressure acts normal to the surface, which in this case is normal to Δy and over the area ΔA. Using the same procedure as before leads us to the solution that

$$p = \gamma \bar{L} \sin \theta \tag{3.36}$$

and since $\bar{L} \sin \theta = \bar{h}$,

$$p = \gamma \bar{h} \tag{3.37}$$

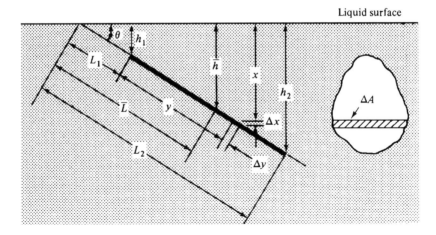

Figure 3.24   Inclined submerged plane.

Note that in equations (3.36) and (3.37) $\bar{L}$ and $\bar{h}$ are the distances to the center of gravity of the plate shown in Figure 3.24. Thus we obtain

$$F_{\text{total}} = \gamma \bar{h} A \qquad\qquad (3.38)$$

It can now be stated as a general result that the total force acting on a submerged plane area is the product of the specific weight of the liquid multiplied by the area and multiplied by the depth of the center of gravity (or centroid) of the area from the liquid surface. This conclusion does not depend on the angle of the plane area.

ILLUSTRATION 3.17   FORCE ON A SUBMERGED SURFACE

A rectangular gate 1 m × 1 m is mounted in the side of a wall having the slope shown in Figure 3.25. Water fills the container until it is 6 m above the top of the gate. What is the total force on the gate?

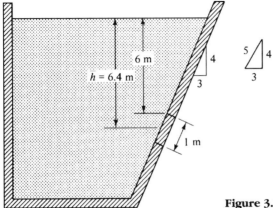

Figure 3.25   Illustrative Problem 3.17.

ILLUSTRATIVE PROBLEM 3.17

Given: Gate 1 m × 1 m; water height = 6 ft over top of gate; gate has slope = $\frac{4}{3}$

Find: F_{total} on gate

Solution: From the figure we have $\bar{h}$ = 6.4 m; therefore,

$$F = \gamma \bar{h} A = 9810 \, \frac{N}{m^3} \times 6.4 \text{ m} \times 1 \text{ m} \times 1 \text{ m} = 62\ 784 \text{ N}$$

ILLUSTRATION 3.18

As shown in Figure 3.26, a valve (plate) is placed in a channel to control the flow. Assume that the valve consists of a plane that makes an angle of 30° with the free surface. If the channel is 5 × 5 ft in cross section and is flowing full, what is the total force on the valve after the flow is cut off? Assume that γ = 62.4 lb/ft³.

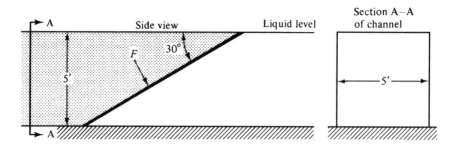

Figure 3.26   Illustrative Problem 3.18.

ILLUSTRATIVE PROBLEM 3.18

Given: θ = 30°, plate = 5 ft × 5 ft, γ = 62.4 lb/ft³

Find: F_{total} on plate

Solution: The area of the valve is (5 × 5)/sin 30 = 50 ft². The centroid of the plate is located $2\frac{1}{2}$ ft below the surface. The total force is therefore

$$F_{total} = \gamma \bar{h} A = 62.4 \frac{lb}{ft^3} \times 2\tfrac{1}{2} \text{ ft} \times 50 \text{ ft}^2 = 7800 \text{ lb}$$

3.6   LOCATION OF THE CENTER OF PRESSURE

In Section 3.5 the discussion was concerned with the magnitude of the force on a plane area. Since force is a vector quantity, a force is completely described by knowing its magnitude, direction, and point of application. Thus far we have determined the magnitude and direction but not the point of application of these vectors (forces). To determine the point of application of each of the forces, let us once more consider the case of a fully submerged vertical plane such as is shown in Figure 3.20. Since the area is in equilibrium, the sum of the moments about any point must be zero. Therefore, by referring to Figure 3.20 and using the notation h_{cp} for the vertical location of the point of application of the resultant force,

$$(F_{total})h_{cp} = \sum \gamma x \Delta A(x) \tag{3.39a}$$

and

$$h_{cp} = \sum \frac{\gamma x^2 \Delta A}{F_{total}} \tag{3.39b}$$

But

$$F_{total} = \gamma \bar{h} A$$

Therefore,

$$h_{cp} = \sum \frac{x^2 \Delta A}{\bar{h} A} \tag{3.40}$$

The summation term $\sum x^2 \Delta A$ in equation (3.40) is the second moment of area, or, more conventionally, it is the moment of inertia (I) of the area A about the axis at the surface of the liquid. Actually, it is more convenient to express equation (3.39) in terms of the moment of inertia about the center of gravity of the area. To do this we utilize the transfer theorem of mechanics, and inserting the results into equation (3.40), we obtain

$$h_{cp} = \frac{\bar{I}}{\bar{h} A} + \bar{h} \tag{3.41}$$

where $\bar{h}$ is the location of the center of gravity and $\bar{I}$ is the moment of inertia of the plane area about a centroidal axis. The location of the resultant force is known as the center of pressure. Table 3.2 gives the properties of some selected areas. Notice that the center of pressure is always below the center of gravity.

The student should note that it is always possible to evaluate either $\bar{h}$ or $\bar{I}$ numerically by subdividing the area in question into many small parts and performing the indicated mathematical operations and summation. Even a relatively coarse subdivision quickly leads to a good numerical approximation of the exact solution.

Table 3.2 Properties of an Area

Section	Area	$\bar{b}$ (location of center of gravity)	$\bar{I}$ (moment of inertia about the center of gravity)
Square	A^2	$\dfrac{1}{2}A$	$\dfrac{1}{12}A^4$
Rectangle	bd	$\dfrac{1}{2}d$	$\dfrac{1}{12}bd^3$
Trapezoid	$\dfrac{1}{2}(B + b)d$	$d\left\{\dfrac{2B + b}{3(B + b)}\right\}$	$\dfrac{d^3(B^2 + 4Bb + b^2)}{36(B + b)}$
Triangle	$\dfrac{1}{2}bd$	$\dfrac{2}{3}d$	$\dfrac{bd^3}{36}$
Circle	πR^2	R	$\dfrac{\pi R^4}{4}$
Ellipse	$\dfrac{\pi bd}{4}$	$\dfrac{1}{2}d$	$\dfrac{\pi bd^3}{64}$
Quarter Circle—Also * for Semicircle	$\dfrac{\pi R^2}{4}$	$0.4244\,R$	$\dfrac{0.055R^4}{8}$
	$\dfrac{{}^*\pi R^2}{2}$	${}^*0.4244R$	$\dfrac{{}^*0.11R^4}{8}$

ILLUSTRATION 3.19   LOCATION OF THE CENTER OF PRESSURE

As shown in Figure 3.27, a sluice gate is 6 ft square. It is hinged at the top and held by two horizontal pins 1 ft from the bottom of the gate. What force must each of these pins resist if the water level is at the top of the gate?

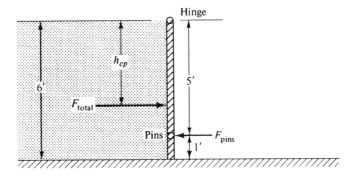

Figure 3.27   Illustrative Problem 3.19.

ILLUSTRATIVE PROBLEM 3.19

Given: Gate 6 ft × 6 ft; pins 1 ft from bottom

Find: F_{pin}

Solution: For the gate to be in equilibrium under the action of the forces shown, it is necessary that the moment of these forces about the hinge be zero. The hydrostatic force, F_{total}, is

$$F_{total} = \tfrac{6}{2} \times 6 \times 6 \times 62.4 = 6740 \text{ lb}$$

The location of h_{cp} is

$$\frac{\bar{I}}{\bar{h}A} + \bar{h} = \frac{bd^3/12}{d/[2(d)(b)]} + \frac{d}{2} = \frac{d}{6} + \frac{d}{2} = 4 \text{ ft}$$

The moment of these forces about the hinge is

$$(F_{total})4 = (F_{pins})5$$

Therefore,

$$F_{pins} = \tfrac{4}{5}(6740) = 5392 \text{ lb}$$

and for one pin

$$F_{pin} = 2696 \text{ lb}$$

ILLUSTRATION 3.20   LOCATION OF THE CENTER OF PRESSURE

Flashboards are installed on top of a dam and are supported by pipes spaced 4 m apart along the dam. If the water is flowing over the flashboard 1 m deep as shown in Figure 3.28, what is the bending moment produced at the bottom of the flashboard on one pipe?

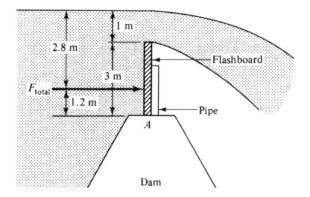

Figure 3.28   Illustrative Problem 3.20.

ILLUSTRATIVE PROBLEM 3.20

Given: Flashboards 3 m × 4 m; water 1 m over flashboards; flashboards supported by pipes at 4 m intervals

Find: $M_{bending}$ on a pipe

Solution: To solve this problem, let us select a repeating section of flashboard 4 m wide. The total force on the flashboard is

$$F_{total} = 9810 \ \frac{N}{m^3} \times 4 \ m \times 3 \ m \times (1 + 1.5) \ m = 294.3 \ kN$$

The location of h_{cp} is

$$h_{cp} = \frac{bd^3/12}{\bar{h} A} + \bar{h}$$

and

$$h_{cp} = \frac{(4 \times 3^3)/12}{(1 + 1.5)(4 \times 3)} + 2.5 = 2.8 \ m \ \text{from the water surface}$$

Taking moments about A, 1.2 × 294.3 = 353.16 kN·m. Since per repeating section there is the equivalent of one pipe, this is the moment at the bottom of a pipe.

For a nonvertical plane, as is shown in Figure 3.24, we can also evaluate the location of the center of pressure in a manner similar to that used for the vertical plane. When this is done using calculus, it can be demonstrated that

$$L_{cp} = \frac{\bar{I}}{\bar{L}A} + \bar{L} \tag{3.42}$$

where L_{cp} denotes the location of the center of pressure along the plane. Thus, the distance from a liquid surface to the center of pressure along the plane is the same regardless of the angle of inclination as long as the angle is not zero. Again we note that the center of pressure is always below the center of gravity.

ILLUSTRATION 3.21   CENTER OF PRESSURE

Determine the center of pressure for the gate of Illustrative Problem 3.17.

ILLUSTRATIVE PROBLEM 3.21

Given: Same as Illustrative Problem 3.17

Find: Location of center of pressure

Solution: Referring to Illustrative Problem 3.17, we have $\bar{h}$ = 6.4 m. Therefore, $\bar{L}$ = 6.4/0.8 = 8 m,

$$\bar{I} = \frac{bd^3}{12} = \frac{1 \times 1^3}{12} = \frac{1}{12}$$

and

$$A = 1 \times 1 = 1 \ m^2$$

From equation (3.42),

$$L_{cp} = \frac{\bar{I}}{\bar{L}A} + \bar{L} = \frac{1/12}{8 \times 1} + 8 = 8.0104 \text{ m along the plane}$$

Vertically, this corresponds to $8.0104 \times 0.8 = 6.408$ m below the surface.

ILLUSTRATION 3.22   CENTER OF PRESSURE

Determine the center of pressure for the valve of Illustrative Problem 3.18.

ILLUSTRATIVE PROBLEM 3.22

Given: Same as Illustrative Problem 3.18

Find: Location of center of pressure

Solution:

$$L_{cp} = \frac{\bar{I}}{\bar{L}A} + \bar{L}$$

For this problem all dimensions will be considered *along* the plane:

$$\bar{I} = \frac{bd^3}{12} = \frac{5 \times (5/\sin 30)^3}{12} = \frac{5 \times (10)^3}{12} = \frac{5000}{12}$$

$$\bar{L} = \frac{2.5}{\sin 30} = 5 \text{ ft}$$

$$A = 5 \times 10 = 50 \text{ ft}$$

Therefore,

$$L_{cp} \text{ along plane } = \frac{5000/12}{5 \times 50} + 5 = 1.67 + 5 = 6.67 \text{ ft}$$

For the vertical height h_{cp},

$$h_{cp} = 6.67 \sin 30 = 3.34 \text{ ft vertically below the surface}$$

3.7   STRESSES IN CYLINDERS AND SPHERES

When a fluid is contained, it exerts forces on the walls of the container that stress the material of the container. As an extension of the principles discussed in this chapter, let us consider the problem of evaluating the tensile stresses in thin cylinders and spheres subjected to an internal pressure. To evaluate these stresses we shall assume that the ratio of the thickness to the diameter of the cylinder or sphere is small and that the stresses developed in these pressure vessels are uniform over the cross section of the vessel. By *thin* we shall mean that the ratio of thickness to diameter is 0.1 or less.

With these assumptions in mind, let us consider the cylinder shown in Figure 3.29. Due to the internal pressure there is a stress in the circumferential direction (S_2) and a stress in

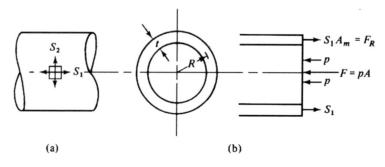

Figure 3.29   Thin cylinder under internal pressure.

the longitudinal direction (S_1), as shown in Figure 3.29a. If the pipe is cut perpendicular to the longitudinal axis (and far from the ends) the resultant free-body diagram will be as shown in Figure 3.29b. The total resisting force in the cylinder will be the stress S_1 multiplied by the area over which the stress acts. Thus, the resisting force is $S_1(2\pi R)(t)$ since the material area is $2\pi Rt$. The applied load is due to the pressure in the tube acting over the tube area. The applied load is therefore $p\pi R^2$. For equilibrium these forces must be equal. Therefore,

$$S_1 2\pi Rt = p\pi R^2$$

or

$$S_1 = \frac{pR}{2t} \qquad (3.43)$$

Let us now consider a diametrical cut through the cylinder as shown in Figure 3.30 on page 104. The resisting force, F_R, equals $S_2 A_m$, where A_m is the metal area. Since there are two metal areas, $A_m = 2tL$ and $F_R = S_2 \times 2tL$. The applied force F consists of the pressure p acting on the projected area, which is $p \times 2RL$. For equilibrium we equate the applied and resisting forces to obtain

$$2tLS_2 = p\,2RL$$

Solving for S_2 yields

$$S_2 = \frac{pR}{t} \qquad (3.44)$$

From equations (3.43) and (3.44) it can be seen that S_2 is twice S_1. S_1 is the tensile stress in the longitudinal seam of the cylinder and S_2 is the transverse stress (circumferential). The student should also note that these stresses were arrived at without having to involve the material of the cylinder. The strength of the material will determine the thickness of the cylinder required for a given internal pressure, and the interested student is referred to any standard text on strength of materials for further discussion of this subject.

Let us now consider a thin sphere subjected to an internal pressure. In this sphere S_1 will be equal to S_2 by symmetry. As can be seen from Figure 3.31 on page 104, the stress S_2 acts over an area equal to $2\pi Rt$. The sum of the force components in the horizontal direction is once again just the product of the pressure multiplied by the projected area, that is, $p\pi R^2$. Therefore,

$$S_2 2\pi Rt = p\pi R^2$$

or

$$S_2 = S_1 = \frac{pR}{2t} \qquad (3.45)$$

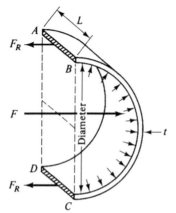

Figure 3.30   Thin cylinder under internal pressure.

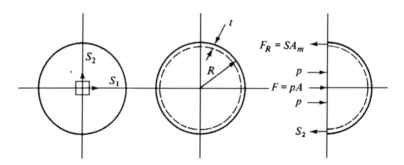

Figure 3.31   Thin sphere under internal pressure.

Equation (3.45) yields the result that S_2 in a thin sphere is half of S_2 in a thin cylinder and equal to S_1 when both are subjected to the same internal pressure. In other words, a thin sphere can be made half as thick as a thin cylinder for the same pressure. Conversely, a thin sphere can be subjected to twice the internal pressure than that of a thin cylinder of the same thickness and diameter. The student is cautioned to consult a text on strength of materials or applicable codes before attempting to design cylinders and spheres for either internal or external pressure.

ILLUSTRATION 3.23   STRESSES IN A PIPE

A 48-in. diameter steel pipe 0.25 in. thick carries water under an internal pressure of 100 psi. Compute the circumferential stress in the steel. If the pressure is raised to 250 psi and the allowable unit stress in the steel is 15,000 psi, what thickness of steel is required?

ILLUSTRATIVE PROBLEM 3.23

Given: (a) $d = 48$ in., $t = 0.25$ in., $p = 100$ psi
           (b) $d = 48$ in., $p = 250$ psi, $S_{\text{allowable}} = 15000$ psi

Find: (a) S_2
          (b) t

Solution: The circumferential stress is S_2. Thus, from equation (3.44),

$$S_2 = \frac{pR}{t} = \frac{100 \times 24}{0.25} = 9600 \text{ psi}$$

The required thickness is

$$t = \frac{pR}{S_2} = \frac{250 \times 24}{15,000} = 0.4 \text{ in.}$$

3.8   REVIEW

In this chapter we have considered fluids that are at rest. After reviewing the pressure–height relationship, the equal level/equal pressure principle was developed and applied to various pressure measurement devices called manometers. For a variable-density fluid, such as air, whose temperature was taken to be constant, we found it possible to develop an analytical expression for the change in pressure as a function of the height of the column.

When a body is submerged in a fluid it experiences a buoyant force that is equal to the weight of the fluid displaced. This is known as Archimedes' principle. Plane surfaces that are placed in a fluid experience forces due to the effect of the hydrostatic pressure variation over the surfaces. By consideration of the forces acting on small sections of the surface it was possible for us to derive a general relation that stated that the total force acting on a submerged plane surface could be obtained by considering that the pressure on the surface was that at the surface's center of gravity. The point of application of the force vector is called the center of pressure, and an analytical expression was developed for determining its location. It was noted that the center of pressure is always below the center of gravity.

Our final application of fluid statics was to calculate the stresses in thin cylinders and spheres due to internal fluid pressure. A thin sphere is a stronger vessel than is a thin cylinder, since it can sustain twice the pressure for a given diameter and a given thickness of material.

KEY TERMS

Terms of importance in this chapter:

Archimedes' principle: the principle that a submerged body is buoyed up by a force equal to the weight of the liquid that it displaces.

Barometer: an absolute pressure measuring device used to measure the pressure of the atmosphere.

Buoyancy: the effect that a fluid has on decreasing the weight of a body when a body is immersed in it.

Center of gravity of an area: the point of an area about which the area will have no net moment—if suspended from this point (in principle), the area will not rotate.

Center of pressure: the point of an area where the hydrostatic force vector acts.

Centroid: *see* center of gravity of an area.

Equal level/equal pressure principle: a statement of the fact that at a given horizontal level in a fluid the pressure must be the same in all directions if the fluid is static.

Manometer: an instrument used to measure pressure or pressure differences in a system.

Moment of inertia: the second moment of an area about an axis.

Standard atmosphere: an agreed-upon defined atmosphere whose properties are accepted for use in performance calculations.

Thin: as used in conjunction with cylinders and spheres, taken to mean a ratio of thickness to diameter of 0.1 or less.

KEY EQUATIONS ·

Pressure–height relation	$p = -\gamma h$	(3.2b)
Pressure–height relation (air at constant temperature)	$\ln \dfrac{b_1}{b_2} = \dfrac{x}{RT}$	(3.8)
Inclined manometer	$p_u - p_A = \gamma h' \sin \theta$	(3.14)
Three-fluid manometer	$p_A - p_B = y(\gamma_3 - \gamma_1) = y(\gamma_3 - \gamma_2)$	(3.16)
Buoyancy	apparent weight of submerged body $= V(\gamma_s - \gamma)$	(3.19)
Force on a submerged vertical surface	$F_{total} = \gamma \bar{h} A$	(3.28)
Force on a general submerged surface	$F_{total} = \gamma \bar{h} A$	(3.38)
Location of center of pressure	$h_{cp} = \dfrac{\bar{I}}{\bar{h} A} + \bar{h}$	(3.41)
Location of center of pressure	$L_{cp} = \dfrac{\bar{I}}{\bar{L} A} + \bar{L}$	(3.42)
Stresses in a thin cylinder	$S_1 = \dfrac{pR}{2t}; S_2 = \dfrac{pR}{t}$	(3.43; 3.44)
Stresses in a thin sphere	$S_1 = S_2 = \dfrac{pR}{2t}$	(3.45)

QUESTIONS

1. Why is there a negative sign in the pressure–height relation?

2. Describe some of the factors that must be considered before using a fluid in a manometer.

3. For the measurement of small pressure differences, would you use a fluid having (a) a high specific gravity, (b) a low specific gravity, or (c) a specific gravity of 1?

4. Why is mercury rather than water a good barometer fluid?

5. What is the problem when mercury is used as a manometer fluid?

6. Manometers can be used to measure (a) absolute pressure, (b) gage pressure, (c) pressure differences, (d) all of these, (e) none of these items.

7. A three-fluid manometer is used (a) to measure large pressure differences, (b) to provide greater sensitivity, (c) to confuse students, (d) to do none of these, (d) to do all of these.

8. What problem would you have if you used a glass manometer to measure small pressure differences in a system whose pressure was variable?

9. Why is it possible for a steel ship to float in water?

10. Explain how a submarine floats on the surface, floats submerged, and returns to the surface after being submerged.

11. In what manner does a blimp resemble a submarine?

12. How does a blimp rise, stay level, and descend?

13. What characteristics should the gas in a blimp have?

14. Hydrogen has some desirable characteristics for use as the gas in blimps. Why is helium used instead in all modern blimps?

15. If a vertical plate is placed in a large pool of water, the total force on it, other than its weight, will be (a) directed to the left, (b) directed to the right, (c) undetermined, or (d) zero?

16. *True or false:* If an area has an axis of symmetry, the center of gravity will always be on this axis.

17. It is possible to have a positive or negative sense in a moment. When determining the moment of inertia of an area, is it possible to have both positive and negative terms?

18. The center of pressure of an area is always (a) at the same point as the center of gravity, (b) above the center of gravity, (c) below the center of gravity, (d) to the right of the center of gravity, or (e) to the left of the center of gravity.

19. Even though a sphere can be made thinner than a cylinder for a given diameter and internal pressure, most pressure vessels are made cylindrical. Why is this so?

20. Quite often cylindrical pressure vessels have hemispherical end caps. How thick would you expect these caps to be?

PROBLEMS

Pressure

3.1. If a column of water is 20 ft high, what is the pressure at its base due to the water only, in psf, psi, and kPa? Use $\gamma = 62.4$ lb/ft^3.

3.2. A column of water is 8 m high. What is the pressure at its base due to the water only in psi, psf, and kPa? Use $\gamma = 9400$ N/m^3.

3.3. If the atmospheric pressure is 14.7 psia, what are the absolute pressures in Problem 3.1?

3.4. If the atmospheric pressure is 100 kPa, what are the absolute pressures in Problem 3.1?

3.5. A column of mercury ($\gamma = 133.5$ kN/m^3) is 25 in. high. What is the pressure at its base due to the mercury only?

3.6. A skin diver descends to a depth of 60 ft in fresh water. What is the total pressure on his body if $\gamma = 62.4$ lb/ft^3?

3.7. The pressure at a level of 100 ft in the Great Salt Lake is 46 psig. What is the specific weight of the lake water if it is constant?

3.8. If the specific weight of air at sea level is taken to be 0.075 lb/ft^3, how high would the atmosphere extend if γ of air is assumed to remain constant?

3.9. If the atmospheric pressure at some unknown elevation is half that at sea level, determine the elevation. Assume the temperature of the air to be constant and equal to 70°F.

3.10. Using equation (3.8), evaluate the height at which the pressure of the atmosphere will become zero. Would you expect this to be the case in the actual atmosphere?

3.11. A barometer reads 755 mm Hg at sea level and 20°C. If it is now taken up a hill that is 1000 m high, what is the expected barometer reading? Assume that the temperature remains constant.

3.12. Table 3.1 for the ICAO Standard Atmosphere gives a pressure of 12.243 psia at 5000 ft. Assuming air to have the constant temperature given at zero altitude (59°F), what pressure would equation (3.8) have given for this altitude? Use $p_0 = 14.696$ psia.

*3.13. An airplane uses a barometer as an altimeter. While approaching an airport, the pilot is told that the local barometer stands at 29.8 in. Hg. Using this value to set zero, the pilot reads 1500 ft altitude. However, the instrument does not compensate for temperature and is only correct at 70°F. If the air is at 0°F, will the plane clear a 1400-ft mountain in the flight path?

3.14. If a barometer is read to be 750 mm Hg, what is the absolute pressure if a pressure gage reads 70 kPa (gage)? Use the specific gravity of mercury to be 13.6 and take $\gamma_w = 62.4$ lb/ft^3.

3.15. A pressure gage reads 90 kPa above atmospheric pressure. If a barometer reads 760 mm Hg (sg = 13.6, $\gamma_w = 9810$ N/m^3), what is the absolute pressure?

3.16. The specific weight for mercury at 32°F is 848.714 lb/ft³ and at 100°F is 842.925 lb/ft³. What maximum percentage error will a barometer have if no temperature compensation is made over this temperature range?

Manometry

3.17. A U-tube mercury manometer that is open at one end is connected to a pressure source as shown in Figure P3.17. What is the unknown pressure in psfa? Assume that the unknown fluid is water, atmospheric pressure is 14.7 psia, and that the mercury has a specific gravity of 13.6.

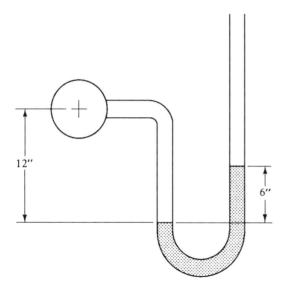

Figure P3.17

3.18. For the inclined manometer shown in Figure P3.18, determine p_A.

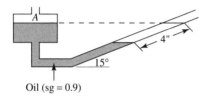

Oil (sg = 0.9)   **Figure P3.18**

3.19. Determine p_A for the manometer shown in Figure P3.19.

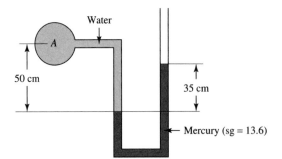

Water

A

50 cm

35 cm

Mercury (sg = 13.6)

Figure P3.19

3.20. Determine *h* in Figure P3.20 if both legs are open to the atmosphere.

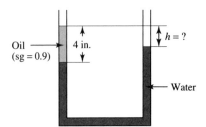

Oil
(sg = 0.9)

4 in.

$h = ?$

Water

Figure P3.20

3.21. Calculate *h* in Figure P3.21.

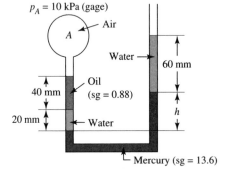

$p_A = 10$ kPa (gage)

Air

A

Water

60 mm

Oil
(sg = 0.88)

40 mm

20 mm

h

Water

Mercury (sg = 13.6)   **Figure P3.21**

3.22. Determine the reading of the pressure gage in Figure P3.22 if the column of water at the side is open to the atmosphere.

3.23. In Figure P3.23, two sources of pressure, *M* and *N*, are connected by a water–mercury differential gage. What is the difference in pressure between *M* and *N* in psi?

3.24. In Figure P3.24, determine the difference in pressure between *A* and *B* if the specific weight of water is 62.4 lb/ft³.

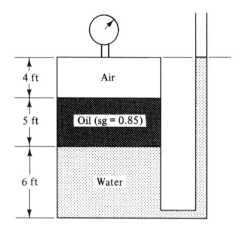

Figure P3.22

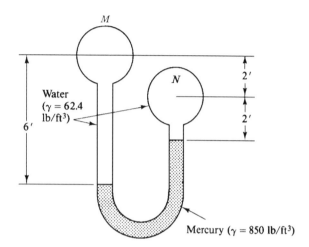

Figure P3.23

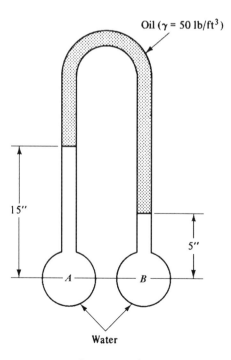

Figure P3.24

3.25. If the liquid in pipe B in Problem 3.24 is carbon tetrachloride, whose specific weight is 99 lb/ft^3, determine the pressure difference between A and B.

3.26. For the arrangement shown in Figure P3.26, determine $p_A - p_B$.

3.27. Two sources of pressure, A and B, are connected by a water–mercury differential gage as shown in Figure P3.27. What is the difference in pressure between A and B in pascals?

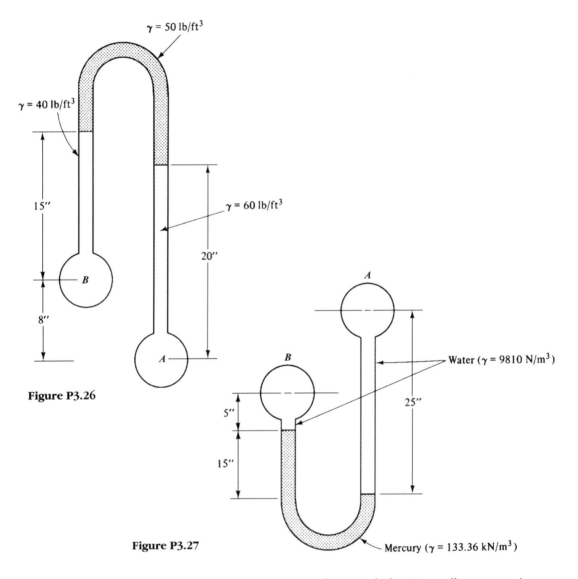

$\gamma = 50$ lb/ft^3

$\gamma = 40$ lb/ft^3

15"

$\gamma = 60$ lb/ft^3

20"

B

8"

A

Figure P3.26

A

B

Water ($\gamma = 9810$ N/m^3)

25"

5"

15"

Figure P3.27

Mercury ($\gamma = 133.36$ kN/m^3)

3.28. In Figure P3.28, a manometer is connected to a tank that is initially open to the atmosphere. If the tank is closed and the pressure in the air space is raised to 70 kPa above atmospheric pressure, what will the manometer reading be?

3.29. What is p_u for the manometer shown in Figure P3.29?

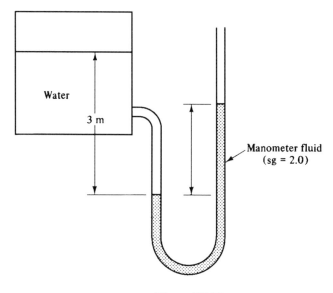

Figure P3.28

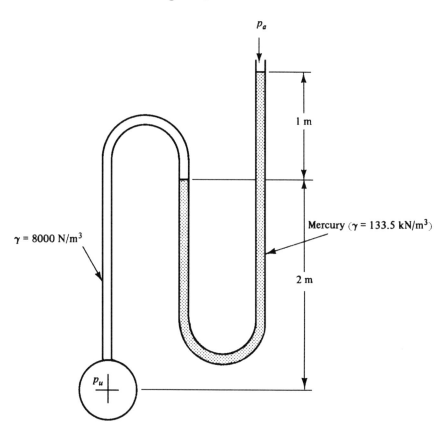

Figure P3.29

3.30. In a U-tube manometer, one end is closed, trapping atmospheric air in the column. The other end is connected to a pressure supply of 5 psig. If the level of mercury in the closed end is 2 in. higher than that in the open end, what is the pressure of the trapped air?

3.31. Calculate the pressure of the trapped air and the reading of the pressure gage for the manometer shown in Figure P3.31. Atmospheric pressure is 100 kPa.

3.32. What pressure will the gage read in Figure P3.32?

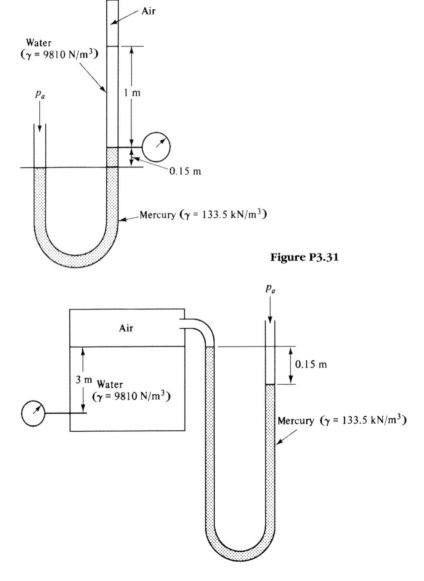

Figure P3.31

Figure P3.32

3.33. In Figure P3.33, what will the gage read if barometric pressure is 100 kPa?

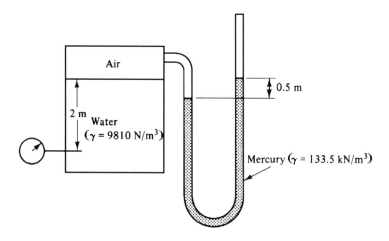

Figure P3.33

3.34. A U-tube manometer is connected to a pressure source as shown in Figure P3.34. What is the unknown pressure if p_a is 100 kPa? Assume the fluid to be air of negligible specific weight.

3.35. Determine the pressure difference between A and B in Figure P3.35.

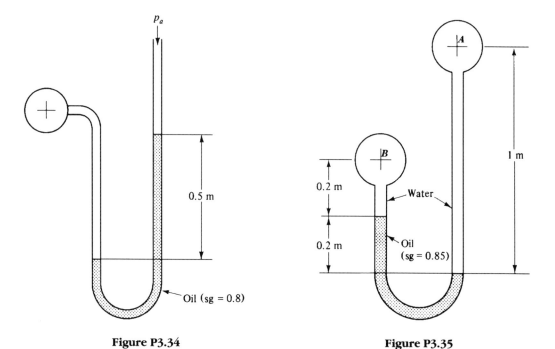

Figure P3.34   **Figure P3.35**

3.36. A manometer is connected to a pipe in which water is flowing as shown in Figure 3.13. Assume that the pipe is 20 in. in diameter and that the specific weight of the water in the pipe is 50 lb/ft³. Because of heat transfer, the specific weight of the water in the manometer leg is 55 lb/ft³. If B is 10 in. above the top of the pipe, h is 5 in., and $\gamma = 850$ lb/ft³, determine the pressure at the center of the pipe.

3.37. In Figure P3.37, what is the pressure in the tank above the water surface?

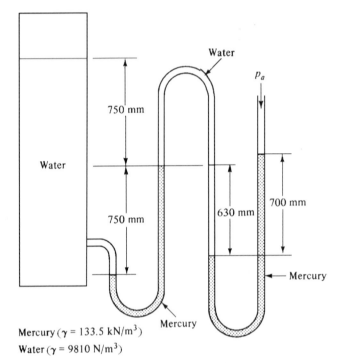

Mercury ($\gamma = 133.5$ kN/m³)

Water ($\gamma = 9810$ N/m³)

Figure P3.37

3.38. In Figure P3.38, what is the pressure in the line A?

*3.39. Determine the pressure in A in Figure P3.39.

*3.40. Determine the pressure difference $p_A - p_B$ for the situation shown in Figure P3.40 on page 118..

*3.41. Determine $p_A - p_B$ for the manometer shown in Figure P3.41 on page 118.

Buoyancy

3.42. A stone weighs 120 lb in air, and when submerged in water weighs 80 lb. Determine the volume and specific gravity of the stone, neglecting the effect of the air.

3.43. A cube of wood having sides of 500 mm and a specific gravity of 0.6 floats in water. What force is required to just submerge the cube?

3.44. Steel weighs 490 lb/ft³. What is the apparent weight of a cube of steel 4 in. on a side when it is suspended in water having $\gamma = 62.0$ lb/ft³?

3.45. A cube of wood has sides equal to 12 in. and a specific gravity of 0.6. If it floats in water, determine the submerged depth of the cube.

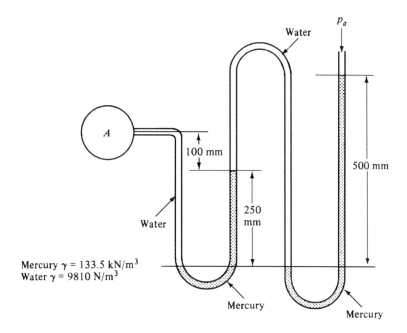

Figure P3.38

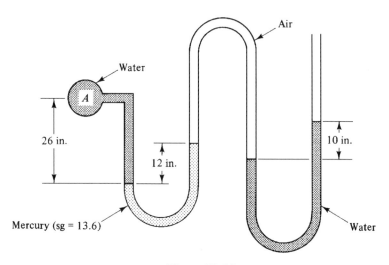

Figure P3.39

3.46. Gold has a specific gravity of 19.3. Would you conclude that you had a gold crown if you (Archimedes) had a crown that weighed 38.6 N in air and 36.6 N in water? Neglect the effect of the air.

3.47. A prism is 12 in. long, 12 in. wide and 24 in. deep, and weighs 150 lb when fully submerged in water. Determine its weight in air and its specific gravity.

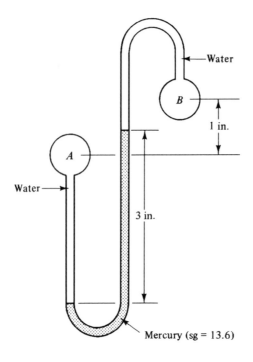

Figure P3.40

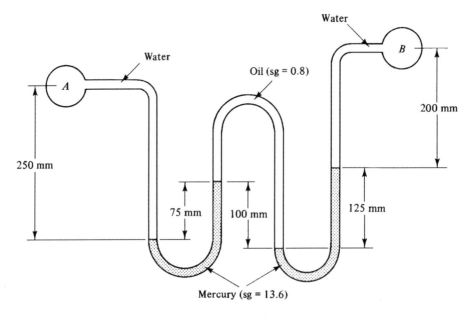

Figure P3.41

3.48. A uniform cylinder is 1 ft in diameter and 1 ft long. The cylinder weighs 30 lb in air. Determine the force necessary to just submerge the cylinder completely in water if (a) the cylinder is vertical and (b) the cylinder is horizontal. Neglect the effect of the air.

3.49. A concrete block having a specific weight of 23.0 kN/m^3 is suspended in water by a cable. If the cable tension is 3.1 kN, determine the volume of the block.

*__3.50.__ A steel cube having sides of 6 in. is to be held in equilibrium under water by attaching a plastic buoy to it. If the specific weight of the steel is 500 lb/ft^3 and the specific weight of the plastic is 5.0 lb/ft^3, calculate the volume of plastic required.

3.51. A cube of steel having sides of 300 mm floats in mercury. If the specific gravity of steel is 8.0 and the specific gravity of the mercury is 13.6, determine the submerged depth of the cube.

3.52. A submerged mine is moored to the ground by a cable. If the mine weighs 150 lb in air and is in the shape of a sphere having a radius of 12 in., what is the tension in the cable when the specific weight of sea water is 64.0 lb/ft^3? Neglect the effect of the air.

3.53. A balloon has a diameter of 30 ft and is filled with helium having a specific weight of 0.01 lb/ft^3. Including its own weight, determine the weight that it can just lift if the specific weight of air is 0.075 lb/ft^3.

3.54. A solid uniform sphere is immersed in fresh water and moored to the bottom by a cable. If the sphere has a radius of 18 in. and the mooring line has a tension of 250 lb, determine the specific weight of the sphere.

3.55. An iceberg floats with one-eighth of its volume above the water surface. Calculate the specific gravity of the iceberg if ocean water has a specific weight of 64.0 lb/ft^3.

3.56. A uniform solid cylinder floats upright in water. If the diameter of the cylinder is 8 in. and its height is 2 ft, what is the specific weight of the cylinder if it floats with three-fourths of its volume below the surface?

3.57. A piece of material is found to weigh 7.848 N in air and 5.00 N in water. What is the specific weight of the material? (γ_{air} = 11.8 N/m^3 and γ_{water} = 9810 N/m^3)

*__3.58.__ A block of material weighs 11.9 N in oil and 10.9 N in water. If the oil has a specific gravity of 0.85, determine the volume of the block.

*__3.59.__ A cylindrical log is 1 ft in diameter and 20 ft long. What weight of iron must be tied to one end of the log to keep it floating in an upright position in seawater with 18 ft of the log submerged? (γ_{log} = 43.7 lb/ft^3, γ_{iron} = 493 lb/ft^3, and $\gamma_{seawater}$ = 64.3 lb/ft^3)

3.60. A ship is placed in a dry dock, and the dry dock is filled until the ship just floats. By measuring the volume of water in the dock with and without the ship in it, it is found that 10,000 ft^3 of seawater is displaced by the ship. If γ is 64.3 lb/ft^3, determine the weight of the ship.

*3.61. Hydrometers are used to determine the specific gravity of liquids. If the hydrometer shown in Figure P3.61 has a weight of 0.05 lb and a stem diameter of 0.25 in. what is the height h when it is placed in a liquid having a specific gravity of 1.2?

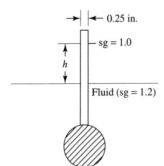

Figure P3.61

3.62. A steel drum is 1.5 ft in diameter and 2.5 ft long. When empty it weighs 20 lb. Steel weights are to be placed inside the drum to make it float submerged in fresh water. What volume of steel is required if steel has a specific weight of 0.285 lb/in.^{3}?

3.63. If the weights in Problem 3.62 are added outside the drum, what volume of steel is required?

3.64. A uniform solid cylinder has a diameter of 500 mm and a height of 1 m. When placed in sea water it floats vertically with 750 mm of its length submerged. Determine the specific weight of the cylinder if the specific weight of sea water is 10.06 kN/m^3.

3.65. A piece of steel, specific gravity = 8.0, floats on a pool of mercury, specific gravity = 13.6. Determine the fraction of the volume of the steel that is above the surface of the mercury.

3.66. A barge has the dimensions 25 ft x 50 ft x 10 ft deep and weighs 10 tons. If a load of 100 tons is placed on the barge, how much deeper will it sink into salt water having a specific weight of 64.0 lb/ft^3?

*3.67. A cylinder is made of two sections. The top section is made of aluminum having a specific weight of 0.1 lb/in.^{3} and the bottom section is made of wood having a specific gravity of 0.6. The cylinder is 3 ft long and has a diameter of 6 in. If the aluminum section is 6 in. long, how much of the length of the aluminum will be out of the water if the cylinder floats vertically with the wood on the bottom?

*3.68. A hollow cylinder is made as shown in Figure P3.68. How deep will it be in water if it floats vertically?

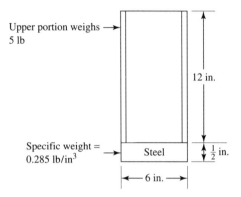

Upper portion weighs → 5 lb

12 in.

Specific weight = 0.285 lb/in.^{3} → Steel $\frac{1}{2}$ in.

6 in.

Figure P3.68

3.69. If the steel in Problem 3.68 is replaced with aluminum having a specific weight of 0.1 lb/in.^{3}, how deep will the cylinder float in water?

3.70. A rectangular barge 25 ft x 8 ft x 7 ft deep weighs 5.0 tons. If it is placed in salt water (specific weight = 64.0 lb/ft^3), how deep will it sink?

3.71. A barrel containing water is placed on a scale and is found to weigh 200 lb. A piece of wood having a cross section of 2 in. × 4 in. × 1 ft is placed vertically in the water. What weight will the scale indicate if the specific gravity of the wood is 0.6?

*3.72. A solid cube is 5 in. on a side and is balanced on an equal arm balance by a 4.5 lb weight when the cube is immersed in water as shown in Figure P3.72. What is the specific weight of the cube?

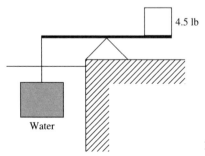

4.5 lb

Water

Figure P3.72

3.73. Two equal-diameter spheres weigh 2 lb and 10 lb respectively. They are connected by a cable and placed in water. Calculate the tension in the cable between the spheres and the cable mooring them to the ocean floor if the spheres each have a radius of 4 in.

*3.74. A steel block has a specific weight of 500 lb/ft^3 and floats as shown in Figure P3.74. Calculate the ratio A/B.

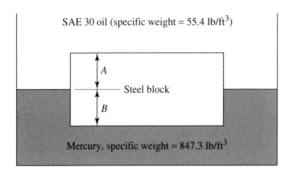

SAE 30 oil (specific weight = 55.4 lb/ft^3)

A

Steel block

B

Mercury, specific weight = 847.3 lb/ft^3

Figure P3.74

3.75. A raft that is 8 ft long and 6 ft wide is constructed by lashing together boards whose finished dimensions are 8 in. wide and 4 in. deep. What weight can the raft support when it becomes completely submerged in water? Take the specific gravity of the wood to be 0.6.

3.76. If the raft in Problem 3.75 is to carry a load of 1/2 ton in addition to its weight, will an empty drum of 55 gallons capacity lashed under the raft be sufficient to just have the top of the raft submerged? Neglect the weight of the drum.

*3.77. A cube having sides of 15 cm is made of steel that has a specific weight of 71 kN/m^3. The cube is suspended in water by a wire so that half of its volume is in water and the other half is in oil having a specific gravity of 0.8. Determine the tension in the wire.

*3.78. When a uniform block of wood floats in water, 2 in. is above the water surface. When the same block floats in oil that has a specific gravity of 0.85, it floats with 1.5 in. above the oil surface. Determine the specific gravity of the wood.

Forces on Surfaces

3.79. A cylindrical tank having flat ends has its long axis vertical. If the top is open to the atmosphere and there is a height of 3 m of water above the base of the tank ($\gamma_w = 9810$ N/m^3), what is the force on the bottom plate due to the water? Assume the tank to have a diameter of 2 m.

*3.80. A cylindrical tank with its long axis horizontal is filled with a fluid to its center. Derive an expression for the force on the end of the cylinder in terms of R, the cylinder radius, and the specific weight of the fluid.

3.81. If a cylindrical tank is placed horizontally and is 10 ft in diameter, determine the force on the end if the water level is at the center of the tank.

3.82. A cylindrical tank is placed horizontally and is filled with oil to its center. If the tank has a 3-m diameter and the oil has a specific gravity of 0.85, what is the force on the end of the cylinder?

3.83. A vertical square plate is used as a dam in a channel, as shown in Figure P3.83. If the plate is 8 ft square and the liquid level on the upstream side is 6 ft high, determine the total force on the plate.

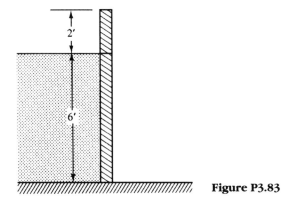

Figure P3.83

3.84. Determine the location of the resultant force in Problem 3.83.

3.85. What is the moment of the force on the plate of Figure P3.83 about its base?

3.86. What is the total force on the rectangular plate shown in Figure P3.86, and where is its center of pressure located? The plate is 5 m wide.

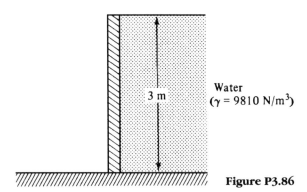

Figure P3.86

3.87. What is the net force on the plate shown in Figure P3.87? It is 6 m wide.

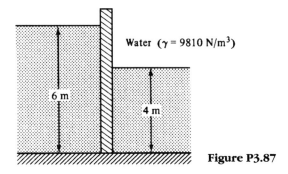

Figure P3.87

3.88. Calculate the force on the wall shown in Figure P3.88 if the wall is 6 m long.

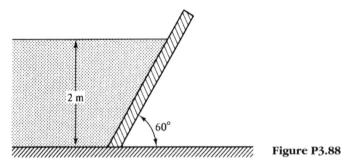

Figure P3.88

3.89. The gate shown in Figure P3.89 is square and is hinged at the top. Determine the total force on the gate if it is 2 m wide.

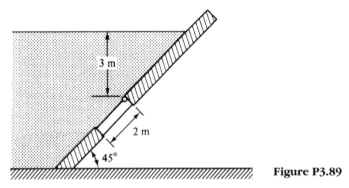

Figure P3.89

3.90. Determine the moment of the hydrostatic force about the hinge in Problem 3.89.

3.91. If the gate of Problem 3.89 is circular, determine the total force on the gate and the moment about the pin.

3.92. If the gate shown in Figure P3.92 is circular, determine the moment of the hydro-static force about the hinge pin.

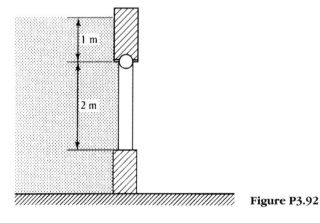

Figure P3.92

3.93. What is the net moment about the base of the plate in Problem 3.87?

*3.94. A trash rack 5 ft wide protects the intake to a power plate as shown in Figure P3.94. In the event that the intake is blocked by ice and debris, what would be the total force acting on the rack, and where would this force be applied?

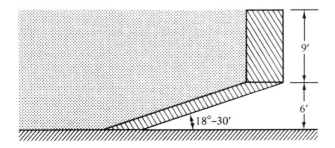

Figure P3.94

*3.95. A vertical plate is used to dam a channel. If the plate has the shape shown in Figure P3.95, determine the total force on the plate. Assume that $\theta_1 = \theta_2$.

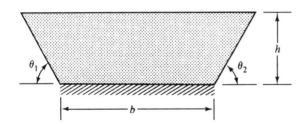

Figure P3.95

*3.96. As shown in Figure P3.96, an aquarium installed a viewing window having a diameter of 1.5 m. Determine the magnitude and location of the hydrostatic force on the window.

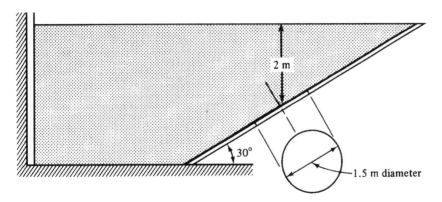

Figure P3.96

Stresses in Cylinders and Spheres

3.97. Determine the thickness of a cylinder subjected to internal pressure if the pressure is 100 psi and the diameter is 4 ft. Assume an allowable stress of 20,000 psi for the material of the cylinder.

3.98. Solve Problem 3.97 if the container is a sphere.

3.99. A pipe 2 ft in diameter has a wall thickness of $\frac{1}{4}$ in. If the allowable stress of the material is 15,000 psi, what internal pressure can it sustain?

3.100. Show that for a given volume and a given internal pressure, the weight of a spherical container will be three-fourths of the weight of a cylindrical container.

4 FLUID DYNAMICS

LEARNING GOALS

1. Define what is meant by *control volume*.
2. State the concept of conservation of mass in a flow system.
3. Express the concept of conservation of mass by the *continuity equation* in its several forms.
4. Explain why the volume rate flow rather than the mass rate flow is used in fluid mechanics applications.
5. Quantitatively evaluate the potential, kinetic, and flow work terms for a system in which there is steady flow.
6. Use the concept of energy conservation to write the Bernoulli equation for an ideal system in which the flow is steady.
7. State the limitations of the Bernoulli equation.
8. Apply the Bernoulli equation to various steady flow situations.

4.1 INTRODUCTION

In Chapter 3 we studied fluids at rest, and by applying the requirement that static systems must be in equilibrium, we were able to analyze many applications that occur in engineering. The next logical extension of the study of fluid mechanics is to those situations in which the fluid is flowing relative to its boundaries or to a fixed datum. A system has already been defined as a grouping of matter taken in any convenient arbitrary manner. When dealing with fluids in motion, however, it is more convenient to utilize the concept of an arbitrary volume in space, known as a *control volume,* that can be bounded by either a real or imaginary surface, known as the *control surface.* By correctly noting all of the forces acting on the fluid within the control volume, the energies crossing the control surface, and the mass crossing the control surface, it is possible to derive mathematical expressions that will evaluate the flow of the fluid relative to the control volume. In this chapter we are concerned with fluids flowing steadily through the control volume, and for this type of system a monitoring station anywhere in the control volume will indicate no change in the fluid properties or energy quantities crossing the control surface with time. These quantities can and will vary from position to position in the control volume.

4.2   CONSERVATION OF MASS—THE CONTINUITY EQUATION

As noted in Section 4.1, both energy and mass can enter and leave a control volume and cross the control surface of a system. Since we are considering steady-flow systems, we can express the fact that the principle of conservation of mass for these systems requires that the mass of the fluid in the control volume at any time be constant. *In turn, this requires that the net mass flowing into the control volume must equal the net mass flowing out of the control volume at any instant of time.* To express these concepts in terms of a given system, let us consider the system shown schematically in Figure 4.1.

Let us assume that at a certain time, fluid starts to enter the control volume by crossing the control surface at section 1 and that after a small interval of time the flowing fluid fills the pipe for a short distance x. If it is further assumed that in this short section of uniform pipe no heat is added or work is exchanged and that the specific weight stays constant, we can evaluate the amount of fluid that flowed in between sections 1 and 2. The weight contained between these sections is equal to the volume contained between the sections multiplied by the specific weight of the fluid. The volume is Ax and the specific weight is γ; therefore, the contained weight is γAx. The distance between stations, x, is simply Vt, where

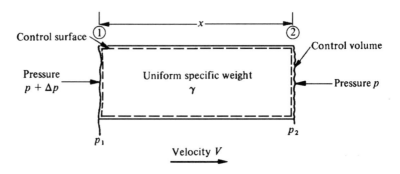

Figure 4.1   Elementary flow system.

V is the velocity of the fluid and t is the flow time required to fill the pipe between sections 1 and 2. Substituting this for x, we have

$$w = \gamma A V t$$

(4.1a)

or

$$\dot{w} = \gamma A V$$

(4.1b)

where $\dot{w}$ is the weight rate of flow per unit time, w/t. Also, we can write equation (4.1) in terms of mass flow rate per unit time:

$$\dot{m} = \rho A V$$

(4.1c)

For the mass within the control volume to remain constant, the amount of mass entering the system must equal the amount of mass leaving the system. Therefore, using the subscripts 1 and 2 to denote any two stations gives us

$$\dot{m}_1 = \dot{m}_2$$

(4.2a)

Equation (4.2a) can also be expressed in the following equivalent forms:

$$\rho_1 A_1 V_1 = \rho_2 A_2 V_2$$

(4.2b)

$$\gamma_1 A_1 V_1 = \gamma_2 A_2 V_2$$

(4.2c)

$$\frac{A_1 V_1}{v_1} = \frac{A_2 V_2}{v_2}$$

(4.2d)

where the specific volume $v = 1/\rho$.

For an incompressible fluid, $\gamma_1 = \gamma_2$ and $\rho_1 = \rho_2$. In each form of equation (4.2) we find the product of AV. Since this combined term appears very frequently in fluid mechanics applications, it is usually given a separate designation,

$$Q = AV$$

(4.3)

where Q is the *volume flow rate*. In English units,

$$Q = \text{ft}^2 \times \frac{\text{ft}}{\text{s}} = \frac{\text{ft}^3}{\text{s}}$$

and in SI units,

$$Q = \text{m}^2 \times \frac{\text{m}}{\text{s}} = \frac{\text{m}^3}{\text{s}}$$

We can obtain the mass and weight flow rates in both systems as

$$\dot{m} = \rho Q$$

(4.4)

$$\dot{w} = \gamma Q$$

where $\dot{m}$ is kg/s or slugs/s and $\dot{w}$ is N/s or lb/s.

ILLUSTRATION 4.1   CONTINUITY EQUATION

One thousand gallons per minute of water flows steadily through a pipe having an inside diameter, d, of 14 in. At a section downstream, the inside diameter reduces to 8 in. Calculate the velocity of flow in each section of the pipe.

ILLUSTRATIVE PROBLEM 4.1

Given: 1000 gpm water, d_1 = 14 in., d_2 = 8 in.

Find: V_1 and V_2

Assumptions: Incompressible, steady flow

Basic Equation: Continuity equation: $A_1 V_1 = A_2 V_2$

Solution: Since a gallon is a measure of volume equal to 231 in.^{3}, the statement of the problem has given us the volume rate of flow as 1000 gallons per minute (gpm), or Q. For our purposes we will need Q in units of cubic feet per second (cfs). Using the definition of a gallon as 231 in.^{3} yields

$$Q = 1000 \, \frac{\text{gal}}{\text{min}} \times 231 \, \frac{\text{in.}^3}{\text{gal}} \times \left(\frac{1}{12 \text{ in./ft}} \right)^3 \times \frac{1}{60 \text{ s/min}}$$

$$= 2.228 \text{ ft}^3/\text{s or } 2.228 \text{ cfs}$$

In order to apply equation (4.3), $Q = AV$, we will now need the cross-sectional area of each section of the pipe. Thus,

$$A_1 = \frac{\pi d_1^2}{4} = \frac{\pi(14 \text{ in./12 in./ft})^2}{4} = 1.069 \text{ ft}^2$$

$$A_2 = \frac{\pi d_2^2}{4} = \frac{\pi(8 \text{ in./12 in./ft})^2}{4} = 0.349 \text{ ft}^2$$

Since equation (4.3) gives us $Q = AV$, $V = Q/A$. Therefore,

$$V_1 = \frac{Q}{A_1} = \frac{(2.228 \text{ ft}^3/\text{s})}{1.069 \text{ ft}^2} = 2.084 \text{ ft/s}$$

$$V_2 = \frac{Q}{A_2} = \frac{(2.228 \text{ ft}^3/\text{s})}{0.349 \text{ ft}^2} = 6.384 \text{ ft/s}$$

We could also have noticed that since $Q_1 = Q_2$, $A_1 V_1 = A_2 V_2$, or

$$V_2 = V_1 \left(\frac{A_1}{A_2} \right) = V_1 \left(\frac{d_1}{d_2} \right)^2$$

Using our calculated value for V_1, we obtain

$$V_2 = 2.084 \frac{\text{ft}}{\text{s}} \times \left(\frac{14 \text{ in.}}{8 \text{ in.}} \right)^2 = 6.382 \text{ ft/s}$$

Within arithmetic round-off error this agrees with our previous calculation.

ILLUSTRATION 4.2 CONTINUITY EQUATION

How many m³/s is flowing in Illustrative Problem 4.1?

ILLUSTRATIVE PROBLEM 4.2

Given: Solution to Illustrative Problem 4.1

Find: Q in m³/s

Assumptions: Same as Illustrative Problem 4.1

Basic equation: Conversion factors

Solution: Using 1 in. = 0.0254 m, we obtain

$$Q = 1000 \frac{\text{gal}}{\text{min}} \times 231 \frac{\text{in.}^3}{\text{gal}} \times \left(0.0254 \frac{\text{m}}{\text{in.}} \right)^3 \times \frac{1}{60 \text{ s/min}}$$

$$Q = 0.063\ 09 \text{ m}^3/\text{s}$$

ILLUSTRATION 4.3 CONTINUITY EQUATION

One thousand liters per minute flows through a 425-mm-diameter pipe. Determine the weight and mass flow rates if the fluid is water whose specific weight is 9.81 kN/m³ and whose density is 1000 kg/m³.

ILLUSTRATIVE PROBLEM 4.3

Given: water: $\gamma = 9.81$ kN/m³, $\rho = 1000$ kg/m³; $d = 425$ mm

Find: $\dot{w}$ and $\dot{m}$

Assumptions: Steady flow

Basic Equations: $\dot{w} = \gamma A V = \gamma Q$; $\dot{m} = \rho A V = \rho Q$

Solution: Since a liter is 10^{-3} m³ (Illustrative Problem 1.3),

$$Q = 1000 \frac{\text{liters}}{\text{min}} \times 10^{-3} \frac{\text{m}^3}{\text{liter}} \times \frac{1}{60 \text{ s/min}} = 0.01667 \text{ m}^3/\text{s}$$

From equations (4.4),

$$\dot{w} = \gamma Q = 9.81 \frac{\text{kN}}{\text{m}^3} \times 0.01667 \frac{\text{m}^3}{\text{s}} = 0.1635 \text{ kN/s}$$

and

$$\dot{m} = \rho Q = 1000 \ \frac{\text{kg}}{\text{m}^3} \times 0.01667 \ \frac{\text{m}^3}{\text{s}} = 16.67 \ \text{kg/s}$$

ILLUSTRATION 4.4   CONTINUITY EQUATION

Water is flowing from one pipe to another as shown in Figure 4.2. Determine the velocity in each section of pipe.

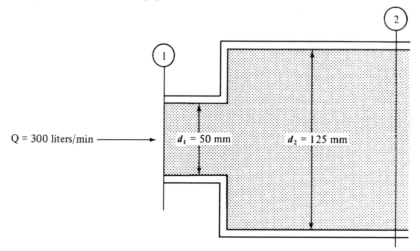

Figure 4.2   Illustrative Problem 4.4.

ILLUSTRATIVE PROBLEM 4.4

Given: Q = 300 L/min of water, d_1 = 50 mm, d_2 = 125 mm
Find: V_1 and V_2
Assumptions: Incompressible, steady flow
Basic Equations: $A_1 V_1 = A_2 V_2$, conversions
Solution: One liter is 1000 cm^3 = 1000 × $(10^{-2})^3$ = 10^{-3} m^3, which is as we found in Illustrative Problem 1.3. Therefore, Q is

$$Q = 300 \ \frac{\text{L}}{\text{min}} \times 10^{-3} \ \frac{\text{m}^3}{\text{L}} \times \frac{1}{60 \ \text{s/min}} = 0.005 \ \text{m}^3/\text{s}$$

The flow areas of each pipe are $\pi d^2/4$, where d is the inside diameter. Therefore,

$$A_1 = \frac{\pi d_1^2}{4} = \frac{\pi (0.050)^2}{4} = 1.963 \times 10^{-3} \ \text{m}^2$$

$$A_2 = \frac{\pi d_2^2}{4} = \frac{\pi (0.125)^2}{4} = 1.227 \times 10^{-2} \ \text{m}^2$$

Since water is incompressible, $\gamma_1 = \gamma_2$ or $\rho_1 = \rho_2$. This gives us the condition that $Q_1 = Q_2 = Q$. Therefore,

$$Q = A_1 V_1 = A_2 V_2$$

Thus,

$$V_1 = \frac{Q}{A_1} = \frac{0.005 \ \text{m}^3/\text{s}}{1.963 \times 10^{-3} \ \text{m}^2} = 2.55 \ \text{m/s}$$

$$V_2 = \frac{Q}{A_2} = \frac{0.005 \ \text{m}^3/\text{s}}{1.227 \times 10^{-2} \ \text{m}^2} = 0.407 \ \text{m/s}$$

4.3   WORK AND ENERGY

We define the *work* done by a force as the product of the displacement of the body multiplied by the component of the force in the direction of the displacement. Thus, in Figure 4.3a the displacement of the body on the horizontal plane is x and the component of the force in the direction of the displacement if $F \cos \theta$. The work done is therefore $(F \cos \theta)(x)$. The constant force $(F \cos \theta)$ is plotted as a function of x in Figure 4.3b, and it will be noted that the resulting figure is a rectangle. The area of this rectangle (shaded) is equal to the work done, since it is $(F \cos \theta)(x)$. If the force varies so that it is a function of the displacement, it is necessary to consider the variation of force with displacement in order to find the work done. Figure 4.3c shows a general plot of force as a function of displacement. If the displacement is subdivided into many small parts, Δx, and for each of these small parts F is assumed to be very nearly constant, it is apparent that the sum of the small areas $(F)(\Delta x)$ will represent the total work done when the body is displaced from x_1 to x_2. Thus, the area under the curve of F as a function of x represents the total work done if F is the force component in the direction of x.

At this point let us define the term *energy* in terms of work. Energy can be described as the capacity to do work. In all instances the observed effects on a system can (ideally) be converted to mechanical work.

4.3.1   Potential Energy

Let us consider the following problem. A body of mass m is in a locality where the local gravitational field is constant and equal to g. A force is applied to the body, and it is raised a distance Z from its initial position as shown in Figure 4.4 on page 134. The force will be assumed to be only infinitesimally greater than the mass. In the absence of electrical, magnetic, and other extraneous effects, determine the work done on the body. The solution to this problem is obtained by noting that the equilibrium of the body requires that a force must be applied to it equal to its weight. The weight of the body is simply w. In moving through a distance Z, the work done by this force will therefore equal

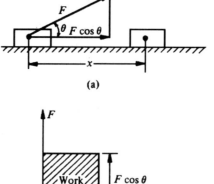

(a)

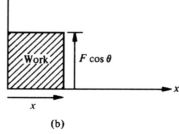

(b)

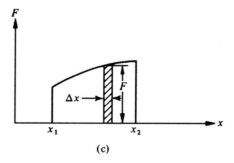

(c)

Figure 4.3

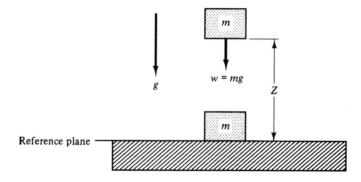

Figure 4.4   Potential energy.

$$\text{work} = wZ \tag{4.5}$$

The work done on the body can be returned to the external environment by simply reversing this process. We therefore conclude that this system has had work done on it equal to wZ and that, in turn, the system has stored in it an amount of energy in excess of the amount it had in its initial position. The energy added to the system in this case is called *potential energy.* Thus,

$$\text{potential energy (P.E.)} = wZ \tag{4.6}$$

or

$$\text{potential energy (P.E.)} = mgZ \tag{4.7}$$

since $w = mg$.

ILLUSTRATION 4.5   POTENTIAL ENERGY

In Figure 4.5, a pump "lifts" water from a well that is 10 m deep. What is the change in the potential energy of the water?

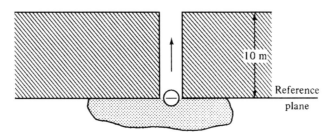

Figure 4.5   Illustrative Problem 4.5.

ILLUSTRATIVE PROBLEM 4.5

Given: Water lifted 10 m

Find: ΔP.E.

Assumptions: negligible velocity

Basic Equation: P.E. $= wZ$

Solution: It is convenient to select a reference plane that is at the lowest point in the system. This helps to avoid negative numbers. Initially, the water is at the reference plane. Consequently, Z_1 is zero and the initial potential energy is zero. Finally, $Z_2 = 10$ m; therefore, for 1 N,

$$(\text{P.E.})_2 = 1 \text{ N} \times 10 \text{ m/N} \quad (\text{since 1 N was assumed})$$

and

$$(\text{P.E.})_2 = 10 \ \text{N·m/N}$$

$$(\text{P.E.})_1 = 10 \ \text{N} \times 0 = 0 \ \text{N·m/N}$$

The change in P.E. is therefore

$$\Delta\text{P.E.} = (\text{P.E.})_2 - (\text{P.E.})_1 = 10 - 0 = 10 \ \text{N·m/N}$$

A feature of importance of potential energy is that a system can be said to possess potential energy only with respect to an arbitrary initial or datum plane.

4.3.2 Kinetic Energy

Let us consider another situation, in which a body of mass m is at rest on a frictionless plane. If a force F is applied to the mass, it will be accelerated in the direction of the force. After moving through a distance S, the velocity of the body will have increased from 0 to V. The only effect of the work done on the body will be an increase in its velocity. Since the force F in Figure 4.6 is constant, the acceleration of the block will be constant.

From Newton's second law we have

$$F = ma \tag{4.8}$$

The work done, $F \times S$, is

$$F \times S = maS \tag{4.9}$$

The distance traveled, S, is equal to the average velocity times time. For constant acceleration the average velocity is the final velocity divided by 2, that is $V/2$. Acceleration, a, is the change in velocity divided by time, and since the initial velocity is zero, is simply V/t. We write these items

$$S = \left(\frac{V}{2}\right)t \tag{4.10a}$$

and

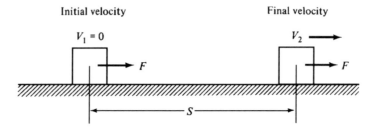

Figure 4.6

$$a = \frac{V}{t} \tag{4.10b}$$

Then substitution into equation (4.9) yields

$$F \times S = m \times \frac{V}{t} \times \left(\frac{V}{2} \right) t \tag{4.10c}$$

or

$$F \times S = \frac{mV^2}{2} \tag{4.11}$$

FS is the work done by the constant force F acting on the body for a displacement of S, and the body is said to possess *kinetic energy* (K.E.), since by virtue of its velocity, it has the ability to do work. With $m = w/g$, we can rewrite equation (4.11) to yield the familiar form for kinetic energy as

$$\text{K.E.} = \frac{wV^2}{2g} \tag{4.12}$$

having units of N·m or ft·lb.

CALCULUS ENRICHMENT

Consider a body of mass m acted upon by a constant force F as shown in Figure A. Initially the body is at rest. After a time t, the body acquires a velocity V and has moved a distance S. At this instant the force F is removed. We wish to know how much work was done on the body by the force and the energy possessed by the body due to its velocity. Using the definitions of velocity and acceleration, we have

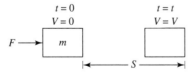

Figure A   Derivation of Kinetic Energy

$$V = \frac{dS}{dt} \tag{a}$$

$$a = \frac{dV}{dt} \tag{b}$$

Eliminating dt in Equations (a) and (b),

$$adS = VdV \qquad\qquad (c)$$

Multiplying both sides of Equation (c) by $\dfrac{w}{g}$,

$$\frac{w}{g}adS = \frac{w}{g}VdV \qquad\qquad (d)$$

From Newton's second law, $F = \dfrac{w}{g}a$. Therefore

$$\int_0^S FdS = \int_0^V \frac{w}{g}VdV \qquad\qquad (e)$$

Integration of Equation (e) yields

$$FS = \text{work} = \frac{w}{g}\frac{V^2}{2} \qquad\qquad (f)$$

The work done by the constant force F increases the energy of the body by an amount equal to $\dfrac{wV^2}{2g}$, that is, the kinetic energy of the body has been increased by $\dfrac{wV^2}{2g}$.

ILLUSTRATION 4.6   KINETIC ENERGY

A body having a mass of 10 kg falls freely from rest (Figure 4.7). After falling 10 m, what will its kinetic energy be and what will its velocity be?

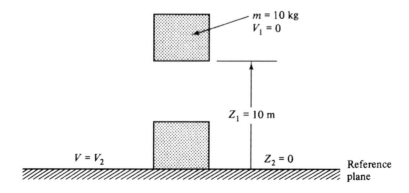

Figure 4.7   Illustrative Problem 4.6.

ILLUSTRATIVE PROBLEM 4.6

Given: $m = 10$ kg, free fall of 10 m

Find: $(K.E.)_2$ and V_2

Assumptions: No energy losses

Basic Equations: P.E. $= mgZ$, K.E. $= \dfrac{mV^2}{2}$

Solution: Since there are no energy losses in the system, we conclude that the sum of the initial potential energy plus kinetic energy must equal the sum of the final potential energy plus kinetic energy. Thus,

$$(P.E.)_1 + (K.E.)_1 = (P.E.)_2 + (K.E.)_2$$

Notice that the selected reference plane and the condition that the body falls from rest gives us $(P.E.)_1 = (K.E.)_2$, that is, potential energy lost equals the kinetic energy gained. Therefore,

$$mgZ = \frac{mV_2^2}{2}$$

$$10 \text{ kg} = 9.81 \,\frac{\text{m}}{\text{s}^2} \times 10 \text{ m} = \frac{10 \text{ kg} \times V_2^2}{2}$$

The left side of this equation is equal to $(K.E.)_2$ and is 981 N·m. Rearranging yields

$$V_2^2 = 2 \times 9.81 \,\frac{\text{m}}{\text{s}^2} \times 10 \text{ m}$$

and

$$V_2 = 14.0 \text{ m/s}$$

4.3.3 Flow Work

When a fluid is caused to flow in a system, it is necessary that somewhere in the system work must have been supplied. At this time let us evaluate the net work required to push the fluid into and out of the system. Consider the system shown in Figure 4.8, where a fluid is flowing steadily across the system boundaries as shown. At the inlet section ①, the pressure is p_1, the area is A_1, the mass flow rate is $\dot{m}$, and the fluid density is ρ_1; at the outlet section ②, the pressure is p_2, the area is A_2, the mass flow rate is still $\dot{m}$, and the fluid density is ρ_2. Let us now consider a plug of fluid of length l_1 entering the system such that the amount of fluid contained in the plug is numerically $\dot{m}$. The force acting on the inlet cross-sectional area A_1 is p_1A_1. In order to push the plug into the system it is necessary for this force to move the plug a distance equal to l_1. In so doing, the work done will be $p_1A_1l_1$. But A_1l_1 is the volume of the plug containing a mass m. Using this, we find the work W to be

$$\text{work} = p_1 A_1 l_1 = p_1 V_1 \tag{4.13}$$

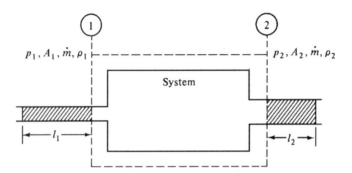

Figure 4.8 Steady-flow system showing flow work.

Because the weight is equal to the specific weight, γ, times the volume, V,

$$w = \gamma V \tag{4.14}$$

Therefore, combining equations (4.13) and (4.14), we obtain

$$\text{work} = w\,\frac{p_1}{\gamma_1} \tag{4.15a}$$

If we now consider the outlet section, using the same reasoning, we have

$$\text{work} = w\,\frac{p_2}{\gamma_2} \tag{4.15b}$$

Each of the right-hand terms in equations (4.15a) and (4.15b) is known as flow work (FW). Thus, the net flow work in the situation depicted in Figure 4.8 is

$$\text{net flow work} = w\left(\frac{p_2}{\gamma_2} - \frac{p_1}{\gamma_1}\right) \tag{4.16a}$$

For $\gamma_1 = \gamma_2 = \gamma$,

$$\text{net flow work} = w\left(\frac{p_2}{\gamma} - \frac{p_1}{\gamma}\right) \tag{4.16b}$$

The units of equations (4.16) are N·m or ft·lb.

4.4   CONSERVATION OF ENERGY—THE BERNOULLI EQUATION

The *First Law of Thermodynamics* is the statement of conservation of energy; it can be expressed by stating that energy can be neither created nor destroyed but only converted from one form to another.

In the derivations made in earlier sections, certain assumptions were made, and these are repeated here for emphasis. The term *steady*, when applied to a flow situation, means that the condition at any section of the system is independent of time. Even though the velocity, specific weight, and elevation of the fluid can vary in an arbitrary manner across the stream, they are not permitted to vary with time. The weight entering the system per unit time must equal the weight leaving the system in the same period of time; otherwise, the system would either store or be depleted of fluid.

To summarize, we have identified three energy terms that apply to various situations. Table 4.1 shows each of these energy terms.

Figure 4.9 shows such a system, where it is assumed that each form of energy can both enter and leave the system. At the entrance, $\dot{w}$ lb of fluid/s enters, and the same amount leaves at the exit. At the entrance the fluid has a pressure of p_1, a specific weight of γ_1, and a velocity of V_1. At the exit we have similar quantities expressed as p_2, γ_2, and V_2 (Table 4.2). The fluid enters and leaves at different elevations. We express the conservation of energy for this steady-flow system by stating that all of the energy entering the system must equal all of the energy leaving the system.

Table 4.1

Item	Value ft·lb	N·m
Potential energy	wZ	wZ
Kinetic energy	$wV^2/2g$	$wV^2/2g$
Flow work	wp/γ	wp/γ

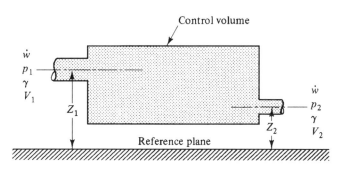

Figure 4.9   Steady-flow system.

Table 4.2

	Energy in ft·lb	J	Energy out ft·lb	J
Potential energy	wZ_1	wZ_1	wZ_2	wZ_2
Kinetic energy	$wV_1^2/2g$	$wV_1^2/2g$	$wV_2^2/2g$	$wV_2^2/2g$
Flow work	wp_1/γ	wp_1/γ	wp_2/γ	wp_2/γ

Equating the energy entering the system to the energy leaving,

$$wZ_1 + \frac{wV_1^2}{2g} + \frac{wp_1}{\gamma} = wZ_2 + \frac{wV_2^2}{2g} + \frac{wp_2}{\gamma} \tag{4.17}$$

Notice that in arriving at equation (4.17) we have assumed that we are dealing with a system in which the flow is steady, there is no change in temperature, no work is done on or by the system, no energy as heat crosses the boundaries of the system, and the fluid is incompressible. We shall further assume that all processes are ideal in the sense that they are frictionless. For this system the energy equation reduces to

$$Z_1 + \frac{V_1^2}{2g} + \frac{p_1}{\gamma} = Z_2 + \frac{V_2^2}{2g} + \frac{p_2}{\gamma} \tag{4.18}$$

Equation (4.18) is the usual form of the Bernoulli equation. Each term of equation (4.18) can represent a height, since the dimension of foot pounds per pound corresponds numerically (and dimensionally) to a height. Hence, the term *head* is frequently used to denote each of the terms in equation (4.18). In SI units, each term is N · m/N. Again, this unit is dimensionally a height or "head." It is more fundamental to interpret each term as the energy per unit weight of fluid. From the derivation of this equation we note that its use is restricted to those situations where the flow is steady, there is no friction, no shaft work is done on or by the fluid, the flow is incompressible, there is no change in internal energy during the process, and there is no heat transfer to or from the system. Although these restrictions severely limit the use of equation (4.18), several modifications have been made to it to apply it to more realistic situations. We shall study these modifications and their reasons in Chapter 5.

CALCULUS ENRICHMENT

We can derive the Bernoulli equation by considering the flow of an element along a streamline as in Figure A.

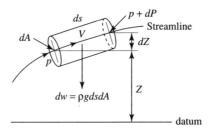

Figure A Derivation of the Bernoulli equation.

Referring to Figure A,

1. Pressure forces at end of cylinder (in the direction of flow)

$$pA - (p + dp)A = -dpdA \tag{a}$$

2. Component of weight in the direction of flow

$$-\rho g ds dA (dZ/ds) = -\rho g dA dZ \qquad \text{(b)}$$

3. The mass being accelerated

$$dm = \rho ds dA \qquad \text{(c)}$$

4. Newton's second law along the streamline

$$dF = (dm)a \qquad \text{(d)}$$

5. Acceleration

$$a = \frac{dV}{dt} \; ; \; V = \frac{ds}{dt} \; ;$$

$$\therefore a = \frac{VdV}{ds} \qquad \text{(e)}$$

Combining Equations (a) through (e)

$$-\frac{dp}{dA} - \rho g dA dZ = (\rho ds dA)\frac{VdV}{ds} \qquad \text{(f)}$$

Equation (g) is known as Euler's equation in one dimension. Dividing Equation (f) by ρdA yields

$$\frac{dp}{\rho} + VdV + gdZ = 0 \qquad \text{(g)}$$

For constant density, dividing Equation (g) by g,

$$d\left(\frac{p}{\gamma}\right) + d\left(\frac{V^2}{2g}\right) + dZ = 0 \qquad \text{(h)}$$

For incompressible flow, γ and g are constant. Therefore

$$\frac{p}{\gamma} + \frac{V^2}{2g} + Z = \text{constant} \qquad \text{(i)}$$

or

$$\frac{p_1}{\gamma} + \frac{V_1^2}{2g} + Z_1 = \frac{p_2}{\gamma} + \frac{V_2^2}{2g} + Z_2 \qquad \text{(j)}$$

Equation (j) is the Bernoulli equation for one-dimensional incompressible flow.

When applying the Bernoulli equation, it is suggested that the following procedure be followed to do the work in a systematic manner.

1. Draw a sketch of the problem, indicating all of the given quantities, paying particular attention to the units of each quantity.

2. Select and indicate the reference plane for potential energy. Usually, this is taken to be the lowest point in the system, to avoid negative heights.

3. Select ① to be at the entrance to the system and ② to be at the exit; that is, your sections should be indicated *in the direction of the flow.*

4. Write the Bernoulli equation.

5. Make sure that all the items that enter the Bernoulli equation have the appropriate units.

6. If necessary, invoke the continuity equation to determine the velocity or velocities required.

7. Solve for the required quantity or quantities.

ILLUSTRATION 4.7   BERNOULLI EQUATION

Water (γ = 9810 N/m³) flows in a pipe. At a section where the inside diameter is 150 mm, the velocity is 3 m/s and the pressure is 350 kPa. At a section located 10 m from the first section, the inside diameter reduces to 75 mm. Calculate the pressure at the second section.

(a) If the pipe is horizontal

(b) If the pipe is vertical and the flow is downward

ILLUSTRATIVE PROBLEM 4.7

Given: γ = 9810 N/m³, d_1 = 150 mm, V_1 = 3 m/s, p_1 = 350 kPa; 10 m from the first section, d_2 = 75 mm

Find: p_2 for pipe, horizontal and vertical

Assumptions: Incompressible, steady, ideal flow with no losses.

Basic Equations: Continuity: $Q_1 = Q_2$; Bernoulli: $Z_1 + \dfrac{V_1^2}{2g} + \dfrac{p_1}{\gamma} = Z_2 + \dfrac{V_2^2}{2g} + \dfrac{p_2}{\gamma}$

Solution: We first draw a sketch of each part. For the pipe horizontal we have Figure 4.10a and for the pipe vertical we have Figure 4.10b. For the horizontal orientation, we select the centerline of the pipe as our reference plane, making $Z_1 = Z_2 = 0$.

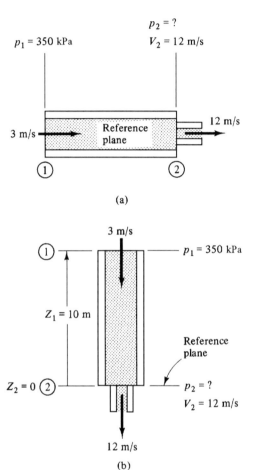

(a)

(b)

Figure 4.10   Illustrative Problem 4.7.

Our first task is to obtain the velocity in the pipe at the second section. Since γ does not change, $Q_1 = Q_2$. Therefore,

$$A_2 V_2 = A_1 V_1$$

$$\frac{\pi}{4} (75 \text{ mm})^2 \times V_2 = \frac{\pi}{4} (150 \text{ mm})^2 \times 3 \text{ m/s}$$

and

$$V_2 = 12 \text{ m/s}$$

Considering the pipe to be horizontal, $Z_1 = Z_2$, and writing equation (4.18) gives us

$$Z_1 + \frac{V_1^2}{2g} + \frac{p_1}{\gamma} = Z_2 + \frac{V_2^2}{2g} + \frac{p_2}{\gamma}$$

Inserting the data, we have

$$0 + \frac{(3 \text{ m/s})^2}{2 \times 9.81 \text{ m/s}^2} + \frac{350 \text{ kPa}}{9.81 \text{ kN/m}^3} = 0 + \frac{(12 \text{ m/s})^2}{2 \times 9.81 \text{ m/s}^2} + \frac{p_2}{9.81 \text{ kN/m}^3}$$

Solving yields

$$p_2 = 282.5 \text{ kPa}$$

For the case of flow being vertically downward, we have (see Figure 4.10b)

$$Z_1 + \frac{V_1^2}{2g} + \frac{p_1}{\gamma} = Z_2 + \frac{V_2^2}{2g} + \frac{p_2}{\gamma}$$

Notice that we have chosen section ① at the entrance, which is the top of the pipe, and section ② at the exit or bottom of the pipe. Our reference plane is taken to be at the bottom of the pipe. With these in mind we have

$$10 \text{ m} + \frac{(3 \text{ m/s})^2}{2 \times 9.81 \text{ m/s}^2} + \frac{350 \text{ kPa}}{9.81 \text{ kN/m}^3} = 0 + \frac{(12 \text{ m/s})^2}{2 \times 9.81 \text{ m/s}^2} + \frac{p_2}{9.81 \text{ kN/m}^3}$$

from which $p_2 = 380.6$ kPa.

     Using the procedure given earlier, we have systematically avoided several possible errors in this problem, especially with the flow being downward.

ILLUSTRATION 4.8   BERNOULLI EQUATION— TORRICELLI'S THEOREM

A large tank contains water and has a well-rounded, small opening in it as shown in Figure 4.11. Determine the velocity of the water flowing from the opening.

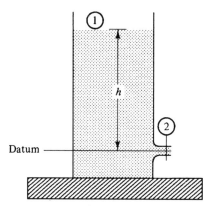

Figure 4.11   Illustrative Problem 4.8.

ILLUSTRATIVE PROBLEM 4.8

Given: Water, height h above opening, discharge from open tank to atmosphere

Find: V_2

Assumptions: Steady, incompressible flow; no losses

Basic Equation: Bernoulli equation

$$Z_1 + \frac{V_1^2}{2g} + \frac{p_1}{\gamma} = Z_2 + \frac{V_2^2}{2g} + \frac{p_2}{\gamma}$$

Solution: Using our procedure, we take ① to be at the free liquid surface, where the pressure is atmospheric and where we will take the velocity to be essentially zero, since the height of the liquid will change very slowly. We now take the free jet centerline to be section ② and also the datum, since it is the lowest point in the system. Notice that the pressure in the free jet will be atmospheric. Writing the Bernoulli equation, we have

$$Z_1 + \frac{V_1^2}{2g} + \frac{p_1}{\gamma} = Z_2 + \frac{V_2^2}{2g} + \frac{p_2}{\gamma}$$

For the conditions of this problem, $Z_1 = h$, $p_1 = 0$, $V_1 = 0$, $Z_2 = 0$, and $p_2 = 0$. Therefore,

$$h = \frac{V_2^2}{2g}$$

and

$$V_2 = \sqrt{2gh}$$

Notice that the result in this problem is simply a statement of the fact that the velocity of the water issuing from the jet equals the velocity that a body in free fall from the surface of the water in the tank to the centerline of the jet would acquire. This result is known as *Torricelli's theorem*.

ILLUSTRATION 4.9   BERNOULLI EQUATION—PIPE FLOW

Water flows in a horizontal pipe as shown in Figure 4.12. At the inlet the pipe has a diameter of 150 mm and the pressure is 420 kPa. At the downstream section of the pipe its diameter is 75 mm and the pressure is 140 kPa. Determine Q in m³/s.

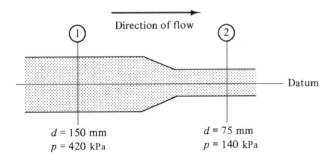

Figure 4.12   Illustrative Problem 4.9.

ILLUSTRATIVE PROBLEM 4.9

Given: $p_1 = 420$ kPa, $d_1 = 150$ mm, $d_2 = 75$ mm, $p_2 = 140$ kPa
      Pipe horizontal, water flowing

Find: Q

Assumptions: Steady, incompressible flow, no losses

Basic Equations: Continuity: $Q_1 = Q_2$; Bernoulli:

$$Z_1 + \frac{V_1^2}{2g} + \frac{p_1}{\gamma} = Z_2 + \frac{V_2^2}{2g} + \frac{p_2}{\gamma}$$

Solution: Writing the Bernoulli equation,

$$Z_1 + \frac{V_1^2}{2g} + \frac{p_1}{\gamma} = Z_2 + \frac{V_2^2}{2g} + \frac{p_2}{\gamma}$$

we note that $Z_1 = Z_2 = 0$ and that neither V_1 nor V_2 are given. To relate V_1 to V_2 we will use the continuity equation:

$$Q_1 = Q_2 = A_1 V_1 = A_2 V_2$$

or

$$V_1 = \left(\frac{A_2}{A_1}\right)V_2 = \left(\frac{d_2}{d_1}\right)^2 V_2$$

From the data given we have

$$V_1 = \left(\frac{75}{150}\right)^2 V_2 = \frac{V_2}{4}$$

Placing these into the Bernoulli equation gives us

$$\frac{(V_2/4 \text{ m/s})^2}{2 \times 9.81 \text{ m/s}^2} + \frac{420 \text{ kPa}}{9.81 \text{ kN/m}^3} = \frac{(V_2 \text{ m/s})^2}{2 \times 9.81 \text{ m/s}^2)} + \frac{140 \text{ kPa}}{9.81 \text{ kN/m}^3}$$

Rearranging gives us

$$\frac{420 - 140}{9.81} = \frac{V_2^2 - V_2^2/16}{2 \times 9.81}$$

and

$$280 = \frac{\frac{15}{16}V_2^2}{2}$$

Solving yields

$$V_2 = 24.4 \text{ m/s}$$

and

$$V_1 = \frac{V_2}{4} = 6.1 \text{ m/s}$$

Finally,

$$Q = A_2 V_2 = \frac{\pi}{4} (0.075 \text{ m})^2 \times 24.4 \frac{\text{m}}{\text{s}} = 0.108 \text{ m}^3/\text{s}$$

ILLUSTRATION 4.10   BERNOULLI EQUATION—VENTURI METER

The device shown in Figure 4.13 is known as a *venturi meter*, which consists of a converging and diverging conical section of a pipe arranged to give an increase in velocity as the pipe converges, causing a measurable drop in pressure. The diverging section is used to reconvert the increased kinetic energy of the fluid stream into pressure energy at the outlet of the device with a minimum of turbulence and friction losses. Show for this meter with no losses that

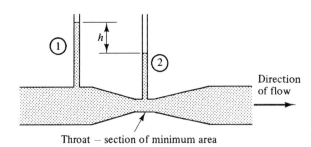

Direction of flow →

Figure 4.13   Simple venturi meter—Illustrative Problem 4.10.

Throat – section of minimum area

ILLUSTRATIVE PROBLEM 4.10

Given: V_1, A_1, A_2, h

Find: Relation of Q to A_1, A_2, and h

Assumptions: Steady, incompressible flow, no losses

Basic Equations: Continuity: $A_1 V_1 = A_2 V_2$; Bernoulli: $Z_1 + \dfrac{V_1^2}{2g} + \dfrac{p_1}{\gamma} = Z_2 + \dfrac{V_2^2}{2g} + \dfrac{p_2}{\gamma}$

Solution: Writing the Bernoulli equation with $Z_1 = Z_2$, we have

$$\frac{V_1^2}{2g} + \frac{p_1}{\gamma} = \frac{V_2^2}{2g} + \frac{p_2}{\gamma}$$

From the continuity equation, $V_2 = V_1(A_1/A_2)$. Using this and the Bernoulli equation gives us

$$\frac{V_1^2}{2g} + \frac{p_1}{\gamma} = \frac{[(V_1(A_1/A_2)]^2}{2g} = \frac{p_2}{\gamma}$$

Transposing and factoring, we obtain

$$\frac{p_1}{\gamma} - \frac{p_2}{\gamma} = \frac{V_1^2}{2g}\left[\left(\frac{A_1}{A_2}\right)^2 - 1\right]$$

But we note that $p_1/\gamma - p_2/\gamma = h$. Therefore,

$$h = \frac{V_1^2}{2g}\left[\left(\frac{A_1}{A_2}\right)^2 - 1\right]$$

Solving for V_1,

$$V_1 = \sqrt{\frac{2gh}{[(A_1/A_2)^2 - 1]}}$$

Since $Q = A_1 V_1$,

$$Q = A_1 \sqrt{\frac{2gh}{[(A_1/A_2)^2 - 1]}}$$

which is the relation that we set out to prove.

ILLUSTRATION 4.11   BERNOULLI EQUATION— VENTURI METER BASIC PROGRAM

Write a BASIC program to solve for Q, given d_1, d_2, and h in English units. Assume that diameters and h are given in inches.

ILLUSTRATIVE PROBLEM 4.11

Given: Venturi meter, V_1, V_2, d_1, d_2, and h in English units

Find: BASIC program for Q

Assumptions: Steady, incompressible flow, no losses

Basic Equations: Continuity: $A_1 V_1 = A_2 V_2$;

Bernoulli: $Z_1 + \dfrac{V_1^2}{2g} + \dfrac{p_1}{\gamma} = Z_2 + \dfrac{V_2^2}{2g} + \dfrac{p_2}{\gamma}$

Solution to Illustrative Problem 4.10: $Q = A_1 \sqrt{\dfrac{2gh}{[(A_1/A_2)^2 - 1]}}$

Solution: The solution to Illustrative Problem 4.10 can be written in terms of pipe diameters as

$$Q = \frac{\pi}{4} d_1^2 \sqrt{\frac{2gh}{(d_1/d_2)^4 - 1}}$$

In English units, $g = 32.17$ ft/s^2 and Q will be in ft^3/s.

$$Q = \frac{\pi}{4} \left(\frac{d_1}{12}\right)^2 \sqrt{\frac{2 \times 32.17 \times h/12}{(d_1/d_2)^4 - 1}}$$

The program is listed below with a numerical test case for $d_1 = 4$ in., $d_2 = 2$ in., and $h = 6$ in.

```
10 INPUT "D one = ";A
20 INPUT "D two = ";B
30 INPUT "h = ";C
40 E = (A/B)
50 D = (3.1416*A*A/576)*(SQR((2*32.17*(C/12))/((E^4) − 1)))
60 LPRINT "D one = ";A
70 LPRINT "D two = ";B
80 LPRINT "h = ";C
90 LPRINT "Q = ";D
100 END
D one = 4
D two = 2
h = 6
Q = .12779924507978
```

ILLUSTRATION 4.12   BERNOULLI EQUATION—VENTURI METER— EXCEL SPREADSHEET

Solve Illustrative Problem 4.11 using the Excel spreadsheet.

ILLUSTRATIVE PROBLEM 4.12

Given: Same inputs as Illustrative Problem 4.11

Find: Excel solution for Q

Assumptions: Same as Illustrative Problem 4.11

Basic Equations: Same as Illustrative Problem 4.11

Solution: Using the solution to Illustrative Problem 4.11,

$$Q = \frac{\pi}{4} \left(\frac{d_1}{12}\right)^2 \sqrt{\frac{2 \times 32.17 \times h/12}{(d_1/d_2)^4 - 1}}$$

The spreadsheet solution proceeds from column to column and row to row. d_1 is entered in box A3, d_2 is entered in box B3, and h is entered in box C3. Box $D3 = \frac{\pi}{4}\left(\frac{d_1}{12}\right)^2$, box E3 $= (2 \times 32.17 \times h/12)\Big/\!\left(\left(\frac{d_1}{d_2}\right)^4 - 1\right)$, and the final solution for Q is obtained in box F3 as D3 × (E3)$^{1/2}$. Notice that no programming language is required, and that the solution could have been obtained in one step. Using several steps makes it easier to detect and correct errors. Note that by changing A3, B3, and C3, the program will calculate Q for the new inputs.

	A	B	C	D	E	F
1			Illustrative Problem 4.12			
2	d1	d2	h	(3.1416/4)*(A3/12)^2	(2*32.17*C3/12)/((A3/B3)^4 − 1)	D3*(E3)^0.5
3	4	2	6	0.087266667	2.144666667	0.127799245

ILLUSTRATION 4.13   BERNOULLI EQUATION—BASIC PROGRAM

Write a BASIC program to solve for any of the individual items in the Bernoulli equation. As an arithmetic test of the program, solve Illustrative Problem 4.7 for the horizontal case.

ILLUSTRATIVE PROBLEM 4.13

Given: Bernoulli equation with any combination of 5 inputs

Find: Sixth input

Assumptions: Steady, incompressible flow; no losses

Basic Equation: Bernoulli equation: $Z_1 + \dfrac{V_1^2}{2g} + \dfrac{p_1}{\gamma} = Z_2 + \dfrac{V_2^2}{2g} + \dfrac{p_2}{\gamma}$

Solution: Based on the statement of the problem and examination of the Bernoulli equation, we see that it will be necessary to solve for any one of six unknowns. Although at first glance this may seem formidable, we will see that it really consists of solving six short programs. We first write the Bernoulli equation and solve for each of the variables. The resulting six equations are

$$p_1 = \gamma\left(\frac{p_2}{\gamma} + \frac{V_2^2}{2g} - \frac{V_1^2}{2g} + Z_2 - Z_1 \right)$$

$$p_2 = \gamma\left(\frac{p_1}{\gamma} + \frac{V_1^2}{2g} - \frac{V_2^2}{2g} + Z_1 - Z_2 \right)$$

$$V_2 = \sqrt{ 2g\left(\frac{p_1}{\gamma} - \frac{p_2}{\gamma} + Z_1 - Z_2 + \frac{V_1^2}{2g} \right) }$$

$$V_1 = \sqrt{ 2g\left(\frac{p_2}{\gamma} - \frac{p_1}{\gamma} + Z_2 - Z_1 + \frac{V_2^2}{2g} \right) }$$

$$Z_1 = \frac{p_2}{\gamma} - \frac{p_1}{\gamma} + \frac{V_2^2}{2g} - \frac{V_1^2}{2g} + Z_2$$

$$Z_2 = \frac{p_1}{\gamma} - \frac{p_2}{\gamma} + \frac{V_1^2}{2g} - \frac{V_2^2}{2g} + Z_1$$

We now use the tactic of a "dummy variable," X, and assign to it an arbitrary value, which in this case is 0.001. If any variable is inputted as 0.001, the program will calculate the value of the variable and replace the dummy variable by the calculated value. If the inputted variable is not equal to the value of the dummy variable, the program then goes to the next variable, and so on. In essence, excluding the input and output statements, the program consists of six, four-step simple statements that include the test, the calculation, a goto statement to the output statement, and an end statement. The program is attached, and the numerical solution of Illustrative Problem 4.7a is also given. If English units are used, G is 32.17 ft/s^2 and units used are feet and seconds. Thus, the program is written generally and can be used with either SI or English units.

```
5 : X = .001
10 : INPUT "P ONE = ";A
20 : INPUT "V ONE = ";B
30 : INPUT "G = ";C
40 : INPUT "Z ONE = ";D
50 : INPUT "GAMMA = ";E
60 : INPUT "P TWO = ";F
70 : INPUT "V TWO = ";H
80 : INPUT "Z TWO = ";I
90 : IF A<> X GOTO 130
100 : A = E*((F/E) + (H*H/(2*C)) − (B*B/(2*C)) + I − D)
110 : GOTO 320
120 : END
130 : IF F<>X GOTO 170
140 : F = E*((A/E) + (B*B/(2*C)) − (H*H/(2*C)) + D − I)
150 : GOTO 320
160 : END
170 : IF B<>X GOTO 210
180 : B = SQR(2*C*(F/E − A/E + I − D + (H*H/(2*C))))
190 : GOTO 320
200 : END
210 : IF H<>X GOTO 250
220 : H = SQR(2*C*(A/E − F/E + D − I + (B*B/(2*C))))
230 : GOTO 320
240 : END
250 : IF D<>X GOTO 290
260 : D = (F/E − A/E + (H*H/(2*C)) − (B*B/(2*C)) + I)
270 : GOTO 320
280 : END
290 : IF I <>X GOTO 320
300 : I = (A/E − F/E + (B*B/(2*C)) −(H*H)/(2*C)) + D)
310 : GOTO 320
315 : END
320 : LPRINT "P ONE = ";A
330 : LPRINT "V ONE = ";B
340 : LPRINT "G = ";C
350 : LPRINT "Z ONE = ";D
360 : LPRINT "GAMMA = ";E
370 : LPRINT "P TWO = ";F
380 : LPRINT "V TWO = ";H
390 : LPRINT "Z TWO = ";I
400 : END


P ONE = 350000
V ONE = 3
G = 9.81
Z ONE = 0
GAMMA = 9810
P TWO = 282500
```

V TWO = 12
Z TWO = 0

ILLUSTRATION 4.14   BERNOULLI EQUATION—EXCEL SPREADSHEET

Solve Illustrative Problem 4.13 using the Excel Spreadsheet

ILLUSTRATIVE PROBLEM 4.14

Given: Same inputs as Illustrative Problem 4.13

Find: Excel Solution

Assumptions: Same as Illustrative Problem 4.13

Basic Equations: Same as Illustrative Problem 4.13

Solution: Illustrative Problem 4.7a has been chosen as a test program, requiring solving for p_2. Boxes A3, B3, C3, D3, E3, F3, and G3 are the inputs. Box H3 is the solution of the equation for p_2. This solution is in SI. It has been made general by using a box for "g" as an input. Notice that box H3 will calculate p_2 for any values of input. The same procedure can be used to solve for any of the variables in the Bernoulli equation. It is emphasized that the use of the spreadsheet eliminates the need for a specialized programming language.

	A	B	C	D	E	F	G	H
1		Illustrative Problem 4.14						
2	p1	gamma	V1	V2	Z1	Z2	g	(B3)*((A3/B3) + (C3)^2/(2*G3) − (D3)^2/(2*G3) + E3 − F3
3	350000	9810	3	12	0	0	9.81	282500

CALCULUS ENRICHMENT

Consider the tank shown in Figure A. It is initially filled with a fluid to a height h_1. At this moment, a plug is removed from a hole in the bottom of the tank. We would like to determine the time for the level of the tank to fall from h_1 to h_2. The cross-sectional area of the tank is A and of the discharging jet is a.

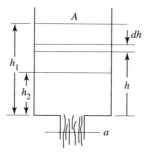

Figure A   Time to empty a tank.

The volume of fluid leaving the tank in a period of time dt is

$$Q = -Adh \qquad\qquad (a)$$

In this time period, the volume flowing out of the jet is, ideally,

$$Q = avdt = a\sqrt{2gh}\,dt \qquad\qquad (b)$$

Equating these two volumetric flows,

$$-Adh = a\sqrt{2gh}\,dt \qquad\qquad (c)$$

Rearranging Equation (c),

$$dt = -\frac{A}{a}\frac{dh}{\sqrt{2gh}} \qquad\qquad (d)$$

Integrating Equation (d)

$$\int_{t_1}^{t_2} dt = -\frac{A}{a\sqrt{2g}} \int_{h_1}^{h_2}\frac{dh}{\sqrt{h}} \qquad\qquad (e)$$

Performing the integration,

$$t_2 - t_1 = \frac{2A}{a\sqrt{2g}}\left(h_1^{1/2} - h_2^{1/2}\right) \qquad\qquad (f)$$

or

$$t_2 - t_1 = \frac{A}{a} = \sqrt{\frac{2}{g}}\left(h_1^{1/2} - h_2^{1/2}\right) \qquad\qquad (g)$$

If the discharge from the tank is not ideal, a coefficient of discharge can be introduced to account for the deviation from ideal. Using equation (g), and measuring the time for the level in a tank to fall from h_1 to h_2 also enables one to calculate the discharge coefficient from the tank.

4.5   REVIEW

This chapter is the first of several chapters that deal with the flow of fluids. After noting that the term "steady" is used to denote effects that do not vary with time, we developed the continuity equation. The continuity equation is a statement of the conservation of mass flow, that is, the rate at which mass enters a system must equal the rate at which mass leaves the system. For incompressible fluids this can be readily shown to yield $Q_{in} = Q_{out}$, where Q is the volume rate of flow.

Energy is the ability to do work, and work is the product of the displacement multiplied by the force causing the displacement. For a fluid system that has no work done on it and in turn does no work, that has a constant temperature, that has no heat transfer, and that is frictionless, we developed an equation expressing the conservation of energy. This equation, known as the Bernoulli equation, equates the potential energy, kinetic energy, and flow work into and out of a system. Quantitative expressions were developed for each of the terms in the Bernoulli equation, and it was noted that each of the terms in this equation has the dimensions of a height of a fluid, which is traditionally called a head.

After setting up a procedure for use of the Bernoulli equation, we applied it to several situations. Also, some BASIC programs were written to show applications of this computer language to problem solving in fluid mechanics. The Excel spreadsheet was also used to solve the same problems. The use of a spreadsheet eliminates the need for special programming languages.

KEY TERMS

Terms of importance in this chapter:

BASIC: *B*eginner's *A*ll-Purpose *S*ymbolic *I*nstruction *C*ode, a computer language.

Bernoulli's equation: a statement of the conservation of energy applied to the steady flow of an ideal incompressible fluid in the absence of work, friction, or heat transfer.

Continuity equation: a statement of the conservation of mass in a steady flow system: the mass flow rate into a system must equal the mass flow rate out of a system. For incompressible fluids, the continuity equation can also be taken to mean that the volume flow rate into a system must equal the volume flow rate out of the system.

Energy: the ability to do work.

Flow work: the work required to cause a fluid to flow into or out of a system.

Head: the terms of the Bernoulli equation have the dimensions of length and can be thought of as being heights of fluid: consequently, they are called heads.

Kinetic energy: the ability of a body to do work as a consequence of its velocity.

Potential energy: the ability of a body to do work as a consequence of its position relative to a reference plane.

Steady flow: the flow of a fluid that is independent of time.

Torricelli's theorem: the statement that the velocity from a jet is equal to the velocity that a body would have if it fell freely from a height equal to the distance from the reservoir surface to the centerline of the jet.

Venturi meter: an arrangement of converging and diverging sections of pipe that is used to measure flow rates in a pipe.

Volumetric flow rate: the product of area multiplied by velocity—expressed as cubic feet per second, cubic feet per minute, cubic meters per second, cubic meters per minute, liters per minute, liters per second, or gallons per minute.

Work: the product of a force and the amount of displacement in the line of action of the force.

KEY EQUATIONS

Continuity equation	$\dot{w} = \gamma AV$	(4.1b)
Continuity equation	$\dot{m} = \rho AV$	(4.1c)
Continuity equation	$\dot{m}_1 = \dot{m}_2$	(4.2a)
Continuity equation	$\rho_1 A_1 V_1 = \rho_2 A_2 V_2$	(4.2b)
Continuity equation	$\gamma_1 A_1 V_1 = \gamma_2 A_2 V_2$	(4.2c)
Continuity equation	$\dfrac{A_1 V_1}{v_1} = \dfrac{A_2 V_2}{v_2}$	(4.2d)
Volume flow rate	$Q = AV$	(4.3)
Mass and weight flow rate	$\dot{m} = \rho Q$	(4.4)
	$\dot{w} = \gamma Q$	
Potential energy (P.E.)	$\text{P.E.} = wZ$	(4.6)
Potential energy (P.E.)	$\text{P.E.} = mgZ$	(4.7)
Kinetic energy (K.E.)	$\text{K.E.} = \dfrac{mV^2}{2}$	(4.11)
Kinetic energy (K.E.)	$\text{K.E.} = \dfrac{wV^2}{2g}$	(4.12)
Flow work	$\dfrac{wp}{\gamma}$	(4.15a or b)
Net Flow Work	$w\left(\dfrac{p_2}{\gamma_2} - \dfrac{p_1}{\gamma_1}\right)$	(4.16a)
Bernoulli equation	$Z_1 + \dfrac{V_1^2}{2g} + \dfrac{p_1}{\gamma} = Z_2 + \dfrac{V_2^2}{2g} + \dfrac{p_2}{\gamma}$	(4.18)

QUESTIONS

1. A fluid flows in a horizontal pipe. How do the velocities at two sections of the pipe compare if the diameters are in the ratio 3:1?

2. A body is dropped freely and strikes a surface that is some distance below it. State what has happened to (a) its potential energy and (b) its kinetic energy.

3. Why is each successive rise of a cyclone roller coaster ride in an amusement park less than the preceding one?

4. If a solid steel ball is dropped onto a flat steel plate, will it rise to its original height? Explain your answer.

5. All fluids have the properties of pressure and specific weight. Does this mean that they all possess flow work?

6. Why would it be inappropriate to apply Bernoulli's equation to a system in which the gas flowing is being cooled?

7. Using the Bernoulli equation, explain in a qualitative manner how a wing can produce a net lifting force.

8. Is it easier to throw a curved pitch with a tennis ball or with a baseball? Use the Bernoulli equation to explain your answer.

9. If two ships move parallel to each other and in the same direction, explain what will occur when they overtake each other.

PROBLEMS

Use $\gamma = 62.4$ lb/ft^3 (9810 N/m^3) for water unless indicated otherwise.

Flow and Continuity

Where necessary use Table B.1 for the properties of pipe. If not stated, all diameters are inside diameters.

4.1. One gallon is a volume measure of 231 in.^{3}. Determine the velocity in a pipe 2 in. in diameter when water flows at the rate of 20 gpm in the pipe.

4.2. What weight of water is flowing in Problem 4.1?

4.3. Solve Problem 4.1 if oil having a specific gravity of 0.85 is flowing.

4.4. If 30 gpm is flowing in a pipe, how many m^3/s is flowing?

4.5. An 8-in. Schedule 40 pipe carries 200 gpm of water. Calculate the average velocity in the pipe.

4.6. Air flows through a 12-in. $\times$ 12-in. duct at the rate of 800 cfm. Determine the mean flow velocity in the duct.

4.7. If 800 liters/min of oil flows through a 4-in. Schedule 80 pipe, determine the average velocity in the pipe.

4.8. Air flows through a rectangular duct 0.5 m $\times$ 0.75 m at the rate of 200 m^3/min. Determine the average velocity of the air.

4.9. Water flows through a 3-in. Schedule 40 pipe with an average velocity of 12 ft/s. Calculate the volume flow rate in cubic feet per second and in gallons per minute, the weight flow rate per second, and the mass flow rate per second.

4.10. A chemical process uses 5000 liters/min. To how many m^3/s does this correspond?

4.11. An American manufacturer can process 100 gpm. In talking to her European factory she finds that she must tell them this rate in liters per second. How many liters per second will she tell them can be processed?

4.12. Twenty liters of fluid per second flows in a pipe whose inside diameter is 7.5 cm. Determine the velocity and weight rate of flow if the fluid is water.

4.13. What is the mass flow rate (kg/s) in Problem 4.12?

4.14. A liquid has a specific gravity of 1.18 and flows in a 6-in. diameter Schedule 40 pipe with an average velocity of 2.0 ft/s. Calculate the flow rate in cfs, gpm, lb/s, and slugs/s.

4.15. Water flows through the pipe shown in Figure P4.15 at the rate of 30 ft³/s. Determine the velocity at sections ① and ②. Also calculate the weight flow rate per second in the pipe at these sections. The diameters shown are internal diameters.

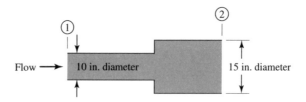

Figure P4.15

4.16. A reservoir contains 1000 gal of water. If a pump can pump at the rate of 4 liters/min, how long will it take for the tank to empty?

4.17. A pipe carrying water at the rate of 100 liters/min is reduced in diameter from 75 mm to 50 mm. What is the velocity of the water in each section of the pipe?

4.18. Water flows at the rate of 50 gallons per minute (gpm) in a pipe that has an internal diameter of 4 in. The diameter of the pipe is reduced to 2 in. Determine the velocity in each section of the pipe.

4.19. When a fluid of constant density flows in a pipe, show that the velocity is inversely proportional to the square of the diameter.

4.20. An 8 in Schedule 40 pipe carries 200 gpm of water. It then branches into two 4 in. Schedule 40 pipes with the flow dividing equally between the branches. Calculate the average velocity in each of the pipes.

4.21. Water flows through the nozzle shown in Figure P4.21 at the rate of 100 lb/s. Calculate the velocity at sections ① and ②.

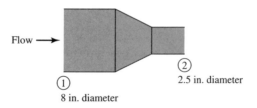

Figure P4.21

4.22. Water flows through a pipe at the rate of 1500 liters/min. At one section the pipe is 250 mm in diameter, and later reduces to a diameter of 100 mm. What is the average velocity in each section of the pipe?

4.23. A pump can fill a tank in 1 min. If the tank has a capacity of 18 gallons, determine the average velocity in the 1.25-in. diameter pipe that is attached to the pump.

4.24. If the water level in the tank in Figure P4.24 remains constant, determine the velocity at section ②.

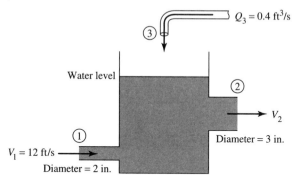

Figure P4.24

4.25. Air having a specific weight of 11.8 N/m³ flows in the pipe shown in Figure P4.25 with a velocity of 10 m/s at section ①. Calculate the weight flow rate per second of the air and its specific weight at section ② if the air has a velocity of 3 m/s at section ②.

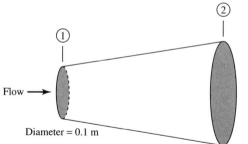

Figure P4.25

4.26. Oil, sg = 0.85, flows through a pipe whose internal diameter is 15 in. at the rate of 1500 gpm. Determine the volume flow rate in cfs, the average velocity, and the mass flow rate per second in the pipe.

4.27. For the liquid rocket shown in Figure P4.27, calculate V_2 for steady operation.

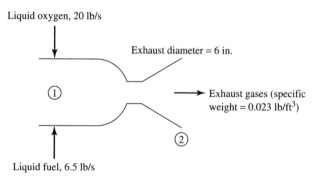

Figure P4.27

4.28. The 25-mm diameter pipe shown in Figure P4.28 supplies two branch pipes each 100 mm in diameter. If 500 gpm flows in the main pipe and each branch carries equal volumes of flow, determine the velocity in each branch pipe. Assume that γ remains constant.

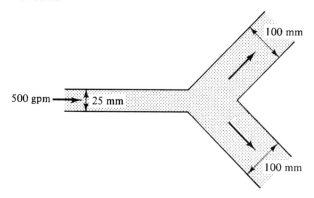

Figure P4.28

Kinetic and Potential Energy

4.29. A body with a mass of 10 slugs is placed 10 ft above an arbitrary plane. If the acceleration of gravity is 10 ft/s², how much work (foot pounds) was done in lifting the body above the plane?

4.30. How much work does it take to lift 10 kg a distance of 3 m?

4.31. A body weighing 100 lb is lifted 10 ft. What velocity will it have after freely falling the 10 ft?

4.32. If the body in Problem 4.30 falls freely, what will its velocity be after it travels the 3 m?

4.33. A body weighing 10 lb is lifted 100 ft. What is its potential energy? What velocity will it possess after falling the 100 ft?

4.34. A body has a mass of 5 kg. If its velocity is 10 m/s, what is its kinetic energy?

4.35. A jet of water issues vertically from a nozzle with a velocity of 25 ft/s. What is the maximum height that it will reach if air resistance is neglected?

4.36. A jet of water issues vertically from a nozzle with a velocity of 10 m/s. What is the maximum height that it will reach if air resistance is neglected?

4.37. A body weighing 10 lb is moving with a velocity of 50 ft/s. From what height would it have to fall to achieve this velocity? What is its kinetic energy?

4.38. Water flows over the top of a dam and falls freely until it reaches the bottom 600 ft below. What is the velocity of the water just before it hits the bottom? What is its kinetic energy per pound at this point?

*4.39. Show that the kinetic energy of a fluid flowing in a pipe varies inversely as the fourth power of the pipe diameter, all other things being equal.

*4.40. The flow in a city water system is being tested by allowing it to flow from a hydrant with an outlet 4 in. in diameter. The outlet is 2.0 ft above the ground, and the issuing stream of water hits the level ground at a distance 9 ft from the outlet. How much water is flowing?

4.41. A hose is 4 in. in diameter and has water flowing in it. If 10 ft³/s is flowing, determine the velocity in the hose. If the pressure in the pipe is 20 psig, determine the maximum height the water can reach.

Flow Work

4.42. A pressure of 5 bars (1 bar = 10^5 Pa) is presented to the piston of a pump, causing it to travel 100 mm. If the cross-sectional area of the piston is 1000 mm² and the pressure is constant, how much work was done by the steam on the piston?

4.43. In a constant-pressure process, steam at 500 psia is presented to the piston of a pump and causes the piston to travel 4 in. If the cross-sectional area of the piston is 5 in.², how much work was done by the steam on the piston?

4.44. Air is compressed steadily until its final volume is half its initial volume. The initial pressure is 35 psia and the final pressure is 100 psia. If the initial volume is 1 ft³, determine the difference in the p/γ terms at entrance and exit in foot pounds per pound if there is 1 pound of air.

4.45. If a fluid flows past a section of pipe with a pressure of 100 kPa and a specific volume of 10^{-3} m³/kg, determine its flow work.

4.46. At the entrance to a steady flow device, the pressure is 350 kPa and the specific volume is 0.04 m³/kg. At the outlet the pressure is 1 MPa and the specific volume is 0.02 m³/kg. Determine the flow work change in this device from entrance to exit.

Bernoulli Equation

4.47. A 75-mm diameter pipe carries water at the rate of 1400 liters/min. The pressure in the pipe is 210 kPa above atmospheric pressure. Determine the velocity, velocity head, and pressure head if a datum is selected 3 m below the center of the pipe. (P_{atmos} = 100 kPa).

4.48. Determine the pressure at section ② of the pipe shown in Figure P4.48 on page 164 if water flows at the rate of 300 gal/min past section ①.

4.49. Water flows into a large tank at the rate of 0.0072 m³/s and leaves the tank through a 4.5 cm diameter hole located as shown in Figure P4.49 on page 164. Determine h for steady flow.

4.50. Using Figure P4.49, determine h in steady flow if water enters the tank at the rate of 0.15 ft³/s and leaves through a 1.5-in. diameter hole at the outlet.

4.51. A nozzle is attached to a hose that has a 2-in. inside diameter as shown in Figure P4.51 on page 164. The nozzle has a discharge diameter of 0.625 in. If the upstream pressure at the inlet to the nozzle is 10 psig, determine the volumetric flow rate through the nozzle when it is horizontal.

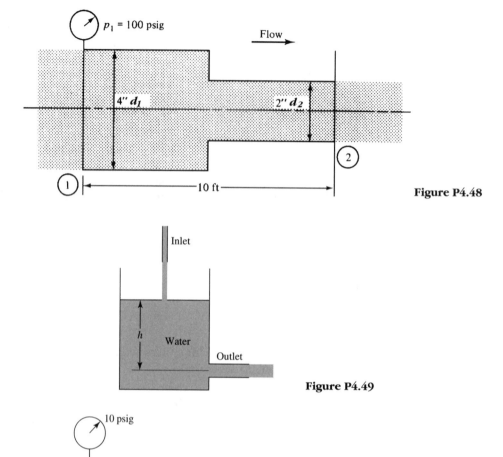

Figure P4.48

Figure P4.49

Figure P4.51

4.52. Solve Problem 4.51 if the nozzle discharges vertically.

4.53. The diameter of a pipe changes slowly from 5 in. at A to 2.5 in. at B. Point A is 30 ft above an arbitrary datum and Point B is 15 ft above this datum. The pressure at A is 20 psig, and the velocity of the water at this point is 10 ft/s. Determine the pressure at B in psi.

4.54. Water flows vertically upward in a pipe. The area of the pipe at the lower section is 30 in.^{2} and at the upper section is 10 in.^{2}. Assuming that the sections are 100 ft apart and that the flow rate is 200 gpm, determine the pressure difference between the two sections.

4.55. If the pipe in Problem 4.48 is turned vertically with the large end (section ①) at the bottom, what is the pressure at section ②? The flow is upward in the pipe.

4.56. A large tank has a 50-mm opening located 4 m below the surface of oil having a specific gravity of 0.85. Determine the volume rate of flow through the opening.

4.57. A tank has a water level of 20 ft above a datum plane. Five feet above the datum plane there exists a 6-in. opening in the tank. If there are no flow losses, determine the velocity of the water leaving the opening and the quantity of water flowing at this instant in gallons per minute.

4.58. Assuming no flow losses, determine the velocity of the water leaving the system shown in Figure P4.58.

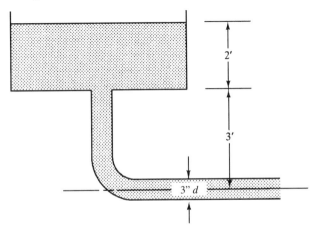

Figure P4.58

4.59. Calculate the quantity of water flowing in Problem 4.58 in gallons per minute.

4.60. Assume that the nozzle shown in Figure P4.60 has no losses. Determine the velocity of the water jet if it discharges to the atmosphere.

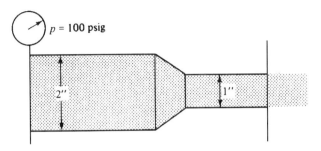

Figure P4.60

4.61. Determine the pressure difference between points ① and ② in Figure P4.61 on page 166 if the fluid flowing is water.

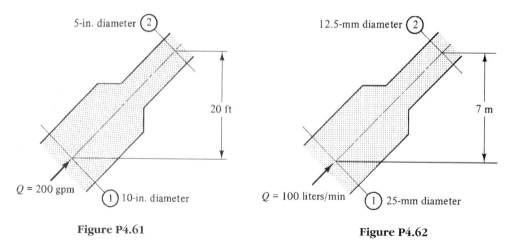

Figure P4.61   **Figure P4.62**

4.62. Determine the pressure difference between points ① and ② in Figure P4.62 if the fluid flowing is water.

4.63. Water is being discharged to the atmosphere from the open tank shown in Figure P4.63. Determine the exit velocity of the water. (Hint: Take 1 to be at the top of the water level and 2 to be at the exit from the tank to the atmosphere.)

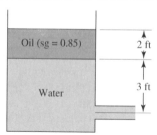

Figure P4.63

4.64. Determine the exit velocity of water from the tank shown in Figure P4.64.

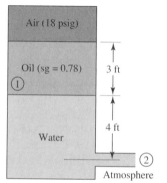

Atmosphere   **Figure P4.64**

4.65. The tank shown in Figure P4.65 discharges water from the opening in the bottom as shown. Determine the height that the water jet will reach assuming that the

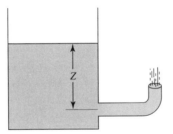

Figure P4.65

tank diameter is much greater than the pipe diameter. The height Z can be taken to be 10 ft.

4.66. If the tank in Problem 4.65 is capped, and there is an air pressure of 10 psig above the water, how high will the jet reach?

4.67. A pipe in the side of a very large closed tank discharges water to the atmosphere. The pressure above the water causes a discharge of 3.0 cfs when the height of the water is 5 ft above the centerline of the pipe. Determine the pressure above the water if the pipe has an inside diameter of 5 in.

4.68. Water flows in a pipe as shown in Figure P4.68. A vertical tube is connected to the pipe. How high will the water rise in the tube?

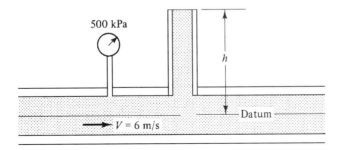

Figure P4.68

4.69. Water flows through the pipe shown in Figure 4.69 at the rate of 2000 liters/min. Determine the water level in tube B.

4.70. Solve Problem 4.69 if the fluid flowing is oil having a specific gravity of 0.9.

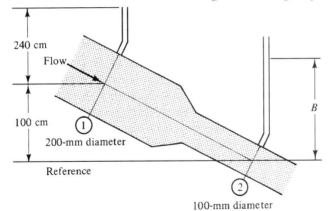

Figure P4.69

4.71. Oil (sg = 0.85) flows at the rate of 6000 liters/min in the pipe shown in Figure P4.71. Calculate the pressure at section ②.

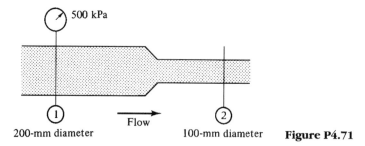

Figure P4.71

4.72. Water flows in the pipe as shown in Figure P4.72 and discharges through the nozzle to the atmosphere. If the outlet velocity is 18 m/s, determine the pressure at section ①.

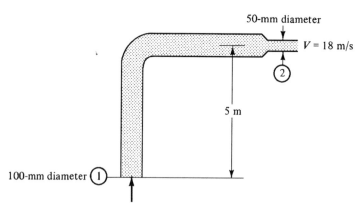

Figure P4.72

4.73. Water flows from a very large tank through a 50-mm diameter tube as shown in Figure P4.73. Determine the velocity of the water in the tube.

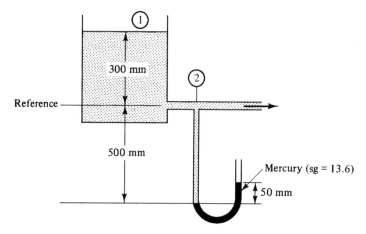

Figure P4.73

4.74. Water flows through the pipe shown in Figure P4.74 from ① to ② at the rate of 500 liters/s. If the pressure at ① is 100 kPa, determine the pressure at ②.

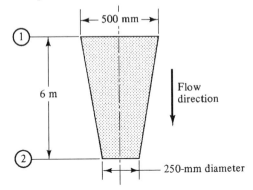

Figure P4.74

4.75. Solve Problem 4.74 if the direction of flow is reversed.

4.76. For the venturi meter shown in Figure P4.76, oil having a specific gravity of 0.9 is flowing. Determine the volumetric flow in this device when $p_1 - p_2 = 3$ psi.

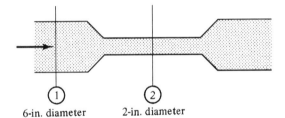

6-in. diameter   2-in. diameter   Figure P4.76

4.77. Determine the discharge from the large tank shown in Figure P4.77.

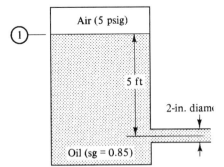

Figure P4.77

4.78. Oil flows through the pipe section shown in Figure P4.78 on page 170. The manometer fluid is water. If the oil has a specific gravity of 0.9, calculate the exit velocity of the oil.

4.79. A venturi meter has a manometer connected to it as shown in Figure P4.79 on page 170. If the velocity of flow of water in the 1-in. diameter section is 25 ft/s, determine the distance y.

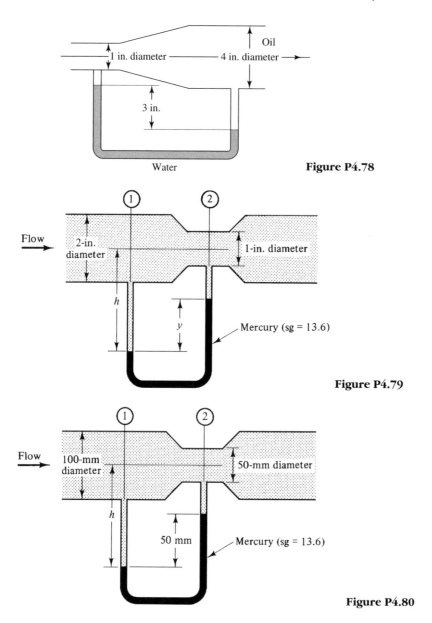

Figure P4.78

Figure P4.79

Figure P4.80

4.80. The venturi meter shown in Figure P4.80 has water flowing in it. If the manometer deflection is 50 mm, determine the volumetric flow.

***4.81.** Water flows from ① to ② in the venturi meter shown in Figure P4.81. Calculate the volumetric flow.

***4.82.** A venturi meter is used to meter the flow of water in a pipe as shown in Figure P4.82. If the velocity of the water at the exit of the device is 3 ft/s, determine the manometer reading h.

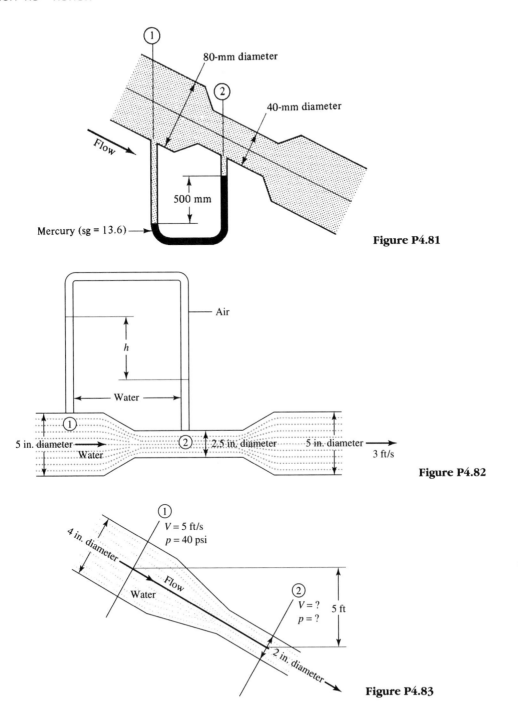

Figure P4.81

Figure P4.82

Figure P4.83

*4.83. Water flows from section ① to section ② in Figure P4.83. Determine the velocity and the pressure at section ②.

*4.84. A 3-in. diameter pipe projects a jet of water vertically into the air. If the rate of flow is 2 cfs, determine the diameter of the jet at a point 9 ft above the end of the pipe, and determine the maximum height that the jet will reach if air resistance is neglected.

*4.85. Sections ① and ② are on a pipe that changes in diameter from 6 in. at section ① to 15 in. at section ②. Section ① is 3 ft lower than section ②. The pressure at section ① is 42 psi, while the pressure at section ② is 50 psi. Determine the flow rate in cfs in the pipe.

*4.86. For the venturi meter shown in Figure P4.86, determine the flow of water when the mercury deflection is 10 in.

*4.87. Assuming that the flow is reversed in Problem 4.86, what will be the pressure difference between points ① and ② $(p_1/\gamma - p_2/\gamma)$ when $Q = 3.1$ cfs?

*4.88. Water flows in the venturi meter shown in Figure P4.88. Determine the volume rate of flow under these conditions.

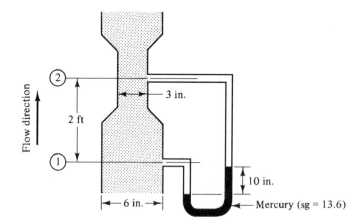

Figure P4.86

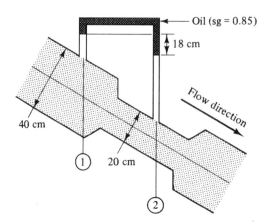

Figure P4.88

5 THE ENERGY EQUATION

LEARNING GOALS

After reading and studying this chapter you should be able to:

1. Restate the limitations of the Bernoulli equation.
2. State the concept of a streamline.
3. Define the terms *head added, head removed, head lost,* and *total head.*
4. State the reasons for the convention adopted for the placement of the head added, head removed, and head lost terms in the energy equation.
5. Use the energy equation to analyze systems in which energy is added, removed, and lost.
6. Express the concept of *power* quantitatively.

5.1   INTRODUCTION

In Chapter 4 the Bernoulli equation was derived and its limitations were discussed. It is interesting to note that the Bernoulli equation was first proposed by Daniel Bernoulli in 1738 and since that time has been used extensively to solve many problems in fluid mechanics. The application of Bernoulli's equation often requires that it be modified to account for deviations from the ideal conditions to which it is strictly applicable. Usually, the equation (or its modifications) yields results that are within the accuracy required for engineering applications. There are several cases, however, where the application of this equation will yield incorrect results, since it is being used beyond the limits of which even its modified forms can be hoped to apply. Several examples of this situation are the high-speed flow of compressible fluids, fluid flow with heat transfer causing large density changes, and fluid flow with large pressure changes due to frictional effects.

This chapter is devoted to the study of those one-dimensional steady flow situations to which the modified Bernoulli equation can be applied to yield reasonable solutions. Many practical cases will be studied and analyzed, but a word of caution must be emphasized at this point. It is quite easy for the student to extend idealizations to cases where they do not apply. It is necessary to be aware of the assumptions being made, and care must be exercised not to attempt to extend these simplified analyses to complex flow situations for which they are not intended.

5.2   THE ENERGY EQUATION

When the Bernoulli equation was considered in Chapter 4, it was very often convenient to select a control volume whose control surface coincided with the physical surface of a system. This is not necessary, however, and frequently we shall wish to utilize a control volume whose control surface is selected for ease of analysis. Any fluid stream consists of molecules of the fluid having a random motion due to collisions with other molecules and in addition having a directed motion along the flow path. A *streamline* is defined as a line with the tangent to it at any point being the direction of its velocity at that point. Streamlines cannot cross, since this would require two molecules having different velocities to be at the same point at the same time. We also note that in steady flow, streamlines do not change with time. Since the streamline coincides with the macroscopic path of the flow, it must be parallel to the surrounding flow, and no flow can cross a streamline. If the flow is accelerating, the streamlines are not parallel to each other and will appear closer together, and if the flow is decelerating, the streamlines will appear farther apart. For the case of uniform steady flow the streamlines are parallel. These concepts are apparent when the ideal airfoil shown in Figure 5.1 is considered. In the undisturbed stream before and after the airfoil, the streamlines are uniformly spaced and parallel. Above the airfoil they are close together, indicating a high-velocity region, and beneath the airfoil they are farther apart, indicating a low-velocity region.

As noted, in steady flow no fluid crosses a streamline, and the flow follows the streamlines. Every streamline is a continuous line that may be considered as starting at an infinite distance upstream and extending to an infinite distance downstream. The streamlines

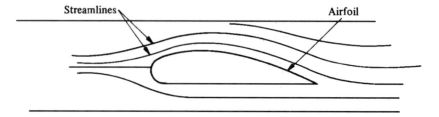

Figure 5.1   Flow over an airfoil.

through the points of a closed curve will result in a closed surface being generated that is known as a *stream tube.* Since the surface of the stream tube (Figure 5.2) consists of streamlines, no flow can enter or leave the stream tube through its lateral surface. When a control volume is selected in which the flow is confined by solid walls, streamlines can be considered to start in a reservoir where the fluid is at rest and to extend to the discharge of the system.

The Bernoulli equation derived in Chapter 4 was

$$\frac{p_1}{\gamma} + \frac{V_1^2}{2g} + Z_1 = \frac{p_2}{\gamma} + \frac{V_2^2}{2g} + Z_2 \tag{5.1}$$

Equation (5.1) can also be written as

$$\frac{p_1}{\gamma} + \frac{V_1^2}{2g} + Z_1 = \text{constant} \tag{5.2}$$

When written in this form we state that the sum of flow energy, the kinetic energy, and the potential energy is a constant *along a streamline.* For another streamline the constant can have a different value, but in many problems all of the streamlines have practically the same total energy. For example, a reservoir that has all the streamlines starting in it will have the same constant for all the streamlines. Thus, for these cases the sum of the energies on the left side of equation (5.1) or (5.2) can be equated to the sum of the corresponding terms between positions in an ideal system regardless of which streamline is being considered.

All real fluid-flow situations are irreversible due to viscous effects giving rise to shear stresses in the fluid. Theoretically, the Bernoulli equation as we have derived it must be modified in order to apply it to those cases in which nonideal effects (friction, turbulence, etc.) are relatively large when compared to the constant of equation (5.2). Let us assume that point 1 is upstream and that point 2 is downstream along a streamline. Assuming that there is no addition of energy into or out of the streamline as either work or heat, we can state that the energy at point 1 equals the energy at point 2 plus all of the flow losses between these two points. Mathematically,

$$\frac{p_1}{\gamma} + \frac{V_1^2}{2g} + Z_1 = \frac{p_2}{\gamma} + \frac{V_2^2}{2g} + Z_2 + \text{losses}_{1 \to 2} \tag{5.3}$$

The loss term on the right-hand side of equation (5.3) can be considered to be a term required to ensure energy conservation. We may also interpret equation (5.3) in an alternative and quite useful manner. The left side of this equation represents the total head entering the system and the right side represents the total head leaving the system. Thus any energy (head) entering will be placed on the left side as an input and any energy (head)

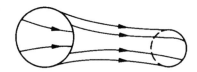

Figure 5.2   Stream tube for steady flow.

leaving the system will be placed on the right side. Using the notation of h_L to represent head lost, we can rewrite equation (5.3) as

$$\frac{p_1}{\gamma} + \frac{V_1^2}{2g} + Z_1 = \frac{p_2}{\gamma} + \frac{V_2^2}{2g} + Z_2 + h_L \tag{5.3a}$$

If a pump adds energy to the streamline between points 1 and 2 or a turbine extracts energy between these points, the Bernoulli equation is usually further modified to account for these energy additions or depletions. Thus, for a pump, with h_p equal to the head added by the pump, we note that this is an *input* and place this term on the left side of equation (5.3a), giving us

$$\frac{p_1}{\gamma} + \frac{V_1^2}{2g} + Z_1 + h_p = \frac{p_2}{\gamma} + \frac{V_2^2}{2g} + Z_2 + h_L \tag{5.4}$$

A turbine removes energy from a system, and consequently we place the head removed by a turbine, h_T, on the right side of equation (5.3a) to give us

$$\frac{p_1}{\gamma} + \frac{V_1^2}{2g} + Z_1 = \frac{p_2}{\gamma} + \frac{V_2^2}{2g} + Z_2 + h_L + h_T \tag{5.5}$$

For a system having both a pump and a turbine, we combine equations (5.4) and (5.5) to obtain

$$\frac{p_1}{\gamma} + \frac{V_1^2}{2g} + Z_1 + h_p = \frac{p_2}{\gamma} + \frac{V_2^2}{2g} + Z_2 + h_L + h_T \tag{5.6}$$

Equation (5.6) will be denoted to be the energy equation, and should be thoroughly understood.

In equations (5.3) through (5.6) the term p/γ is referred to as the *pressure head,* the term $V^2/2g$ is referred to as the *velocity head,* and the term Z is referred to as the *potential head*. Using the usual engineering units it will be found that each of these terms has the units of feet; their sum is called the *total head* or *total energy*. These terms are more readily understood by referring to Figure 5.3, on which each term is shown for sections along an irregular tube. In Figure 5.3a the total head or total energy is a constant for flow without friction, and is represented by the total head or total energy line as a horizontal line at a constant distance from the datum plane. With friction this is not true, and the total head line in Figure 5.3b is shown as a dashed curved line dropping off to the right. It will be noticed that when the velocity in the tube increases, the sum of the potential and pressure heads must decrease, and that for a decreased velocity, the sum of the potential and pressure heads increases. For steady flow, $V_3^2/2g$ is a constant, requiring p_3/γ to be constant for the case of no friction (Figure 5.3a); in the case of friction (Figure 5.3b), p_3/γ must decrease, since $V_3^2/2g$ is constant.

Figure 5.4 on page 178 shows Hoover Dam and Lake Mead. This dam is used to provide flood protection, river control, and water storage for irrigation; for municipal and industrial use; and for the generation of low-cost energy. Figure 5.5 on page 178 shows a view of the first generating unit that went on-line at Hoover Dam.

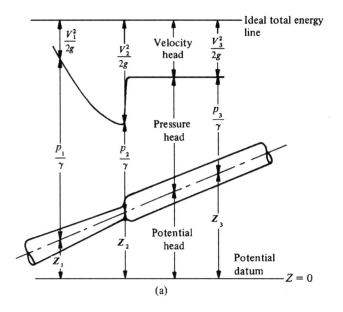

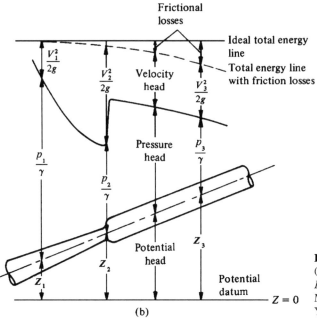

Figure 5.3   Head variation in a tube. (Adapted with permission from *Basic Fluid Mechanics* by J. L. Robinson, McGraw-Hill Book Company, New York, 1963, p. 49.)

Figure 5.6 on page 179 shows a pumped storage plant located on Lake Michigan. The pumped storage plant is used to store water in an upper reservoir at times of low power usage and to generate power at times of high power demand. Figure 5.7 on page 179 shows the principle of operation of this type of plant.

Figure 5.4   Hoover Dam and Lake Mead. (Courtesy of the United States Department of the Interior, Bureau of Reclamation.)

Figure 5.5   Generator at Hoover Dam (Note the size of the man.) (Courtesy of the United States Department of the Interior, Bureau of Reclamation.)

Figure 5.6   Pumped storage hydroelectric power plant. (Courtesy of Detroit Edison Co.)

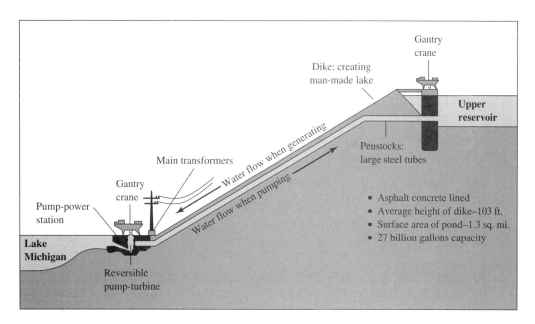

Figure 5.7   Principle of operation of a pumped storage hydroelectric power plant. Pumped storage hydroelectric plants of this type consist of an upper and lower reservoir connected by steel tubes or penstocks each equipped with a reversible pump-turbine. Essentially an adaptation of the principle of waterfalls, the plant generates energy by releasing water from the upper reservoir. As the water courses down the penstocks, it turns the plant's hydraulic turbines, which spin the generators, producing energy. In the second phase of the cycle, the process is reversed. The generators become motors, driving the turbines as pumps. Water is drawn from the plant's lower reservoir (in this case, Lake Michigan) and is pumped up the penstocks, refilling the upper reservoir and recharging the plant (like a storage battery). (Courtesy of Detroit Edison Co.)

ILLUSTRATION 5.1   THE ENERGY EQUATION—FRICTION LOSSES

Water is to be delivered from an open tank through a pipeline to a lower elevation. Flow rate is to be 100 gal/min of 60°F water (γ = 62.4 lb/ft³), and the inside diameter of the pipeline is 2 in. At the second section the desired pressure is to be 4 psi above atmospheric pressure. If the friction drop in the pipeline is estimated to be 38 ft, determine the vertical distance required between the level in the supply tank and the point of water discharge.

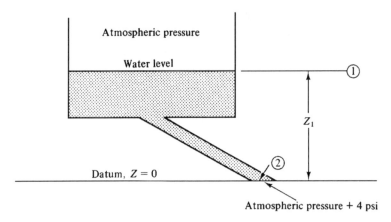

Figure 5.8   Illustrative Problem 5.1

ILLUSTRATIVE PROBLEM 5.1

Given: p_1 = 0, p_2 = 4 psig, Z_2 = 0, γ = 62.4 lb/ft³, h_L = 38 ft

Find: Z_1

Assumptions: Steady, incompressible flow; losses

Basic Equations: Continuity: $A_1 V_1 = A_2 V_2$

$$\text{Energy:} \qquad \frac{p_1}{\gamma} + \frac{V_1^2}{2g} + Z_1 = \frac{p_2}{\gamma} + \frac{V_2^2}{2g} + Z_2 + h_L$$

Solution: Referring to Figure 5.8, we shall select two stations, one at the water-air free surface in the tank and the other in the pipe. Also, we shall select the potential head datum at the pipe outlet as shown. For the conditions of this problem we note that all streamlines have the same total energy at the water level in the tank and that we can therefore properly apply the energy equation at all positions in the flow regardless of the streamline involved. Using equation (5.3a), we obtain

$$\frac{p_1}{\gamma} + \frac{V_1^2}{2g} + Z_1 = \frac{p_2}{\gamma} + \frac{V_2^2}{2g} + Z_2 + h_L$$

The selection of the datum as shown automatically makes Z_2 = 0, and it will be assumed that V_1 is very small compared to V_2, since the tank is presumably large compared to the pipe. Thus, for this problem we have

$$\frac{p_1}{\gamma} + Z_1 = \frac{p_2}{\gamma} + \frac{V_2^{\,2}}{2g} + h_L$$

or

$$Z_1 = \frac{p_2}{\gamma} - \frac{p_1}{\gamma} + \frac{V_2^{\,2}}{2g} + h_L$$

But p_2 exceeds p_1 by 4 psi. Therefore,

$$\frac{p_2}{\gamma} - \frac{p_1}{\gamma} = \frac{4 \text{ lb/in.}^2 \times 144 \text{ in.}^2/\text{ft}^2}{62.4 \text{ lb/ft}^3} = 9.23 \text{ ft}$$

and

$$Z_1 = \left(\frac{p_2}{\gamma} - \frac{p_1}{\gamma}\right) + \frac{V_2^{\,2}}{2g} + h_L = 9.23 + \frac{V_2^{\,2}}{2g} + 38$$

By definition, a gallon is a volume measure equal to 231 in.^{3}. Therefore,

$$100 \, \frac{\text{gal}}{\text{min}} = \frac{100 \text{ gal}}{60 \text{ s}} = \frac{100 \times 231 \text{ in.}^3}{60 \times 1728 \text{ in.}^3/\text{ft}^3} = 0.2228 \, \frac{\text{ft}^3}{\text{s}}$$

Since volume flow equals AV,

$$0.2228 = \left[\frac{(\pi/4)(2)^2}{144}\right] V_2 \quad \text{and} \quad V_2 = 10.21 \text{ ft/s}$$

Using this value of V_2, we obtain

$$Z_1 = 9.23 + \frac{(10.21)^2}{2 \times g} + 38 = 48.85 \text{ ft or about 49 ft}$$

ILLUSTRATION 5.2   THE ENERGY EQUATION—FRICTION LOSSES

A pipe is connected to a large reservoir as shown in Figure 5.9 and discharges to the atmosphere. If 10 liters/s flows in the pipe, determine the energy losses in the system.

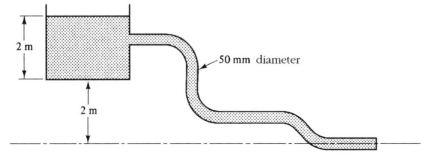

Figure 5.9   Illustrative Problem 5.2.

ILLUSTRATIVE PROBLEM 5.2

Given: $Q = 10$ liters/s, $p_1 = 0$, $p_2 = 0$, $Z_1 = 4$ m, $Z_2 = 0$, $V_1 = 0$

Find: h_L

Assumptions: Steady, incompressible flow; losses

Basic Equations: Continuity: $A_1 V_1 = A_2 V_2$

$$\text{Energy: } \frac{p_1}{\gamma} + \frac{V_1^2}{2g} + Z_1 = \frac{p_2}{\gamma} + \frac{V_2^2}{2g} + Z_2 + h_L$$

Solution: Taking both the surface of the liquid in the tank and the outlet of the pipe as atmospheric pressure, we have $p_1 = p_2 = p_a$. If the tank is considered to be large, the velocity of the fluid at its surface, V_1, may be taken as zero. The velocity in the pipe, which is also the velocity leaving the pipe, is found from the continuity equation as

$$Q = AV = \frac{10 \text{ liters}}{s} \times \frac{1000 \text{ cm}^3}{\text{liter}} \times 10^{-6} \frac{m^3}{cm^3} = 10 \times 10^{-3} \, m^3/s$$

Therefore,

$$\frac{\pi}{4} (0.05)^2 V_2 = 10 \times 10^{-3}$$

$$V_2 = 5.09 \text{ m/s}$$

Writing the energy equation, we obtain

$$\frac{p_1}{\gamma} + \frac{V_1^2}{2g} + Z_1 = \frac{p_2}{\gamma} + \frac{V_2^2}{2g} + Z_2 + h_L$$

Since $p_1 = p_2$ and $V_1 = 0$,

$$4 = \frac{V_2^2}{2g} + 0 + h_L$$

The loss in energy is therefore

$$h_L = 4 - \frac{(5.09)^2}{2 \times 9.81} = 2.68 \text{ m or } 2.68 \text{ N·m/N}$$

ILLUSTRATION 5.3   THE ENERGY EQUATION—PUMP

Calculate the work required for a pump to pump water from a well to ground level 125 m above the bottom of the well (see Figure 5.10). At the inlet to the pump the pressure is 96.5 kPa, and at the system outlet it is 103.4 kPa. Assume constant pipe diameter. Use $\gamma = 9810$ N/m^3, and assume it to be constant. Neglect any flow losses in the system.

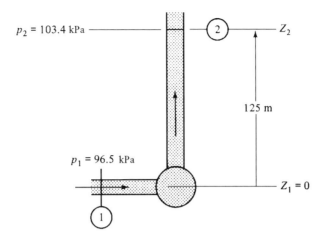

Figure 5.10   Illustrative Problem 5.3.

ILLUSTRATIVE PROBLEM 5.3

Given: $Z_1 = 0$, $Z_2 = 125$ m, $p_1 = 96.5$ kPa, $p_2 = 103.4$ kPa, $h_L = 0$, $d_1 = d_2$

Find: h_p

Assumptions: Steady, incompressible flow, no losses

Basic Equations: Continuity: $A_1V_1 = A_2V_2$

$$\text{Energy: } \frac{p_1}{\gamma} + \frac{V_1^2}{2g} + Z_1 + h_p = \frac{p_2}{\gamma} + \frac{V_2^2}{2g} + Z_2 + h_L$$

Solution: Write the energy equation,

$$\frac{p_1}{\gamma} + \frac{V_1^2}{2g} + Z_1 + h_p = \frac{p_2}{\gamma} + \frac{V_2^2}{2g} + Z_2 + h_L$$

Note that $V_1 = V_2$ and $h_L = 0$; thus,

$$h_p = \frac{p_2}{\gamma} - \frac{p_1}{\gamma} + Z_2 - Z_1$$

With $Z_1 = 0$,

$$h_p = \frac{103.4 \text{ kPa}}{9.81 \text{ kN/m}^3} - \frac{96.5 \text{ kPa}}{9.81 \text{ kN/m}^3} + 125 \text{ m}$$

and

$$h_p = 125.7 \text{ m} = 125.7 \text{ N·m/N}$$

ILLUSTRATION 5.4   THE ENERGY EQUATION—PUMP

Solve Illustrative Problem 5.3 if there is friction in the system whose total head loss equals 12.5 m

ILLUSTRATIVE PROBLEM 5.4

Given: $Z_1 = 0$, $Z_2 = 125$ m, $p_1 = 96.5$ kPa, $p_2 = 103.4$ kPa, $h_L = 12.5$ m, $d_1 = d_2$

Find: h_p

Assumptions: Steady, incompressible flow; losses

Basic Equations: Continuity: $A_1 V_1 = A_2 V_2$

$$\text{Energy:} \quad \frac{p_1}{\gamma} + \frac{V_1^2}{2g} + Z_1 + h_p = \frac{p_2}{\gamma} + \frac{V_2^2}{2g} + Z_2 + h_L$$

Solution: Write the energy equation,

$$\frac{p_1}{\gamma} + \frac{V_1^2}{2g} + Z_1 + h_p = \frac{p_2}{\gamma} + \frac{V_2^2}{2g} + Z_2 + h_L$$

As before, $V_1 = V_2$, but $h_L = 12.5$ m. Therefore,

$$h_p = \frac{p_2}{\gamma} - \frac{p_1}{\gamma} + Z_2 + h_L$$

and

$$h_p = 125.7 \text{ m} + 12.5 \text{ m} = 138.2 \text{ m}$$

or

$$h_p = 138.2 \text{ N·m/N}$$

Notice that the pump is required to overcome the additional friction head loss, and for the same flow this requires more pump work. The additional pump work is equal to the head loss.

ILLUSTRATION 5.5   THE ENERGY EQUATION—
TURBINE WITH LOSSES

A turbine in a hydroelectric plant has a flow of 800 cfs at a pressure at inlet of 60 psig. If the turbine discharges the water to the atmosphere, determine the work out of the turbine. Assume that the inlet and outlet pipes have the same diameter and that the losses in the turbine equal 5 ft.

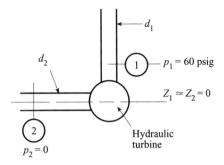

Figure 5.11   Illustrative Problem 5.5.

ILLUSTRATIVE PROBLEM 5.5

Given: $d_1 = d_2$, $p_1 = 60$ psig, $Z_1 \simeq Z_2$, $p_2 = 0$, $h_L = 5$ ft

Find: h_T

Assumptions: Steady, incompressible flow, losses

Basic Equations: Continuity: $A_1 V_1 = A_2 V_2$

$$\text{Energy:}\quad \frac{p_1}{\gamma} + \frac{V_1^2}{2g} + Z_1 = \frac{p_2}{\gamma} + \frac{V_2^2}{2g} + Z_2 + h_L + h_T$$

Solution: Writing the energy equation,

$$\frac{p_1}{\gamma} + \frac{V_1^2}{2g} + Z_1 = \frac{p_2}{\gamma} + \frac{V_2^2}{2g} + Z_2 + h_L + h_T$$

Note that $p_2 = 0$, $V_1 = V_2$, and $Z_1 = Z_2$, which gives us

$$h_T = \frac{p_1}{\gamma} - h_L$$

For this problem,

$$h_T = \frac{60 \text{ lb/in.}^2 \times 144 \text{ in.}^2/\text{ft}^2}{62.4 \text{ lb/ft}^3} - 5 \text{ ft}$$

and

$$h_T = 138.5 \frac{\text{ft·lb}}{\text{lb}} - \frac{5 \text{ ft·lb}}{\text{lb}}$$

$$h_T = 133.5 \frac{\text{ft·lb}}{\text{lb}}$$

Notice that the losses subtract from the ideally possible output of the turbine.

ILLUSTRATION 5.6   THE ENERGY EQUATION—BASIC PROGRAM

Write a BASIC program to solve the energy equation. Solve Illustrative Problem 5.4 as a numerical check on your solution.

ILLUSTRATIVE PROBLEM 5.6

Given: The energy equation

Find: BASIC program

Assumptions: All inherent to the energy equation

Basic Equations: Energy: $\dfrac{p_1}{\gamma} + \dfrac{V_1^2}{2g} + Z_1 + h_p = \dfrac{p_2}{\gamma} + \dfrac{V_2^2}{2g} + Z_2 + h_L + h_T$

Solution: In Chapter 4 we programmed the Bernoulli equation in BASIC by solving for each of the variables successively. We can use the same procedure for this problem to program the energy equation (5.6). However, we now need nine equations, since we have added three terms to the Bernoulli equation. The equations needed become

$$p_1 = \gamma\left(\frac{p_2}{\gamma} + \frac{V_2^2}{2g} - \frac{V_1^2}{2g} + Z_2 - Z_1 + h_T + h_L - h_p\right)$$

$$p_2 = \gamma\left(\frac{p_1}{\gamma} + \frac{V_1^2}{2g} - \frac{V_2^2}{2g} + Z_1 - Z_2 + h_p - h_T - h_L\right)$$

$$V_2 = \sqrt{2g\left(\frac{p_1}{\gamma} - \frac{p_2}{\gamma} + Z_1 - Z_2 + \frac{V_1^2}{2g} + h_p - h_T - h_L\right)}$$

$$V_1 = \sqrt{2g\left(\frac{p_2}{\gamma} - \frac{p_1}{\gamma} + Z_2 - Z_1 + \frac{V_2^2}{2g} + h_T + h_L - h_p\right)}$$

$$Z_1 = \frac{p_2}{\gamma} - \frac{p_1}{\gamma} + \frac{V_2^2}{2g} - \frac{V_1^2}{2g} + Z_2 + h_T + h_L - h_p$$

$$Z_2 = \frac{p_1}{\gamma} - \frac{p_2}{\gamma} + \frac{V_1^2}{2g} - \frac{V_2^2}{2g} + Z_1 + h_p - h_T - h_L$$

$$h_p = \frac{p_2}{\gamma} - \frac{p_1}{\gamma} + Z_2 - Z_1 + \frac{V_2^2}{2g} - \frac{V_1^2}{2g} + h_T + h_L$$

$$h_T = \frac{p_1}{\gamma} - \frac{p_2}{\gamma} + \frac{V_1^2}{2g} - \frac{V_2^2}{2g} + Z_1 - Z_2 + h_p - h_L$$

$$h_L = \frac{p_1}{\gamma} - \frac{p_2}{\gamma} + \frac{V_1^2}{2g} - \frac{V_2^2}{2g} + Z_1 - Z_2 + h_p - h_T$$

The program and numerical check are given below, using the same method of reasoning that was used in Chapter 4. Since the two velocities are equal, a value of 10 (arbitrary) was entered for each, and a value of 0.001 was entered for the pump head.

```
5 : X = .001
10 : Input "P one = ";A
20 : Input "V One = ";B
30 : Input "G = ";C
40 : Input "Z One = ";D
50 : Input "Gamma = ";E
60 : Input "P Two = ";F
70 : Input "V Two = ";H
80 : Input "Z Two = ";I
92 : Input "Pump Head = ";L
94 : Input "Turbine Head = ";J
96 : Input "Lost Head = ";K
98 : IF A<> X GOTO 130
100 : A = E*((F/E) + (H*H/(2*C)) − (B*B/(2*C)) + I − D + J + K − L)
110 : GOTO 330
120 : END
130 : IF F<>X GOTO 170
140 : F = E*((A/E) + (B*B/(2*C)) − (H*H/(2*C)) + D − I + L − J − K)
150 : GOTO 330
160 : End
170 : IF B<>X GOTO 210
180 : B = SQR(2*C*(F/E−A/E + I−D + J + K − L + (H*H/(2*C))))
190 : GOTO 330
200 : END
210 : IF H<>X GOTO 250
220 : H = SQR (2*C*(A/E − F/E + D−I + L − J−K + (B*B/(2*C))))
230 : GOTO 330
240 : END
250 : IF D<>X GOTO 290
260 : D = (F/E − A/E + (H*H/(2*C)) − (B*B/(2*C)) + I + J + K − L)
270 : GOTO 330
280 : END
290 : IF I<>X GOTO 316
300 : I = (A/E − F/E + (B*B/(2*C)) − (H*H/(2*C)) + D + L − J − K)
310 : GOTO 330
315 : END
316 : IF L<>X GOTO 320
317 : L = (F/E − A/E − (B*B/(2*C)) + (H*H/(2*C)) + I − D + J + K)
318 : GOTO 330
319 : END
320 : IF J<>X GOTO 324
321 : J = (A/E − F/E + L − K + D + (B*B/(2*C)) − (H*H/(2*C)))
322 : GOTO 330
323 : END
324 : IF K<>X GOTO 330
325 : K = (A/E − F/E + L − J + D + (B*B/(2*C)) − (H*H/(2*C)))
326 : GOTO 330
327 : END
330 : LPRINT "V ONE = ";B
335 : LPRINT "P ONE = ";A
340 : LPRINT "G = ";C
350 : LPRINT "Z ONE = ";D
360 : LPRINT "GAMMA = ";E
370 : LPRINT "P TWO = "; F
```

```
380 : LPRINT "V TWO = "; H
390 : LPRINT "Z TWO = ";I
392 : LPRINT "PUMP HEAD = "; L
394 : LPRINT "TURBINE HEAD = "; J
396 : LPRINT "LOST HEAD = "; K
400 : END

V ONE = 10
P ONE = 96.5
G = 9.81
Z ONE = 0
GAMMA = 9.81
P TWO = 103.4
V TWO = 10
Z TWO = 125
PUMP HEAD = 138.20336391437
TURBINE HEAD = 0
LOST HEAD = 12.5
```

ILLUSTRATION 5.7   THE ENERGY EQUATION—
SPREADSHEET SOLUTION

Solve Illustrative Problem 5.4 using the Excel spreadsheet.

ILLUSTRATIVE PROBLEM 5.7

Given: Illustrative Problem 5.4

Find: Solution using Excel spreadsheet

Assumptions: Same as Illustrative Problem 5.4

Basic Equations: Same as Illustrative Problem 5.4

Solution: From Illustrative Problem 5.6,

$$h_p = \frac{p_2}{\gamma} - \frac{p_1}{\gamma} + Z_2 - Z_1 + \frac{V_2^2}{2g} - \frac{V_1^2}{2g} + h_T + h_L$$

The attached program is almost self explanatory. For γ, G has been used. The answer in box M3 is seen to consist of the terms in the energy equation. With g and γ as inputs, this procedure can be used for English units, too. A similar spreadsheet can be set up for the other equations in Illustrative Problem 5.6.

	A	B	C	D	E	F	G	H	I	J	K	L	M
1						Illustrative Problem 5.7							
2	p2/G	p1/G	Z2	Z1	V2	V1	hT	hL	g	V2^2/2g	V1^2/2g	Gamma(G)	hp
3	10.54	9.84	125	0	0	0	0	12.5	9.81	0	0	9.81	138.2
	K3 = F3^2/2*I3												
	J3 = E3^2/2*I3												
	M3 = A3 − B3 + C3 − D3 + J3 − K3 + G3 + H3												

5.3   POWER

In mechanics, *work* is defined as the product of a force times the displacement (distance moved) in the direction of the line of action of force. The student will note that we have already used this definition in Chapter 4. At this point we will define the term *power* to be the rate at which work is being done. Since work is expressed as either ft·lb or N·m, power is expressed as ft·lb/s or N·m/s. The term *horsepower* (hp) is also a defined term that equals 550 ft·lb/s in the English system or 746 N·m/s or 746 W in the SI system.

To determine the horsepower required or delivered in a given situation, it will first be necessary to define the system. As our first example, let us consider a force F acting on a solid body moving with a velocity of V ft/s or V m/s, as shown in Figure 5.12. The rate of doing work by the applied force will be the product of the force and the displacement per unit time, which equals FV. The horsepower being supplied by the applied force is given by

$$\text{hp} = \frac{FV}{550} \qquad \text{(English units)} \tag{5.7}$$

or

$$\text{hp} = \frac{FV}{746} \qquad \text{(SI units)} \tag{5.8}$$

Equations (5.7) and (5.8) refer to a constant force acting on a body that is moving with a constant velocity. Note that this power is needed to replenish the power lost by friction between the wheels and the ground.

Let us now consider a pump that is required to pump a quantity of fluid against a given "head." This terminology is commonly used, but it really should be worded: "How much energy must be imparted to a fluid to pump it from one location to another?" If we review Chapter 4 at this point, we will see that the weight flow rate multiplied by the total head gives us power in ft·lb/s or N·m/s. The weight flow rate is given in units of ft·lb/s or N·m/s, and in either set of units equals the volume flow rate, Q, multiplied by the specific weight. *Therefore, power is the product of head multiplied by the volume flow rate and the specific weight of the fluid.* In terms of horsepower,

$$\text{hp} = \frac{(h_p \text{ or } h_T) \times Q \times \gamma}{550} \qquad \text{(English units)} \tag{5.9}$$

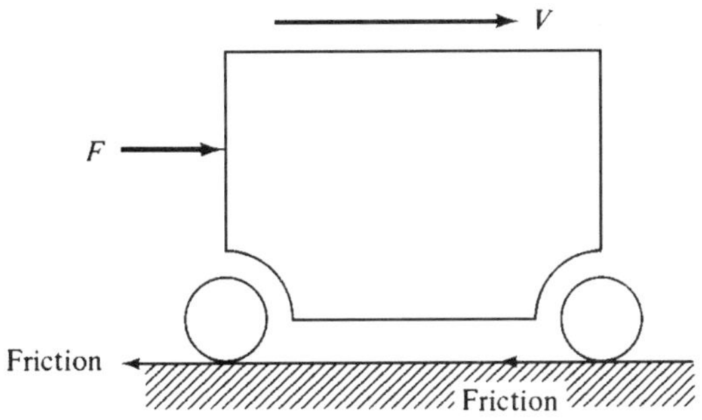

Figure 5.12   Work.

or

$$\text{hp} = \frac{(h_p \text{ or } h_r) \times Q \times \gamma}{746} \quad \text{(SI units)} \tag{5.10}$$

where h_p and h_r are defined by equations (5.3) through (5.6).

ILLUSTRATION 5.8 POWER

A pump is required to pump water from a large reservoir to a point located 20 m above the reservoir (Figure 5.13). If 0.05 m³/s of water having a density of 1000 kg/m³ is pumped through a 50-mm pipe, how much horsepower is required to be delivered to the water by the pump? Neglect all flow losses in the pipe.

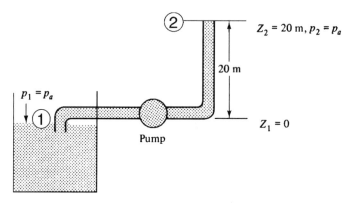

Figure 5.13 Illustrative Problem 5.8.

ILLUSTRATIVE PROBLEM 5.8

Given: Pump, $Z_2 = 20$ m, $Z_1 = 0$, $p_1 = 0$, $p_2 = 0$, $Q = 0.05$ m³/s, $\rho = 1000$ kg/m³, $d = 50$ mm, $h_L = 0$, $V_1 = 0$

Find: Required pump horsepower

Assumptions: Steady, incompressible flow; no losses

Basic Equations: Continuity: $Q = A_1V_1 = A_2V_2$

$$\text{Energy:} \frac{p_1}{\gamma} + \frac{V_1^2}{2g} + Z_1 + h_p = \frac{p_2}{\gamma} + \frac{V_2^2}{2g} + Z_2$$

$$\text{Power: hp} = \frac{h_p \times Q \times \gamma}{746}$$

Solution: Let us first write the energy equation between sections ① and ②,

$$\frac{p_1}{\gamma} + \frac{V_1^2}{2g} + Z_1 + h_p = \frac{p_2}{\gamma} + \frac{V_2^2}{2g} + Z_2$$

since there are no losses in the pipe. Therefore, h_p is

$$h_p = \frac{p_2 - p_1}{\gamma} + \frac{V_2^2 - V_1^2}{2g} + (Z_2 - Z_1)$$

but $p_2 = p_1 = p_a$, $V_1 = 0$, and $Z_1 = 0$. h_p is thus given by

$$h_p = \frac{V_2^2}{2g} + Z_2$$

Since $Q = AV$, we have

$$V = \frac{0.05}{\pi[(0.05)^2/4]} = 25.5 \text{ m/s}$$

Using this value of V gives us

$$h_p = \frac{(25.5)^2}{2 \times 9.81} + 20 = 53.1 \text{ m}$$

We can now obtain the horsepower required as follows:

$$hp = \frac{h_p \times Q \times \gamma}{746}$$

$$= \frac{53.1 \times 0.05 \times 1000 \times 9.81}{746} = 34.9 \text{ hp}$$

ILLUSTRATION 5.9   TURBINE

A hydraulic turbine is connected as shown in Figure 5.14. How much horsepower will it develop? Use 1000 kg/m³ for the density of water. Neglect the flow losses in the system.

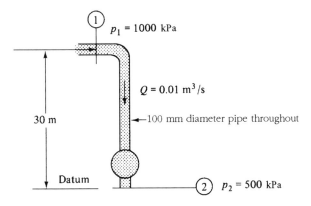

Figure 5.14   Illustrative Problem 5.9.

ILLUSTRATIVE PROBLEM 5.9

Given: Turbine, $p_1 = 1000$ kPa, $Q = 0.01$ m³/s, $d_1 = d_2 = 100$ mm, $Z_1 = 30$ m, $Z_2 = 0$, $p_2 = 500$ kPa, $h_L = 0$

Find: Turbine horsepower

Assumptions: Steady, incompressible flow, no losses

Basic Equations: Continuity: $Q = A_1 V_1 = A_2 V_2$

$$\text{Energy: } \frac{p_1}{\gamma} + \frac{V_1^2}{2g} + Z_1 = \frac{p_2}{\gamma} + \frac{V_2^2}{2g} + Z_2 + h_T$$

$$\text{Power: } hp = \frac{h_T Q \gamma}{746}$$

Solution: Again let us write the energy equation, but this time for a turbine.

$$\frac{p_1}{\gamma} + \frac{V_1^2}{2g} + Z_1 = \frac{p_2}{\gamma} + \frac{V_2^2}{2g} + Z_2 + h_T$$

Since the pipe diameter is constant, $V_1 = V_2$, $Z_2 = 0$, and $Z_1 = 30$ m. Therefore, h_T is found as

$$h_T = \frac{p_1 - p_2}{\gamma} + (Z_1 - Z_2)$$

For the data given,

$$h_T = \frac{(1000 - 500)1000}{1000 \times 9.81} + (30 - 0) = 80.97 \text{ m}$$

and the horsepower is given by

$$hp = \frac{h_T Q \gamma}{746}$$

$$= \frac{80.97 \times 0.01 \times 1000 \times 9.81}{746} = 10.65 \text{ hp}$$

5.4 REVIEW

Chapter 5 is a continuation of the material that was studied in Chapter 4. We have at this time taken the Bernoulli equation and modified it to account for pumps, turbines, and losses in a system. This modification produced an equation that we now call the energy equation. The energy equation is applicable to many real systems in which incompressible fluids flow. It is not applicable to systems in which there is heat transfer, phase changes, high-speed flow, or systems in which compressible fluids flow.

We applied the energy equation to those systems in which a fluid was flowing and which contained pumps and turbines, and had losses. A BASIC program was written to solve the energy equation, and a spreadsheet was used to solve the same problem. Since power is the rate of doing work, we were also interested in the power produced, expended, or needed in systems. The term "horsepower" is used to define a given rate of doing work. Using the defined quantities of 550 ft·lb/s or 746 N·m/s, we finally arrived at analytical expressions for the horsepower used or produced in specific situations.

KEY TERMS

Terms of importance in this chapter are:

Energy equation: the modified Bernoulli equation that takes into account pumps, turbines, and losses in systems in which fluids are flowing.

Horsepower: a defined rate of doing work equal to 550 ft·lb/s or 746 N·m/s.

Streamline: a line with the tangent to it at any point being the direction of the velocity at that point.

Total energy: *see* total head.

Total head: the sum of the velocity head, pressure head, and potential head in a system.

KEY EQUATIONS

Complete energy equation

$$\frac{p_1}{\gamma} + \frac{V_1^2}{2g} + Z_1 + h_p = \frac{p_2}{\gamma} + \frac{V_2^2}{2g} + Z_2 + h_L + h_T \quad (5.6)$$

Horsepower—constant force

$$hp = \frac{FV}{550} \text{ (English units)} \quad (5.7)$$

Horsepower—constant force

$$hp = \frac{FV}{746} \text{ (SI units)} \quad (5.8)$$

Horsepower—fluid flow

$$hp = \frac{(h_p \text{ or } h_T) \times Q \times \gamma}{550} \quad (5.9)$$

Horsepower—fluid flow

$$hp = \frac{(h_p \text{ or } h_T) \times Q \times \gamma}{746} \quad (5.10)$$

QUESTIONS

1. Explain the need to modify the Bernoulli equation.
2. Does the total head in a system always remain constant?

3. Why do we call equations (5.3) through (5.6) energy equations when each term has the dimension of length?

4. To what type of system does the energy equation apply?

5. Give some examples of systems where the energy equation that we derived in this chapter does not apply.

6. How much power does a weight lifter exert while holding a 200-lb weight over her head?

7. Does a student running up a flight of steps from one floor to the next use more power than another student running from one building to another to avoid being late for class? Explain how you arrived at your answer.

8. If two cars have the same weight, the same tires, and the same horsepower, will they have the same acceleration from a standing start?

9. What is meant by the statement that a person has power?

10. Explain in your own words the concept of a pumped hydroelectric plant.

11. Are the production of power and flood control in a dam mutually compatible? Why?

PROBLEMS

Use $\gamma = 62.4$ lb/ft³ (9810 N/m³) for water unless indicated otherwise. If not stated, all diameters are inside diameters.

Head

5.1. A 4-in. pipe carries water at the rate of 300 gal/min. If the pressure in the pipe is 25 psig, determine the velocity in the pipe, the velocity head, the pressure head, and the total head referenced to a datum 10 ft below the center of the pipe.

5.2. A 75-mm pipe carries water at the rate of 1400 liters/min. The pressure in the pipe is 210 kPa above atmospheric pressure. Determine the velocity, velocity head, pressure head, and total head if a datum is selected 3 m below the center of the pipe. ($p_{atmos} = 100$ kPa)

5.3. Oil (specific weight 50 lb/ft³) flows in a 4-in. pipe. If the pressure in the pipe is 30 psia and the total head with respect to a reference plane 5 ft below the center of the pipe is 200 ft of oil, determine the velocity in the pipe and the volume rate of flow in gallons per minute.

5.4. A stream flows over a waterfall 150 ft high. What is the velocity with which it strikes the base of the fall if air resistance is negligible?

Energy Equation

5.5. Determine the pressure at section ② of the pipe shown in Figure P5.5 if water flows at the rate of 300 gal/min past section ①. Losses in this system equal 10 ft.

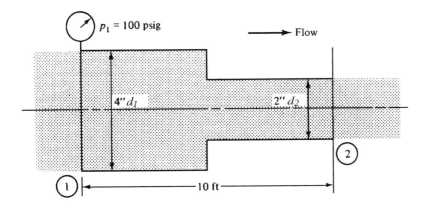

Figure P5.5

5.6. If the pipe in Problem 5.5 is turned vertically with the large end (section ①) at the bottom, what is the pressure at section ②? Include losses. Flow is upward.

5.7. A tank has a water level of 20 ft above a datum plane. Five feet above the datum plane there exists a 6-in. opening in the tank. If there are flow losses, determine the velocity of the water leaving the opening and the quantity of water flowing at this instant in gallons per minute. The flow losses in the opening equal 3 ft.

5.8. Assuming flow losses equal to 0.75 ft, determine the velocity of the water leaving the system shown in Figure P5.8.

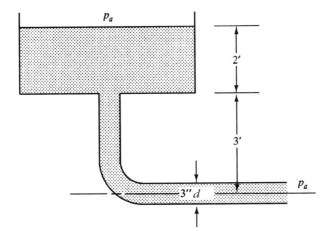

Figure P5.8

5.9. A test is conducted to determine the pressure loss in the horizontal section of pipe shown in Figure P5.9 on page 197. If the pipe carries water, determine the head loss.

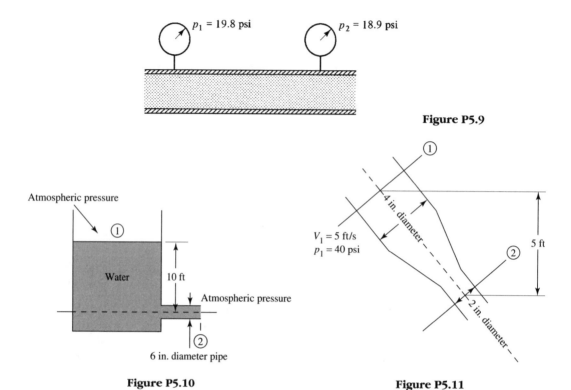

Figure P5.9

Figure P5.10

Figure P5.11

5.10. Determine the velocity and volume flow rate of water being discharged from the pipe in Figure P5.10 if the total losses between the water surface and the water jet are 4.0 ft.

5.11. Water flows in a venturi meter as shown in Figure P5.11. If the losses from section ① to ② equal 9 ft, determine the pressure and velocity at section ②.

*5.12. A pipe carries oil having a specific gravity of 0.9. At section *A* it has an 8 in. diameter and at section *B* it has a 12 in. diameter. Section *A* is 10 ft lower than section *B*. The pressure at section *A* is 12.0 psi and at section *B* the pressure is 9.0 psi. If the pipe carries 4.2 cfs of oil, determine the head loss and the direction of flow.

5.13. Water flows from a tank and discharges into air as shown in Figure P5.13. Calculate the air pressure needed for 0.5 ft³/s of water to flow if the head loss from ① to ② equals 4.0 ft of water.

5.14. Water flows up the pipe shown in Figure P5.14. At point *A* the velocity is 30 ft/s and the pressure is 40 psi. Calculate the pressure at *B* if the head loss between *A* and *B* is 20 ft.

5.15. Calculate the pressure at *B* if the flow direction is reversed in Problem 5.14.

5.16. Water flows from a reservoir and discharges to the atmosphere from a pipe whose inside diameter is 4 in., as shown in Figure P5.16. Calculate the energy losses in the pipe if 2.8 ft³/s discharges from the pipe.

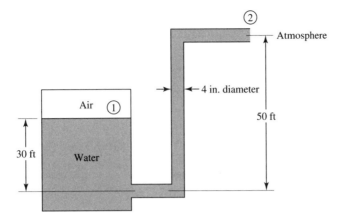

Figure P5.13

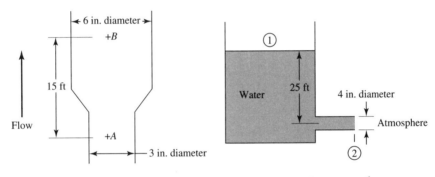

Figure P5.14   **Figure P5.16**

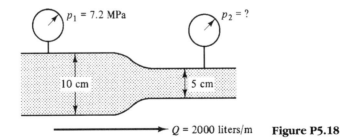

$Q = 2000$ liters/m   **Figure P5.18**

5.17. Determine the volumetric flow out of the 4-in. diameter pipe in Problem 5.16 assuming that the losses equal a head of 6.6 ft.

5.18. Water flows at the rate of 2000 liters/min. If losses in the system shown in Figure P5.18 on page 198 equal 4.5 m, determine p_2.

5.19. A valve is tested in the setup shown in Figure P5.19. If 20 cfs of water is flowing, determine the head loss in the valve.

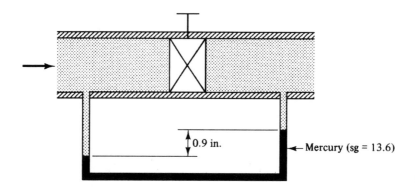

0.9 in.   ← Mercury (sg = 13.6)

Figure P5.19

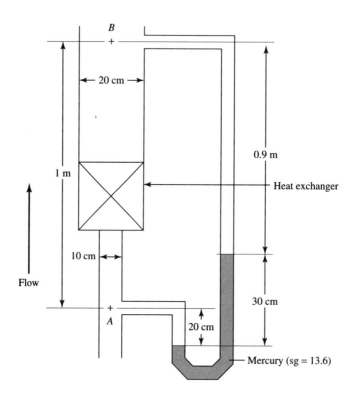

Figure P5.21

5.20. Water is piped from the ground floor of a building up 4 m to the second floor. At the ground floor the pipe is 10 cm in diameter and the pressure is 198 kPa. At the second floor the pipe diameter is 8 cm. Flow losses in the pipe are estimated to be 0.5 m. Assuming that the velocity in the pipe on the ground floor is 12 m/s, determine the pressure at the second floor.

5.21. Water flows through a heat exchanger as shown in Figure P5.21. Calculate the head loss between points A and B if the flow rate through the exchanger is 0.012 m³/s.

*5.22. Calculate the velocity of the jet of water in Figure P5.22 if the pump delivers 8 hp to the water. Assume losses equal to a head of 2 feet.

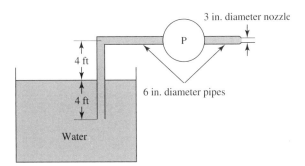

Figure P5.22

Power

5.23. The total head of a stream of water whose flow rate is 200 ft^3/s is 20 ft. Determine the theoretical horsepower available from the stream.

5.24. The total head of a stream of water whose flow rate is 2000 liters/s is 10 m. Calculate the theoretical horsepower available from the stream.

5.25. A pumped storage hydroelectric plant (see Figure 5.7) is proposed using a reversible pump-turbine. In the pumping mode, each pump is to handle 50,000 cfm of water and pump this volume to a reservoir that is 350 ft above the pump. If the total number of installed pumps is 10, determine the total theoretical horsepower of the plant.

5.26. A turbine delivers 500 hp when 35.0 cfs of water flows through it. What is the theoretical head acting on the turbine?

5.27. A turbine at a hydroelectric plant accepts 30 m^3/s of water at a gage pressure of 400 kPa and discharges the water to the atmosphere. Determine the maximum power output of the turbine.

5.28. A pump increases the pressure of water from 400 to 4200 kPa. What horsepower does the pump impart to the water for a flow rate of 0.15 m^3/s? Assume that the inlet pipe to the pump has a diameter of 50 mm and the outlet pipe has a diameter of 100 mm.

5.29. What horsepower must the pump shown in Figure P5.29 on page 200 deliver to a fluid having $\gamma = 60$ lb/ft^3? The pump delivers 40 gal/min. Assume no flow losses.

5.30. If there are flow losses in Problem 5.29 that are equivalent to 3 ft (ft·lb/lb), what power must the pump deliver?

5.31. If the system shown in Problem 5.29 is a turbine with flow downward, what power will the fluid deliver to the turbine?

5.32. Solve Problem 5.30 if the pump is replaced by a turbine and the flow losses remain at 3 ft.

5.33. Water flows through a turbine as shown in Figure P5.33 on page 200. If the flow rate is 200 liters/s and the pressures at ① and ② are 160 kPa and ⁻15 kPa respectively, determine the horsepower delivered to the turbine by the water.

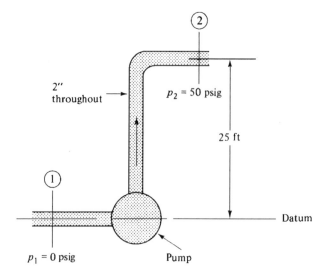

$p_2 = 50$ psig

25 ft

2" throughout

$p_1 = 0$ psig   Pump   Datum

Figure P5.29

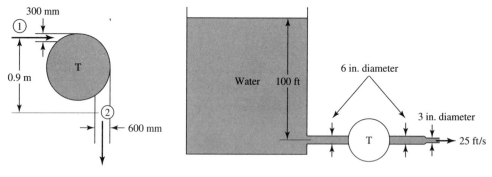

300 mm

0.9 m

T

600 mm

Figure P5.33

6 in. diameter

Water   100 ft

3 in. diameter

T   25 ft/s

Figure P5.34

*5.34. Calculate the power developed by the turbine shown in Figure P5.34. Assume that the head losses up to the turbine inlet equal 7 ft and that the losses from the turbine outlet to the atmosphere equal 2 ft.

5.35. Water flows in a channel with a velocity of 1.5 m/s. The channel has a cross section that is 1.0 × 2.0 m and leads to a dam. The dam is capable of developing a head of 2.5 m above the outlet of a turbine. What is the maximum power output of the turbine?

5.36. The pump shown in Figure P5.36 delivers water from the lower tank to the upper tank. If 0.2 m³/s is flowing, determine the power delivered to the fluid by the pump. Assume no flow losses.

5.37. If there is a loss in energy between the inlet and outlet of the system shown in Problem 5.36 of 0.75 m, what power must the pump deliver?

*5.38. For the system shown in Figure P5.38 in which water is flowing, determine the turbine output. Losses in the system equal 8.5 ft.

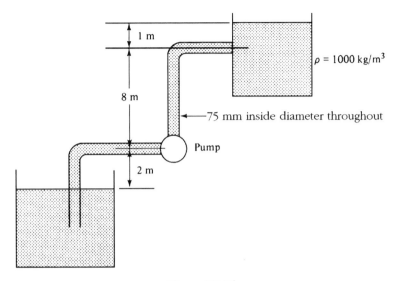

Figure P5.36

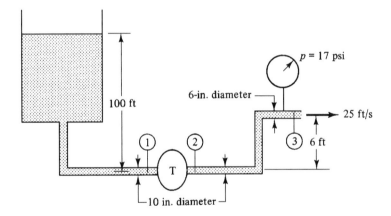

Figure P5.38

*5.39. If the outlet nozzle of the pump shown in Figure P5.39 on page 202 is to produce a jet of water that reaches an elevation of 15 m, determine the pump power required. Flow losses can be taken to equal 0.5 m.

*5.40. The pump shown in Figure P5.40 on page 202 has a 5-in. diameter inlet and a 3-in. diameter discharge. The pressure gages are 2 ft apart and read 5 psi and 30 psi, respectively. If the velocity of the water is 12 ft/s at the outlet of the system, determine the horsepower required by the pump.

*5.41. As shown in Figure P5.41 on page 202, a deep-well pump is used to pump water up to an elevation that is 100 ft above its inlet. The pump must deliver 180 gpm through a 4-in. diameter pipe. At the discharge the pressure is 12 psig. Determine the pump power if the head loss is 3.45 ft.

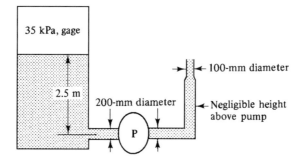

Figure P5.39

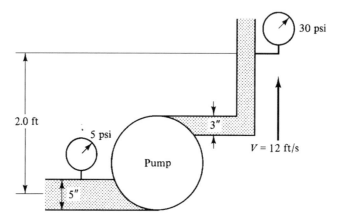

Figure P5.40

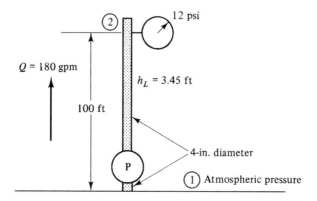

Figure P5.41

*5.42. In the system shown in Figure P5.42 the pump delivers 60 liters/s of water. The loss in the inlet pipe to the pump is 0.5 m and the loss in the discharge pipe is 4 m. Determine the power required by the pump.

*5.43 Oil, having a specific gravity of 0.85, is pumped from a storage tank to the top of a nearby hill, which is 265 ft above the tank, as shown in Figure P5.43. The pump is required to maintain a flow rate of 25 cfs with a head loss in the pipe of 18 ft. Determine the horsepower required by the pump.

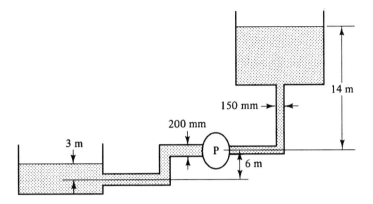

150 mm

200 mm

14 m

3 m

P

6 m

Figure P5.42

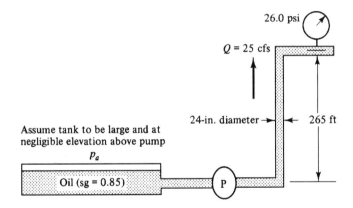

26.0 psi

$Q = 25$ cfs

Assume tank to be large and at
negligible elevation above pump

p_a

Oil (sg = 0.85)

P

24-in. diameter

265 ft

Figure P5.43

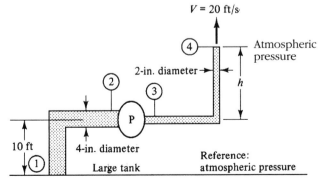

$V = 20$ ft/s

Atmospheric
pressure

2-in. diameter

h

2

3

P

10 ft

4-in. diameter

Large tank

Reference:
atmospheric pressure

Figure P5.44

*5.44. In the system in Figure P5.44, the pump imparts 20 hp to the water. If the loss at
the inlet from ① to ② is 1.5 ft and at the outlet from ③ to ④ is 12 ft, determine
the pressure at ② and the height h to which the pump can pump the water.

6 STEADY FLOW OF INCOMPRESSIBLE FLUIDS IN PIPES

LEARNING GOALS

1. Characterize laminar, transition, and turbulent flow.
2. Define and calculate the Reynolds number.
3. Determine the flow regime based on the Reynolds number.
4. Calculate the friction factor in laminar flow from the Reynolds number.
5. Calculate the head loss in laminar flow using the Darcy-Weisbach equation.
6. Use the Moody chart to determine the friction factor in turbulent flow from the Reynolds number and the relative roughness.
7. Calculate the head losses in pipe systems due to valves, fittings, and so on.
8. Apply the Darcy-Weisbach formula to noncircular pipes using the concept of equivalent diameter.
9. Show how to calculate the pressure drop and flow in pipes that are connected in parallel.

6.1   INTRODUCTION

Fluids are most commonly transported from one location to another by forcing them through pipes and tubes. The Bernoulli equation that was developed in Chapter 4 and applied extensively in Chapter 5 is applicable to the flow of a frictionless fluid, and with modification can be applied (judiciously) to the flow of real fluids by inserting a term in the equation to account for these real effects (losses). In this chapter we consider the problem of the steady flow of incompressible fluids in pipes, define the flow regimes encountered and the general form of the friction equation, and give expressions for entrance, exit, enlargement, contraction, bend, and valve losses. The literature on this subject is quite extensive, and the data relating to these effects have been presented in many different (and many times inconsistent) forms. One of the most comprehensive, consistent correlations of available data has been carried out by the Hydraulic Institute and has been published as the Pipe Friction section of the *Standards of the Hydraulic Institute*. Several of the tables and charts in this chapter are taken from this standard with permission of the Hydraulic Institute, and for more complete and extensive literature references and details, the reader is specifically referred to this publication in addition to the other general references given at the end of the book.

6.2   CHARACTER OF FLOW IN PIPES—LAMINAR AND TURBULENT

To visualize the character of a fluid flowing in a pipe, Osborne Reynolds devised a simple experiment, shown schematically in Figure 6.1. Basically, the dye is introduced into the glass tube by injecting it from a fine piece of tubing into the well-rounded entrance of the glass tube. The velocity of the test fluid is controlled by varying the height of the fluid in the glass tank and by changing the setting of the valve in the downstream portion of the glass tube.

    At small average velocities it is found that the dye filament appears as a straight unbroken line parallel to the axis of the tube. This type of flow is known as *laminar, viscous,* or *streamline flow* and is composed of concentric cylindrical layers flowing past each other in a manner determined by the viscosity of the fluid. The fluid particles are retained in layers, and their motion is along parallel paths. As the flow rate is increased by changing the valve setting, it is found that the dye line will remain straight until a velocity is reached that causes it to waver and break into diffused patterns. The velocity at which the dye streak wavers and breaks is known as the *critical velocity*. At velocities greater than the critical velocity the colored dye filament becomes completely diffused in the main body of the fluid a short distance downstream from its point of injection. At velocities greater than the critical velocity the flow is termed *turbulent,* and the particles have a random motion that is transverse to the main flow direction and that causes the particles to intermingle in a random manner. In

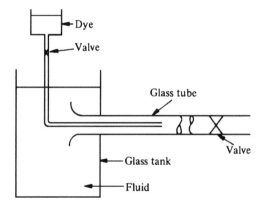

Figure 6.1   Schematic of Reynolds' apparatus.

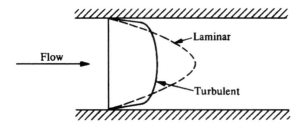

Figure 6.2   Velocity profiles in a pipe.

laminar flow the velocity of the fluid is maximum at the pipe axis and decreases to zero at the wall of the pipe, while in turbulent flow the velocity distribution is more uniform across the pipe diameter, as shown in Figure 6.2. By taking the average velocity as the characteristic velocity and the pipe diameter as the characteristic length, Reynolds showed that the character of the flow of a fluid in a pipe was dependent on the pipe diameter, the velocity of the flow, the density of the fluid, and its viscosity. The combination of these four variables yields a dimensionless parameter known as the Reynolds number, $DV\rho/\mu$.

During the course of his experiments, Reynolds was able to obtain the change from laminar to turbulent flow at Reynolds numbers as low as 1200 and higher than 40,000. The conditions under which these high Reynolds numbers were obtained, however, are not ordinarily found in commercial installations. These numbers at which the transition in character from laminar to turbulent flow occur are known as the Reynolds upper critical numbers and are not usually significant in the analysis of normal pipe flow. If, however, the flow is initially turbulent and the velocity of the fluid is decreased, it will be found that all initial disturbances will disappear and that the flow will become laminar. The value of this Reynolds number, known as the Reynolds lower critical number, is generally agreed to be approximately 2000. The usual piping installation will be found to undergo a transition from laminar to turbulent flow at Reynolds numbers from 2000 to 4000, with the flow always laminar below a Reynolds number of 2000 and always turbulent above a Reynolds number of 4000. Between these two values there is a region known as the *transition region* in which the flow may be either laminar or turbulent. In the transition region a disturbance will cause the character of the flow to change from laminar to turbulent.

6.3   EVALUATION OF REYNOLDS NUMBER

We have noted earlier that the Reynolds number is a dimensionless parameter involving diameter, velocity, density, and viscosity. Mathematically, the relationship is expressed as

$$\text{Re} = \frac{DV\rho}{\mu} \tag{6.1}$$

Analysis of the Reynolds number shows that it is really the ratio of the inertia forces to the viscous forces involved in the flow. Depending on the value of the Reynolds number, either the viscous forces will predominate, yielding laminar flow, or the inertia forces will predominate, yielding turbulent flow.

Since the kinematic viscosity was defined in Chapter 2 to be the ratio of the absolute viscosity to the density, we can also write equation (6.1) as

$$\mathrm{Re} = \frac{DV}{\nu}, \qquad \text{since } \nu = \frac{\mu}{\rho} \qquad (6.2)$$

The evaluation of the Reynolds number from either equation (6.1) or (6.2) requires that care be exercised, since the technical literature abounds with units of μ and ν that are not necessarily consistent. The following illustrative problems will help to clarify this point.

ILLUSTRATION 6.1   REYNOLDS NUMBER

Water flows in a pipe of inside diameter 50 mm with a velocity of 3 m/s. If the water is at 40°C, what is its Reynolds number? Use the data of Table 2.5 on page 45.

ILLUSTRATIVE PROBLEM 6.1

Given: Water at 40°C, $V = 3$ m/s, $D = 50$ mm

Find: Re

Assumptions: Data for water given in Table 2.5

Basic Equations:

$$\mathrm{Re} = \frac{DV\rho}{\mu} = \frac{DV}{\nu}$$

Solution: From Table 2.5 at 40°C, $\rho = 992.2$ kg/m³ and $\mu = 0.656 \times 10^{-3}$ N·s/m². Using equation (6.1) gives us

$$\mathrm{Re} = \frac{DV\rho}{\mu}$$

$$= \frac{0.050 \text{ m} \times 3 \text{ m/s} \times 992.2 \text{ kg/m}^3}{0.656 \times 10^{-3} \text{ N·s/m}^2} = 226\,875 \; \frac{\text{kg·m}}{\text{N·s}^2}$$

Since a Newton has the units kg · m/s²,

$$\mathrm{Re} = 226\,875 \; \frac{\text{kg·m}}{\text{N·s}^2} = 226\,875 \; \frac{\text{kg·m}}{\text{s}^2} \times \frac{1}{\text{kg·m/s}^2} = 226\,875$$

Note that the Reynolds number is dimensionless, *as it must be.*

We can also solve this problem using the kinematic viscosity for water at 40°C from Table 2.5. $\nu = 0.661 \times 10^{-6}$ m²/s. Thus,

$$\mathrm{Re} = \frac{DV}{\nu}$$

and

$$\mathrm{Re} = \frac{0.050 \text{ m} \times 3 \text{ m/s}}{0.661 \times 10^{-6} \text{ m}^2/\text{s}} = 226\,929$$

For all purposes, both methods yield the same result. The use of the kinematic viscosity makes the calculation easier and the units can readily be seen to cancel to yield a dimensionless Reynolds number.

ILLUSTRATION 6.2   REYNOLDS NUMBER

Water flows in a pipe of inside diameter 2 in. at a velocity of 30 ft/s. If the water is at 100°F, find its Reynolds number. Use the data of Table 2.4.

ILLUSTRATIVE PROBLEM 6.2

Given: Water at 100°F, $V = 30$ ft/s, $D = 2$ in.

Find: Re

Assumptions: Data for water given in Table 2.4

Basic Equations: $\mathrm{Re} = \dfrac{DV\rho}{\mu} = \dfrac{DV}{\nu}$

Solution: From Table 2.4 at 100°F, $\rho = 1.927$ slugs/ft³, $\mu = 1.424 \times 10^{-5}$ lb·s/ft². Using equation (6.1) with all dimensions in feet yields

$$\mathrm{Re} = \frac{DV\rho}{\mu} = \frac{2 \text{ in.}/(12 \text{ in./ft}) \times 30 \text{ ft/s} \times 1.927 \text{ slugs/ft}^3}{1.424 \times 10^{-5} \text{ lb·s/ft}}$$

$$= 676{,}615 \, \frac{\text{slug·ft}^2/\text{s·ft}^3}{\text{lb·s/ft}^2}$$

At first glance our answer would seem to be far from dimensionless. If, however, we note that the slug has the units of lb·sec²/ft, we obtain

$$\mathrm{Re} = 676{,}615 \, \frac{\text{lb·s}^2/\text{ft} \times \text{ft}^2/\text{s·ft}^3}{\text{lb·s/ft}^2} = 676{,}615$$

which is indeed dimensionless.

We can also use the kinematic viscosity of water at 100°F as 0.739×10^{-5} ft²/s. Since Reynolds number, Re, also equals $\dfrac{DV}{\nu}$,

$$\mathrm{Re} = \frac{(2 \text{ in.}/(12 \text{ in./ft})) \times 30 \text{ ft/s}}{0.739 \times 10^{-5} \text{ ft}^2/\text{s}} = 676{,}590$$

It is easy to see that the Reynolds number is dimensionless using the kinematic viscosity.

ILLUSTRATION 6.3   REYNOLDS NUMBER

An oil has a specific weight of 50 lb/ft³. If its viscosity is 5 cP, determine its Reynolds number if it is flowing in a 1-in. pipe of inside diameter 1 in. with an average velocity of 1 ft/s.

ILLUSTRATIVE PROBLEM 6.3

Given: $\gamma = 50$ lb/ft³, $\mu = 5$ cP, $D = 1$ in., $V = 1$ ft/s

Find: Re

Assumptions: Steady flow

Basic Equations:

$$\text{Re} = \frac{DV\rho}{\mu} \; ; \rho = \frac{\gamma}{g}$$

Solution:

Using the conversion from Table 2.6a, 1 cP $= 2.09 \times 10^{-5}$ lb·s/ft², we have a viscosity of $5 \times 2.09 \times 10^{-5}$ lb·s/ft², which equals 1.045×10^{-4} lb·s/ft². To convert the specific weight to density requires that we divide the specific weight by g, where g also has the units of lb/slug.

$$\rho = \frac{50 \text{ lb/ft}^3}{32.17 \text{ lb/slug}}$$

$$\rho = 1.55 \text{ slugs/ft}^3$$

Using the data given, we obtain

$$\text{Re} = \frac{DV\rho}{\mu} = \frac{1 \text{ in.}/(12 \text{ in./ft}) \times 1 \text{ ft/s} \times 1.55 \text{ slugs/ft}^3}{1.045 \times 10^{-4} \text{ lb·s/ft}^2}$$

$$\text{Re} = 1236$$

Using slugs $=$ lb·s²/ft, we find Re to be dimensionless.

ILLUSTRATION 6.4   LAMINAR AND TURBULENT FLOW

Discuss the character of the flow in Illustrative Problems 6.1 through 6.3.

ILLUSTRATIVE PROBLEM 6.4

Given: Re from Illustrative Problems 6.1 through 6.3

Find: Character of each flow, that is, turbulent or laminar

Assumptions: Character of flow only a function of Reynolds number

Basic Equations: Re ≤ 2000 for laminar flow; Re ≥ 4000 for turbulent flow

> *Solution:* Illustrative Problem 6.1: Re = 226 875. Since this is greater than 4000, the flow is turbulent.
>
> Illustrative Problem 6.2: Re = 676,615. Since this is greater than 4000, the flow is turbulent.
>
> Illustrative Problem 6.3: Re = 1236. Since this is less than 2000, the flow is laminar.

6.4   LAMINAR FLOW IN PIPES

In circular pipes where the Reynolds number is less than 2000, the flow is said to be laminar and we can characterize the flow pattern as consisting of a series of thin shells that are sliding over one another. At the center the velocity of the fluid is the greatest and at the wall the velocity is zero. This type of flow is depicted in Figure 6.3 and can readily be analyzed mathematically. When this is done, we obtain an equation relating the pressure drop (head loss) to the other variables involved. Equation (6.3) gives us this expression, which is known as the *Hagen-Poiseuille equation:*

$$\Delta p = \frac{128\mu LQ}{\pi D^4} \tag{6.3}$$

where Δp = pressure difference (pressure drop)
   μ = viscosity
   D = inside diameter of the pipe
   L = length of the pipe
   Q = volume rate of flow (since there is a wide variation in velocity, Q is taken to be equal to $A\bar{V}$, where $\bar{V}$ is the average velocity; for laminar flow the velocity varies parabolically from zero at the wall to a maximum value at the center of the pipe; the average velocity for this case equals one-half of the maximum velocity)

Equation (6.3) has been verified against experiments and found to give excellent agreement with the experimental results. It is important to note that the pressure drop in laminar flow is independent of the character (roughness) of the pipe wall. Also, because of the excellent agreement between measured and calculated results in laminar flow, the Hagen-Poiseuille relation has been used as the basis for measuring the viscosity of fluids in commercial viscosimeters.

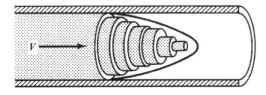

Figure 6.3   Laminar flow in a pipe.

It is sometimes more convenient to express equation (6.3) in terms of the average velocity, $\bar{V}$. Since $Q = A\bar{V} = (\pi D^2/4)\bar{V}$, we have

$$\Delta p = \frac{128\mu L A\bar{V}}{\pi D^4} = \frac{128\mu L\,(\pi D^2/4)\bar{V}}{\pi D^4} = \frac{32\mu L\bar{V}}{D^2} \qquad (6.4)$$

In terms of a "head loss," and using $p_1 - p_2 = \Delta p$, we obtain

$$\frac{p_1 - p_2}{\gamma} = h_f = \frac{32\mu L\bar{V}}{\gamma D^2} \qquad (6.5)$$

where h_f is the friction "head loss."

For laminar flow, it can also be shown that the velocity profile is a parabola given by

$$V = 2\bar{V}(1 - (r/r_0)^2) \qquad (6.6)$$

where V is the velocity at radius r, $\bar{V}$ is the average velocity determined from the continuity equation, r is any radius measured from the centerline of the pipe, and r_0 is the pipe radius. From Equation (6.6) it can be seen that the velocity at the centerline of the pipe ($r = 0$) is twice the average velocity. From Figure 6.4 and Equation (6.6) we can also see that the average velocity occurs when $r = \dfrac{\sqrt{2}}{2} r_0$.

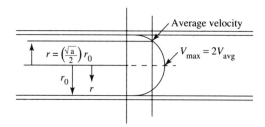

Figure 6.4   Velocity profile in laminar flow.

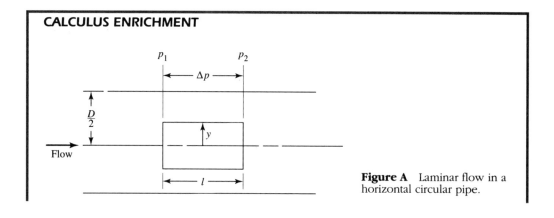

Figure A   Laminar flow in a horizontal circular pipe.

Consider the cylindrical mass having a radius y and a length l. The pressure p_1 acts on one end of the cylinder and p_2 acts on the other. The net force on the ends due to these pressures equals $-\Delta p (\pi y^2)$. In steady flow this force is opposed by the friction on the external cylindrical surface which is $2\pi y l \tau$, where τ is the shear stress due to the viscous action of the fluid. Since the flow is steady, we equate these two forces to obtain

$$\tau = \frac{-\Delta p}{l}\left(\frac{y}{2}\right) \tag{a}$$

From Chapter 2, for laminar flow, the shear stress can be written as

$$\tau = \mu\left(\frac{dV}{dy}\right) \tag{b}$$

From Figure A, it is apparent that the velocity at the center of the pipe is maximum and that it is zero at the walls. Therefore, combining Equations (a) and (b) yields

$$\frac{dV}{dy} = \frac{-\Delta p}{2\mu l}\,y \tag{c}$$

As an integral,

$$\int_V^0 dV = \frac{-\Delta p}{2\mu l}\int_y^{D/2} ydy \tag{d}$$

Therefore,

$$V = \frac{\Delta p}{4\mu l}\left(\frac{D^2}{4} - y^2\right) \tag{e}$$

Equation (e) is the equation of a parabola with the maximum velocity at the center, where $y = 0$. Using this,

$$V_{max} = \frac{\Delta p D^2}{16\mu l} \tag{f}$$

The volumetric flow, Q, is obtained from $Q = AV$. Therefore,

$$dQ = \frac{\Delta p}{4\mu l}\left(\frac{D^2}{4} - y^2\right)2\pi ydy \tag{g}$$

The area in Equation (g) is an annulus located at a distance y from the center and dy thick.

$$Q = \frac{\Delta p \pi}{2\mu l}\int_0^{D/2}\left(\frac{D^2}{4} - y^2\right)ydy \tag{h}$$

Integrating and combining terms yields

$$Q = \frac{\Delta p \pi D^4}{128\mu l} \tag{i}$$

In order to obtain the average velocity across the flow, we define

$$Q = \frac{\pi}{4} D^2 \overline{V} \tag{j}$$

Equating Q from Equations (i) and (j),

$$\frac{\pi}{4} D^2 \overline{V} = \frac{\Delta p \pi D^4}{128 \mu l} \tag{k}$$

Thus

$$\overline{V} = \frac{\Delta p D^2}{32 \mu l} \tag{l}$$

When compared to Equation (f) for the maximum velocity, we obtain

$$\overline{V} = \frac{V_{\text{max}}}{2} \tag{m}$$

ILLUSTRATION 6.5   PRESSURE DROP IN LAMINAR FLOW

What is the pressure drop in a straight, horizontal run of 20 ft of pipe for the flow of Illustrative Problem 6.3?

ILLUSTRATIVE PROBLEM 6.5

Given $\gamma = 50$ lb/ft³, $\mu = 5$ cP, $D = 1$ in., $V = 1$ ft/s, Re $= 1236$; also,
$\mu = 1.045 \times 10^{-4}$ lb·s/ft²

Find: h_f

Assumptions: Equation (6.5) applicable

Basic Equations: $h_f = \dfrac{32 \mu L \overline{V}}{\gamma D^2}$

Solution: Since Re is less than 2000, the flow is laminar and we can use equation (6.5). The use of equation (6.5) requires care that all units be consistent, since the equation has dimensions. Proceeding with the data given, we have

$$h_f = \frac{32 \mu L \overline{V}}{\gamma D^2} = \frac{32 \times 1.045 \times 10^{-4} \text{ lb·s/ft}^2 \times 20 \text{ ft} \times 1 \text{ ft/s}}{50 \text{ lb/ft}^3 \times [1 \text{ in.}/(12 \text{ in./ft})]^2}$$

and

$$h_f = 0.1926, \text{ or about } 0.193 \text{ ft}$$

or

$$h_f = 0.193 \text{ ft·lb/lb}$$

It will be found later that the general form of equation most used to calculate the pressure loss in a pipe is known as the *Darcy-Weisbach equation*, given by

$$\frac{p_1 - p_2}{\gamma} = h_f = f \frac{L}{D} \frac{V^2}{2g} \tag{6.7}$$

where h_f is the head loss due to friction in feet or meters of fluid flowing, f is known as the *friction factor*, and $\overline{V}$ is the average velocity. If we equate equations (6.6) and (6.7), we obtain

$$f = \frac{L}{D} \frac{V^2}{2g} = \frac{32\mu L\overline{V}}{\gamma D^2} \tag{6.8}$$

Solving for f yields

$$f = \frac{64}{DV\gamma/\mu g} = \frac{64}{DV\rho/\mu} = \frac{64}{Re} \tag{6.9}$$

where Re is based on the *average* velocity. We can therefore conclude that the friction factor in laminar flow is simply 64 divided by the Reynolds number *and is independent of the pipe roughness.*

ILLUSTRATION 6.6   PRESSURE DROP IN LAMINAR FLOW

Solve Illustrative Problem 6.5 using equation (6.9).

ILLUSTRATIVE PROBLEM 6.6

Given: Same inputs as Illustrative Problem 6.5

Find: h_f using equation (6.9)

Assumptions: Equations (6.9) and (6.7) applicable

Basic Equations: Equation (6.9): $f = \dfrac{64}{Re}$

Equation (6.7): $h_f = f \dfrac{L}{D} \dfrac{V^2}{2g}$

Solution: Earlier we had to find the Reynolds number to determine that the flow is laminar. From Illustrative Problem 6.3, Re = 1236. Therefore,

$$f = \frac{64}{Re} = \frac{64}{1236} = 0.052$$

and

$$h_f = f \frac{L}{D} \frac{V^2}{2g} = 0.052 \times \frac{20 \text{ ft}}{1 \text{ in.}/(12 \text{ in./ft})} \times \frac{(1 \text{ ft/s})^2}{2 \times 32.17 \text{ ft/s}^2}$$

and

$$h_f = 0.194 \text{ ft·lb/lb}$$

which agrees with the result of Illustrative Problem 6.5.

ILLUSTRATION 6.7   BASIC PROGRAM FOR LAMINAR FLOW

In laminar flow we have shown that $f = 64/Re$. Devise a BASIC program to solve for the head loss due to friction in laminar flow if D, V, ρ, μ, L, D, and g are given. Print out Re, f, the head loss, and the input variables. Use Illustrative Problems 6.3 through 6.6 as a numerical check of your program.

ILLUSTRATIVE PROBLEM 6.7

Given: Same inputs as Illustrative Problems 6.3 through 6.6

Find: BASIC program to solve for head loss in laminar flow

Assumptions: Same as Illustrative Problems 6.3 through 6.6

Basic Equations: Same as Illustrative Problems 6.3 through 6.6

Solution: The head loss is given by the Darcy-Weisbach equation, which is $h_f = f$ $(L/D)(V^2/2g)$, and the friction factor in laminar flow is given by $f = 64/Re$. The program listed below uses the inputs to calculate Re, f, and the head loss. The numerical check with the Illustrative Problems cited is excellent. Note that the diameter is entered directly in units of feet to maintain dimensional consistency.

```
10 INPUT "diameter = "; D
20 INPUT " velocity = "; V
30 INPUT "density = "; R
40 INPUT "viscosity = "; M
50 INPUT "length = "; L
60 INPUT "gravity = "; G
70 N = (D*V*R)/M
80 F = 64/N
90 H = F*(L/D)*(V*V/(2*G))
100 LPRINT "diameter = "; D
110 LPRINT "density = "; R
120 LPRINT "viscosity = "; M
130 LPRINT "length = "; L
140 LPRINT "gravity = "; G
145 LPRINT "velocity = "; V
150 LPRINT "Reynolds number = "; N
160 LPRINT "friction factor = "; F
170 LPRINT "head loss = "; H
180 END


diameter = .08333
density = 1.55
viscosity = .0001045
length = 20
```

gravity = 32.17
velocity = 1
Reynolds number = 1235.995215311
friction factor = .051780135721558
head loss = .19315712903887

ILLUSTRATION 6.8   EXCEL PROGRAM FOR LAMINAR FLOW

Use the Excel spreadsheet to solve Illustrative Problem 6.7.

ILLUSTRATIVE PROBLEM 6.8

Given: Same as Illustrative Problem 6.3

Find: Excel spreadsheet to solve for losses in laminar flow

Assumptions: Same as Illustrative Problems 6.3 through 6.6

Basic Equations: Same as Illustrative Problems 6.3 through 6.6

Solution: The inputs are placed in boxes A3 through F3. The Reynolds number is given in box G3 as = A3*F3*B3/C3. The friction factor is given in box H3 as = 64/G3. The head loss due to friction is given in box I3 as = H3*(D3/A3)*(F3^2/(*E3)). The values of f, Re, and h_f agree with the previous solutions.

	A	B	C	D	E	F	G	H	I
1			Illustrative Problem 6.8						
2	D	Rho	Mu	L	g	V	Re	f	hf
3	0.0833	1.55	0.0001045	20	32.17	1	1235.55024	0.05179878	0.19329628
	G3 = A3*F3*B3/C3								
	H3 = 64/G3								
	I3 = H3*(D3/A3)*(F3^2/(2*E3))								

6.5 BOUNDARY LAYER[1]

The parabolic velocity profile of fluid flowing in laminar flow inside a pipe is due partly to the viscosity of the fluid and partly to the adhesive force between the liquid and the pipe wall. As the flow remains laminar, the velocity profile remains parabolic. When the flow becomes turbulent, the velocity decreases only very slightly from the axis of the pipe to the vicinity of the pipe wall. There is, however, a steep velocity gradient in the thin layer of liquid next to the stationary boundary. This is called the *boundary layer*. Although the main body of fluid is turbulent, there still exists a thin laminar layer immediately next to the pipe wall. Some typical velocity distributions are shown in Figure 6.5.

Let us consider the flow of fluid over a flat plate held parallel to the direction of flow (see Figure 6.6). The vertical scale is purposely enlarged to show the detail of the flow

[1]The material in this section is developed with permission of the Van Nostrand Reinhold Company from *Engineering Heat Transfer* by S. T. Hsu, Litton Educational Publishing, Inc., New York, 1963, pp. 209–211.

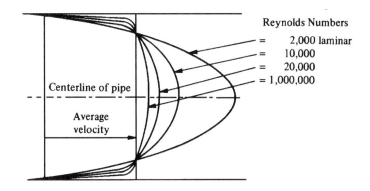

Figure 6.5   Velocity distributions in a smooth pipe.

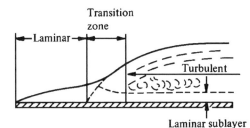

Laminar sublayer   **Figure 6.6**   Boundary-layer transition.

pattern. When the fluid passes the leading edge of the plate, the velocity gradient and the viscous boundary shear are high. The fluid is moving in the laminar state, and the boundary layer is thin. This is called the *laminar boundary layer.* As the fluid travels farther down the stream along the plate, the retardation of fluid flow increases due to shearing force, and the boundary layer also grows in thickness. As a result, the velocity gradient gradually decreases and concurrently the boundary shear is reduced as the thickness increases. When the boundary layer becomes thick enough, the particles begin to move out of the smooth layers, or laminae, the laminar motion becomes unstable, and finally, the flow becomes turbulent. Under the turbulent boundary layer, however, there is still a thin layer of fluid immediately next to the solid boundary, and this thin layer is flowing in the laminar regime. This layer is called the *laminar sublayer* of the turbulent boundary layer. The layer of transition from the laminar sublayer to the turbulent layer is called the *buffer layer.* Once the boundary layer is in the turbulent state, velocity distribution becomes more nearly constant in a lateral direction. A laminar boundary layer cannot change suddenly into a turbulent one, and a transition zone is found to exist between the turbulent and laminar zones. The turbulent motion is accompanied by an increase of the boundary shear and by an expansion of the thickness of the boundary layers. For a submerged plate, the transition takes place in the range of Reynolds numbers from 500,000 to 1 million, where the characteristic dimension is the distance from the leading edge of the plate. The number depends on the initial state of flow and also on the shape of the front edge and the roughness of the plate.

The thickness of the boundary layer is defined as the distance from the boundary to the point where the velocity reaches the value of the main stream velocity. It is extremely difficult actually to measure the thickness of the boundary layer, especially in turbulent flow. But the thickness of the boundary layer can be predicted by analytical methods or from the experimental results of velocity and temperature measurements. In laminar boundary layers the velocity profiles join the outside velocity curve asymptotically.

The velocity profile inside the boundary layer may be approximated by an analytical method based on the theory of velocity distribution in pipes, and the drag force due to the boundary layer may also be derived by an analytical method. A complete discussion of these methods, however, is too advanced to fall within the scope of this book, and the reader is referred to the references at the end of the book for further information on this subject.

6.6   FRICTION PRESSURE LOSSES IN TURBULENT PIPE FLOW

In Chapter 5 it was indicated that the energy equation can be used when a real fluid flows in a pipe. It will be recalled that a term was added to the right side of the equation to account for "head losses" that occur in actual flow situations. Specifically, we wrote

$$\frac{p_1}{\gamma} + \frac{V_1^2}{2g} + Z_1 = \frac{p_2}{\gamma} + \frac{V_2^2}{2g} + Z_2 + h_L \tag{6.10}$$

For a horizontal pipe of constant diameter,

$$h_L = \frac{p_1 - p_2}{\gamma} \tag{6.11}$$

That is, the loss between two sections of a horizontal pipe in which an incompressible fluid is flowing is simply the difference in the static pressures that exist at each of the sections.

When an incompressible fluid flows in a pipe and the flow is turbulent, it has been found experimentally that the head loss is a function of the length of the pipe, the diameter of the pipe, the surface roughness of the pipe wall, the velocity of the fluid, the density of the fluid, and the viscosity of the fluid. As noted earlier, the Darcy-Weisbach equation (6.7) is most generally used to calculate the frictional losses in pipes. It is repeated here for convenience:

$$h_f = f\frac{L}{D}\frac{V^2}{2g} \tag{6.7}$$

Unfortunately, in the turbulent flow regime it is not possible to obtain an analytical solution for the friction factor as we just did for the case of laminar flow. Most of the usable data available for evaluating the friction factor in turbulent flow have been derived from experiments. Prior to 1933, most of these results were scattered throughout the technical literature. In 1933, J. Nikuradse published his work on pipes whose walls had been artificially roughened using glued sand grains of different diameters. He termed the diameter of the sand grains (e) the absolute roughness and the ratio of the sand grain diameter to the inside pipe diameter (e/D) the relative roughness. The results of Nikuradse's work are shown graphically in Figure 6.7. In the laminar zone (Re $\leq$ ~2100) the friction factor is independent

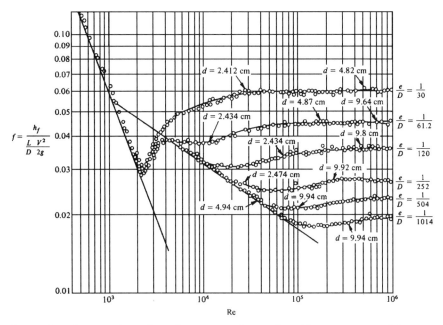

Figure 6.7   Nikuradse's sand-roughened pipe tests. (Reproduced by permission of the Van Nostrand Reinhold Company from *Engineering Heat Transfer* by S. T. Hsu, Litton Educational Publish-

of the relative or absolute roughness. In the turbulent zone the friction factor is seen to be a function of the relative roughness and the Reynolds number, and it will be seen that as the Reynolds number increases, the friction factor becomes constant for each given value of the relative roughness, and consequently independent of the Reynolds number.

In the same year, 1933, R. J. S. Piggott and E. Kemler published papers in which much of the existing data on friction in pipes were correlated and plotted in a diagram similar to Figure 6.7. This type of diagram, in which the logarithm of the friction factor is plotted as a function of the logarithm of the Reynolds number, is known as a Stanton diagram and is a very convenient portrayal of the data. In 1944, L. F. Moody published his paper on friction factors in pipe flow and in this paper gave his results in the form of a Stanton diagram, which had proved to be the most widely used source of friction factors for new or clean commercial pipes. This figure is reproduced as Figure 6.8 on page 220. Qualitatively, the trends in Nikuradse's data shown in Figure 6.7 and the Moody diagram shown in Figure 6.8 are in good agreement. Detailed differences exist in the critical and transition zones, as might be expected when comparing artificially roughened pipes with commercial pipes. Also, the values of the absolute roughness of pipe used by Moody and given in Table 6.1 are arbitrary and not comparable to the sand grain diameters used by Nikuradse. The dashed line on the Moody diagram indicates the value of the Reynolds number at which the flow becomes independent of Reynolds number and becomes a function of e/D only. This is noted to be the zone of complete turbulence, rough pipes. Figure 6.9 on page 222 gives the values of e/D for various types of pipe and also the friction factor for complete turbulence, rough pipes. It will be noted that the friction factor in the turbulent zone decreases with increasing Reynolds number until the limiting value of complete turbulence, rough pipes, is reached.

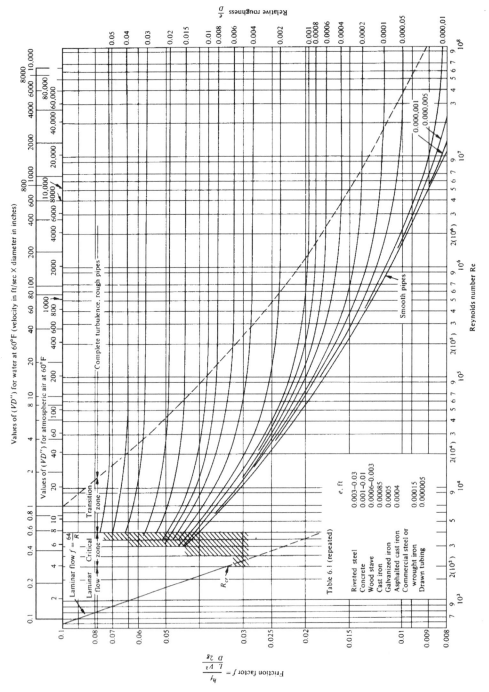

Figure 6.8 Friction factors for any type and size of pipe. (Reproduced with permission from *Pipe Friction Manual*, 3rd ed., Hydraulic Institute, New York, 1961.)

Table 6.1 Absolute Roughness of Pipes

| Material | Absolute roughness, e | |
	ft	m
Glass, new commercial pipe surfaces, drawn tubing (brass, copper, lead)	0.000005	1.52×10^{-6}
Commercial steel or wrought iron	0.00015	4.57×10^{-5}
Asphalted cast iron	0.0004	1.22×10^{-4}
Galvanized iron	0.0005	1.52×10^{-4}
Cast iron	0.00085	2.59×10^{-4}
Wood stave	0.0006−0.0003	$1.83 \times 10^{-4} - 9.14 \times 10^{-5}$
Concrete	0.001−0.01	$3.05 \times 10^{-4} - 3.05 \times 10^{-3}$
Riveted steel	0.003−0.03	$9.14 \times 10^{-4} - 9.14 \times 10^{-3}$

The values of e/D used on the Moody diagram apply to commercial, clean pipes and are not comparable to Nikuradse's artificially roughened values. For convenience, Figure 6.10 is included on page 223, since it is a handy presentation of both the kinematic viscosity and Reynolds number for many liquids and gases. Also, note that across the top of Figure 6.8 there are two auxiliary scales that are very useful. One is for water at 60°F and the other for air at 60°F. For a given fluid at a given temperature, Re is simply proportional to the product of $V \times D$. The scales along the top of Figure 6.8 have the product of V in ft/s and D in inches. Simply by going vertically downward, one reads the value of Re. Actually, the value of Re is not needed, since f can be obtained from the "VD" value and e/D.

ILLUSTRATION 6.9   FRICTION PRESSURE DROP

Water at 50°F is supplied from a reservoir to a 1000-ft-long, horizontal, round, concrete pipe 36 in. in diameter. Determine the frictional pressure drop in the pipe if the flow is 20,000 gal/min. Neglect minor losses.

ILLUSTRATIVE PROBLEM 6.9

Given: Water at 50°F, $Q = 20{,}000$ gpm, $L = 1000$ ft, $D = 36$ in., concrete pipe

Find: h_f

Assumptions: Incompressible flow, no losses other than friction.

Basic Equations: Continuity: $Q = AV$

$$\text{Re} = \frac{DV}{\nu}\, ; \, h_f = f \frac{L}{D} \frac{V^2}{2g}$$

Solution: Using 231 in.^{3} in 1 gal, we obtain

$$Q = \left(\frac{20{,}000}{60}\right)\left(\frac{231}{1728}\right) = 44.6 \text{ ft}^3/\text{s}$$
(continues on page 224)

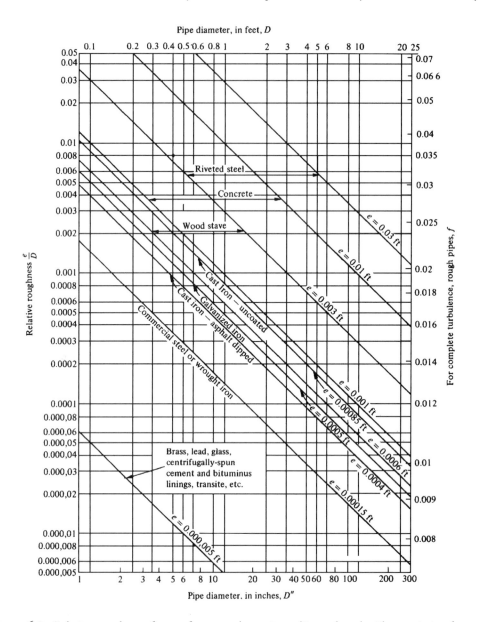

Figure 6.9 Relative roughness factors for new, clean pipes. (Reproduced with permission from *Pipe Friction Manual,* 3rd ed., Hydraulic Institute, New York, 1961.) *f* is for complete turbulence, rough pipes.

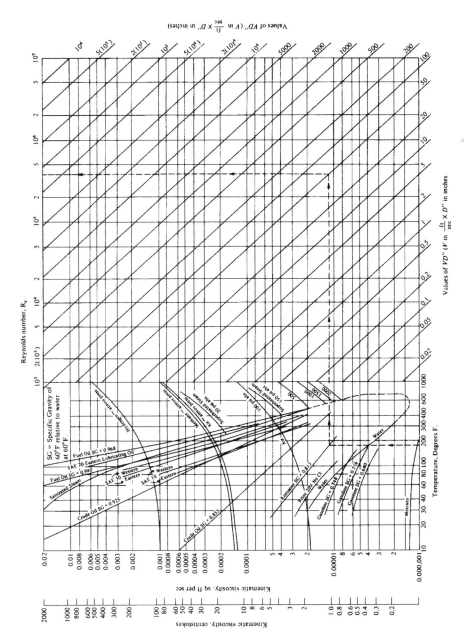

Figure 6.10 Kinematic viscosity and Reynolds number chart. (Reproduced with permission from *Pipe Friction Manual*, 3rd ed., Hydraulic Institute, New York, 1961.)

Since $Q = AV$,

$$V = \frac{Q}{A} = \frac{44.6}{(\pi/4)[(36)^2/144]} = 6.3 \text{ ft/s}$$

From Table 2.4, $\nu = 1.41 \times 10^{-5}$ ft²/s. Therefore,

$$\text{Re} = \frac{DV}{\nu} = \frac{\left(\dfrac{36 \text{ in.}}{12 \text{ in./ft}}\right) \times 6.3 \text{ ft/s}}{1.41 \times 10^{-5} \text{ ft}^2/\text{s}} = 1.34 \times 10^6$$

Also, the product of (VD) is $6.3 \times 36 = 227$. Using Figure 6.8 or 6.10, we can evaluate the Reynolds number directly:

$$\text{Re} \simeq 1.6 \times 10^6$$

Had we used 60°F for ν, we would have calculated Re $= 1.55 \times 10^6$. In most cases this difference will lead to a negligible error, and the use of (VD) is usually justified for temperatures not greatly different from 60°F.

From Table 6.1, e for concrete varies from 0.001 to 0.01 ft. Using the arithmetic average of 0.005 gives us

$$\frac{e}{D} = \frac{0.005 \times 12}{36} = 0.0017$$

From Figure 6.8, $f = 0.023$.
Continuing the problem, the pressure drop due to friction is

$$h_f = f\frac{L}{D}\frac{V^2}{2g} = 0.023\left(\frac{1000}{36/12}\right)\frac{(6.3)^2}{2g} = 4.73 \text{ ft of water}$$

or

$$h_f = 4.73 \times \frac{62.4}{144} = 2.05 \text{ psi}$$

ILLUSTRATION 6.10   FRICTION PRESSURE DROP

If the pipe of Illustrative Problem 6.1 is steel and 50 m long, what is the pressure drop due only to friction?

ILLUSTRATIVE PROBLEM 6.10

Given: water at 40°C, $V = 3$ m/s, $D = 50$ mm, Re $= 226{,}900$.

Find: h_f for $L = 50$ m

Assumptions: See Illustrative Problem 6.1 and the use of the Moody chart.

Basic Equations: Darcy: $h_f = f\dfrac{L}{D}\dfrac{V^2}{2g}$

Solution: In Illustrative Problem 6.1 we found that Re = 226,900. For commercial steel pipe, $e = 4.57 \times 10^{-5}$ m. Therefore,

$$\frac{e}{D} = \frac{4.57 \times 10^{-5}}{0.050} = 0.0009$$

From Figure 6.8, $f = 0.02$ and

$$h_f = f\frac{L}{D}\frac{V^2}{2g} = 0.02\left(\frac{50}{0.05}\right)\frac{(3)^2}{2 \times 9.81} = 9.17 \text{ m} = 9.17 \text{ N·m/N}$$

or

$$h_f = 9.17 \text{ m} \times 9810 \frac{N}{m^3} = 90 \text{ kPa}$$

ILLUSTRATION 6.11   FRICTION PRESSURE DROP

Water flows at 60°F in a commercial steel pipe whose inside diameter is 2.067 in. If the pipe is 200 ft long and a pump can furnish a head of 50 ft to overcome friction losses, determine the flow in the pipe.

ILLUSTRATIVE PROBLEM 6.11

Given: Water at 60°F, $D = 2.067$ in., $L = 200$ ft, $h_f = 50$ ft, commercial steel pipe
Find: Q
Assumptions: Incompressible flow, friction losses only
Basic Equations: Continuity: $Q = AV$

$$\text{Darcy: } h_f = f\frac{L}{D}\frac{V^2}{2g}$$

Solution: In this problem it is not possible to obtain a direct solution, and it will be necessary to use successive iterations (trial an error). The procedure is to assume a value of friction factor and to solve the problem based on this assumption. The friction factor is then evaluated using the solution based on the initial assumption. If the calculated and assumed values agree, this is the solution. If they do not agree, a new assumption is made and the procedure is repeated.

As a first trial, assume fully developed turbulent flow. From Figure 6.9, for 2.067-inch diameter, commercial steel pipe, $f = 0.019$. This is a good first guess. Therefore

$$50 = 0.019\left(\frac{200}{2.067/12}\right)\frac{V^2}{2 \times 32.17}$$

$$V = 12.1 \text{ ft/s}$$

$$VD = 12.1 \times 2.067 = 25.0. \text{ From Table 6.1, } e = 0.00015 \text{ ft.}$$

Therefore,

$$\frac{e}{D} = \frac{0.00015 \text{ ft}}{2.067 \text{ in.}/(12 \text{ in./ft})} = 0.00087$$

From Figure 6.8, $f = 0.021$. Using this value of f,

$$50 = 0.021\left(\frac{200}{2.067/12}\right)\frac{V^2}{2 \times 32.17}$$

$$V = 11.5 \text{ ft/s}$$

$VD = 11.5 \times 2.067 = 23.8$. From Figure 6.8, $f = 0.021$. This is a satisfactory agreement, since we assumed $f = 0.021$ and obtained $f = 0.021$. Using $V = 11.5$ ft/s,

$$Q = AV = 11.5 \times \frac{\pi}{4}\left(\frac{2.067}{12}\right)^2 = 0.268 \text{ ft}^3/\text{s}$$

and in terms of gallons per minute,

$$Q = \frac{0.268 \text{ ft}^3/\text{s} \times 1728 \text{ in.}^3/\text{ft.}^3 \times 60 \text{ s/min}}{231 \text{ in.}^3/\text{gal}}$$

$$Q = 120.3 \text{ gallons/minute}$$

While Figure 6.8 is convenient when solving problems in pipe friction, it is often desirable to have an explicit formula expressing friction factor in turbulent flow as a function of the relative roughness and the Reynolds number. In 1947, Moody published such a formulation, which is sufficiently accurate for most engineering purposes, namely,

$$f \simeq 0.0055\left[1 + \left(20,000\,\frac{e}{D} + \frac{10^6}{\text{Re}}\right)^{1/3}\right] \tag{6.12}$$

Equation (6.12) is useful when problems involving pipe friction are programmed on a digital computer, since an explicit solution can be obtained. It is sometimes more convenient to write it in the following form:

$$f \simeq 0.0055 + 0.0055\left(20,000\,\frac{e}{D} + \frac{10^6}{\text{Re}}\right)^{1/3} \tag{6.13}$$

ILLUSTRATION 6.12   BASIC PROGRAM FOR TURBULENT FLOW

Program in BASIC the head loss due to friction when the fluid flowing is turbulent. Assume that e, D, V, ρ, μ, and L are the given inputs. Use equation (6.12) and the Darcy-Weisbach equation (6.7). Print out the Reynolds number and the friction factor. Use Illustrative Problem (6.10) as a numerical check of your program.

ILLUSTRATIVE PROBLEM 6.12

Given: Equation (6.7) and equation (6.12)

Find: Solution using a BASIC program

Assumptions: Those inherent in equations (6.7) and (6.12)

Basic Equations: Darcy: $h_f = f \dfrac{L}{D} \dfrac{V^2}{2g}$; $f \simeq 0.0055\left[1 + \left(20{,}000 \dfrac{e}{D} + \dfrac{10^6}{Re}\right)^{1/3}\right]$

Solution: The program is given below and proceeds in a straightforward manner. One additional input, *g*, is needed depending on the system of units used. Comparison of the printout to Illustrative Problem 6.10 shows excellent agreement with the Reynolds number and agreement to within 5% for the friction factor and the head loss. This is satisfactory, since equation (6.12) is approximate to within this value.

```
10 INPUT "absolute roughness = ";E
20 INPUT "diameter = "; D
30 INPUT " velocity = "; V
40 INPUT "density = "; R
50 INPUT "viscosity = "; M
60 INPUT "length = "; L
70 INPUT "gravity = "; G
80 N=(D*V*R)/M
90 F = .0055*( 1 + (20000*(E/D) + (1000000#/N))^(1/3))
100 H = F*((L/D))*(V*V/(2*G))
110 LPRINT "absolute roughness = "; E
120 LPRINT "diameter = "; D
130 LPRINT "velocity = "; V
140 LPRINT "density = "; R
150 LPRINT "viscosity = "; M
160 LPRINT "length = "; L
170 LPRINT "gravity = "; G
180 LPRINT "Reynolds number = "; N
190 LPRINT "friction factor = "; F
200 LPRINT "friction head loss = "; H
210 END
```

```
absolute roughness = .0000457
diameter = .05
velocity = 3
density = 992.2
viscosity = .000656
length = 50
gravity = 9.81
Reynolds number  = 226875
friction factor = .021070154878497
friction head loss = 9.6652086598611
```

ILLUSTRATION 6.13   EXCEL SPREADSHEET FOR TURBULENT FLOW

Use the Excel spreadsheet to determine the head loss due to friction when the fluid flowing is turbulent. Solve using the same parameters as Illustrative Problem 6.12 and use Illustrative Problem 6.10 as a numerical check of the program.

ILLUSTRATIVE PROBLEM 6.13

Given: Same inputs as Illustrative Problem 6.12

Find: Solution using an Excel spreadsheet

Assumptions: Same as Illustrative Problem 6.12

Basic Equations: Same as Illustrative Problem 6.12

Solution: The inputs are found in boxes A3 through G3. Box H3 calculates Re. Since equation (6.12) is somewhat lengthy, it has been solved in parts. Thus box A4 = $\frac{10^6}{Re}$ and box B4 = (20,000 e/D). I3 = $0.0055 \left[\left(\frac{10^6}{Re} + 20,000\,\frac{e}{D}\right)^{1/3}\right] + 1$, which is *f*. Finally, J3 = h_f. The spreadsheet is shown below.

	A	B	C	D	E	F	G	H	I	J
1			Illustrative Problem 6.13							
2	e	D	V	rho	mu	L	g	Re	f	hf
3	4.57E−05	0.05	3	992.2	0.000656	50	9.81	226875	0.0211	9.66520866
4	4.407713	18.28								
5	H3 = B3*C3*D3/E3									
6	A4 = 10^6/H3									
7	B4 = 20000*A3/B3									
8	I3 = 0.0055*((A4 + B4)^(1/3) + 1)									
9	J3 = (I3*F3/B3)*C3^2/(2*G3)									

ILLUSTRATION 6.14   FRICTION IN TURBULENT FLOW

Water at 60°F flows in a commercial steel pipe whose inside diameter is 75 mm. If the pipe is 100 m long and a pump can furnish a head of 20 m to overcome friction losses, determine the flow in the pipe.

ILLUSTRATIVE PROBLEM 6.14

Given: Commercial steel pipe, D = 75 mm, L = 100, water at 60°F, h_f = 20 m

Find: Q

Assumptions: Turbulent flow, friction only

Basic Equations: Continuity: Q = AV

Darcy: $h_f = f\dfrac{L}{D}\dfrac{V^2}{2g}$

Solution: As in Illustrative Problem 6.11, we cannot get an explicit solution to this problem. We will assume a friction factor of 0.02. Therefore,

$$20 = 0.02 \times \frac{100}{0.075}\left(\frac{V^2}{2 \times 9.81}\right)$$

and

$$V = 3.84 \text{ m/s}$$

Although we could fundamentally obtain Re, it is simpler first to obtain (VD). Therefore,

$$(VD) = (3.84 \times 3.281 \text{ ft/m}) \left(\frac{75}{25.4 \text{ mm/in.}} \right) = 37.2$$

$$\frac{e}{D} = \frac{4.57 \times 10^{-5}}{0.075} = 0.000\ 61$$

Entering Figure 6.8, we read $f = 0.019$. If we now use $f = 0.019$ as our second trial, $V = 3.94$ m/s and $(VD) = 38.1$. Entering Figure 6.8, we read $f = 0.019$. Using this value, $V = 3.94$ m/s and

$$Q = AV = \frac{\pi}{4} (0.075)^2 \times (3.94) = 1.74 \times 10^{-2} \text{ m}^3/\text{s}$$

It is apparent from Illustrative Problems 6.11 and 6.14 that the work of solving problems that require several iterations is made easier if the first "guess" is close to the final solution. While Figure 6.9 can be used for this purpose, it is also possible to obtain the value of f for fully developed turbulent flow by noting that Re is usually large when this occurs. If we use equation (6.12) with Re very large, the $10^6/\text{Re}$ term becomes small relative to the 20,000 e/D term. Since their sum is raised to the one-third power, neglecting the $10^6/\text{Re}$ term leads to a small percentage error in f calculated from equation (6.12). Using the Excel spreadsheet, f has been determined for various diameters for drawn tubing, commercial steel pipe, concrete, and asphalted cash iron pipe. Table 6.2 on page 230 shows these calculations and Figure 6.11 on page 231 shows f as a function of the pipe diameter for fully developed turbulent flow.

The problem of evaluating friction factors for old pipes and allowing for the deterioration of new pipes is particularly difficult, since pipe deterioration is a function of the fluid and pipe chemical properties. Thus it is recommended that the Moody curve and the data presented in this chapter not be used when estimating the pressure losses in old pipes or to allow for the aging of new pipes. If at all possible, experience with an existing installation will yield the best data for estimating the pressure drop in old or aging pipes. Data in the engineering literature are not satisfactory and may yield results that are in large error when applied to a specific situation.

6.7   OTHER LOSSES

In addition to the pressure drop incurred due to friction as a fluid flows in a pipe, other losses in pressure occur. These other losses occur at the entrance to a pipe, at abrupt changes in section in a pipe, at valves, at fittings, at bends, and at the pipe exit; they are often called minor losses. The data available in the literature on the losses in valves and fittings have been

Table 6.2   Excel Spreadsheet Calculations of Approximate Values of f for Fully Developed Turbulent Flow

		f, Drawn tubing		
Diameter D Inches	1.2/D	(1.2/D)^(1/3)	.0055 ∗ (1.2/D)^(1/3)	.0055 + .0055 ∗ (1.2/D)^(1/3)
1	1.2	1.062658569	0.005844622	0.01134462
2	0.6	0.843432665	0.00463888	0.01013888
4	0.3	0.66943295	0.003681881	0.00918188
10	0.12	0.493242415	0.002712833	0.00821283
20	0.06	0.391486764	0.002153177	0.00765318
40	0.03	0.310723251	0.001708978	0.00720898
48	0.025	0.292401774	0.00160821	0.00710821

		f, Commercial steel pipe		
Diameter D Inches	36/D	(36/D)^(1/3)	.0055 ∗ (36/D)^(1/3)	.0055 + .0055 ∗ (36/D)^(1/3)
1	36	3.301927249	0.0181606	0.0236606
2	18	2.620741394	0.014414078	0.019914078
4	9	2.080083823	0.011440461	0.016940461
10	3.6	1.532618865	0.008429404	0.013929404
20	1.8	1.216440399	0.006690422	0.012190422
40	0.9	0.965489385	0.005310192	0.010810192
48	0.75	0.908560296	0.004997082	0.010497082

		f, Concrete		
Diameter D Inches	1200/D	(1200/D)^(1/3)	.0055 ∗ (1200/D)^(1/3)	.0055 + .0055 ∗ (1200/D)^(1/3)
1	1200	10.62658569	0.058446221	0.063946221
2	600	8.434326653	0.046388797	0.051888797
4	300	6.694329501	0.036818812	0.042318812
10	120	4.932424149	0.027128333	0.032628333
20	60	3.914867641	0.021531772	0.027031772
40	30	3.107232506	0.017089779	0.022589779
48	25	2.924017738	0.016082098	0.021582098

		f, Asphalted cast iron		
Diameter D Inches	96/D	(96/D)^(1/3)	.0055 ∗ (96/D)^(1/3)	.0055 + .0055 ∗ (96/D)^(1/3)
1	96	4.57885697	0.025183713	0.030683713
2	48	3.634241186	0.019988327	0.025488327
4	24	2.884499141	0.015864745	0.021364745
10	9.6	2.125317138	0.011689244	0.017189244
20	4.8	1.686865331	0.009277759	0.014777759
40	2.4	1.3388659	0.007363762	0.012863762
48	2	1.25992105	0.006929566	0.012429566

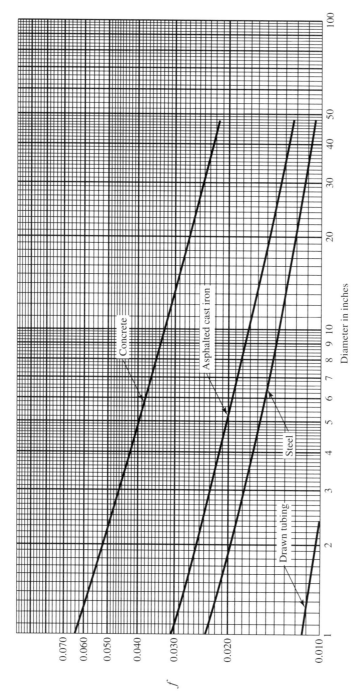

Figure 6.11 Approximate value of f for fully developed turbulent flow in pipes.

presented in two different forms. The first form expresses the fact that these losses can be expressed in terms of velocity heads of fluid flowing; that is,

$$h = K \frac{V^2}{2g} \tag{6.14}$$

The value of K increases with increasing roughness and decreases with increasing Reynolds numbers but depends primarily on the geometric shape of the valve or fitting. Figure 6.12 gives the resistance coefficients for pipe fittings and Figure 6.13 gives the resistance coefficients for valves, couplings, and unions. To appreciate that these values are approximate and to appreciate the wide variations that are found in specific fittings, Table 6.3 has been included on page 235. It is apparent from this table that, at best, the calculation of these losses is approximate and that the values given in Figures 6.12 and 6.13 on pages 233–234 for the resistance coefficients are just good approximations. To illustrate further the difficulty of estimating the resistance coefficients for gate valves that are partially open, it is noted that a gate valve 3/4 open can have a resistance coefficient that is 5 times that of a fully open gate valve and a gate valve that is 3/4 closed can have a resistance coefficient that is as much as 120 times that of a fully open gate valve. Estimating how open or closed a valve is positioned is difficult, making the accurate estimate of K almost impossible. Table 6.4 on page 236 shows the range of losses in partially open valves.

The second method of presenting these losses is to express them as equivalent lengths of pipe, L_{eq}, that have the same head loss for the same discharge. Therefore,

$$f \frac{L_{eq}}{D} \frac{V^2}{2g} = K \frac{V^2}{2g} \tag{6.15}$$

and

$$L_{eq} = \frac{KD}{f} \tag{6.16}$$

These minor losses are in excess of the friction loss produced in a straight pipe whose length is equal to the length of the axis of the pipe (see Table 6.5 on page 239).

ILLUSTRATION 6.15   EQUIVALENT LENGTH

A commercial steel pipeline has an inside diameter of 6 in., and the flow is completely turbulent. If the sum of all the loss coefficients in the pipe (K) is 12, determine the equivalent length that must be added to the actual pipe length to cause the same resistance to flow.

ILLUSTRATIVE PROBLEM 6.15

Given: Commercial steel pipe, D = 6 in., turbulent flow, K = 12

Find: L_{eq}

Assumptions: Fully developed turbulent flow, clean pipe

Basic Equations: Figure 6.11, Equation (6.16): $L_{eq} = \dfrac{KD}{f}$

Solution: For steel pipe, $e = 0.00015$ and $e/D = 0.00015/(6/12) = 0.00030$.
For completely turbulent flow Figure 6.11 gives us $f = 0.015$. Therefore,

$$L_{eq} = \frac{12 \times (6/12)}{0.015} = 400 \text{ ft to be added}$$

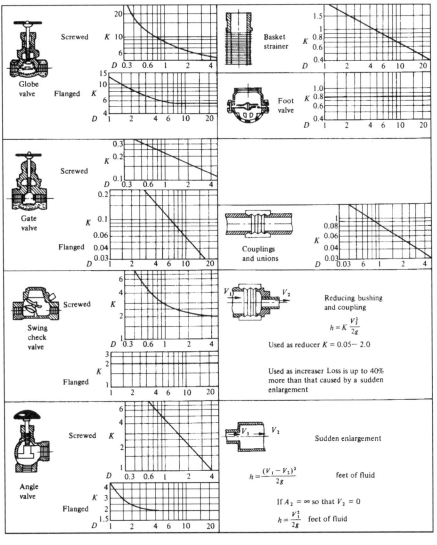

$$h = K \frac{V^2}{2g} \text{ Feet of fluid}$$

Figure 6.12   Resistance coefficients for valves and fittings (see Figure 6.13). (Reproduced with permission from *Pipe Friction Manual,* 3rd ed., Hydraulic Institute, New York, 1961.)

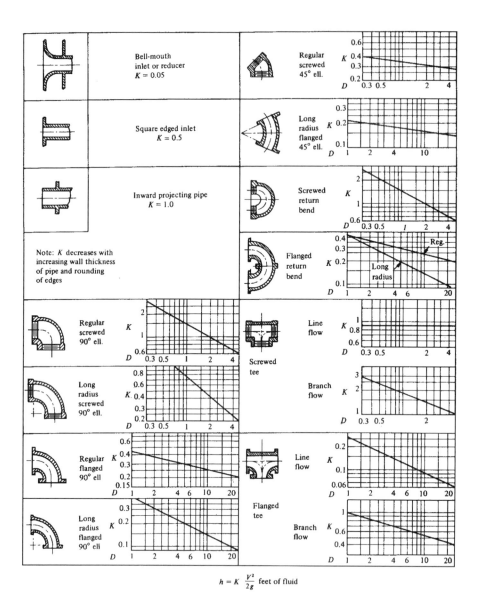

$$h = K \frac{V^2}{2g} \text{ feet of fluid}$$

Figure 6.13 Resistance coefficients for valves and fittings. The value of D is nominal iron pipe size. For velocities below 15 ft/s, check valves and foot valves will be only partially open and will exhibit higher values of K than that shown. (Reproduced with permission from *Pipe Friction Manual,* 3rd ed., Hydraulic Institute, New York, 1961.)

Table 6.3   Resistance Coefficients for Valves and Fittings

	Fitting	Range of variation
90-deg elbow	Regular screwed	±20% above 2-in. size
	Regular screwed	±40% below 2-in. size
	Long radius, screwed	±25%
	Regular flanged	±35%
	Long radius, flanged	±30%
45-deg elbow	Regular screwed	±10%
	Long radius, flange	±10%
180-deg bend	Regular screwed	±25%
	Regular flanged	±35%
	Long radius, flanged	±30%
Tee	Screwed, line, or branch flow	±25%
	Flanged, line, or branch flow	±35%
Globe valve	Screwed	±25%
	Flanged	±25%
Gate valve	Screwed	±25%
	Flanged	±50%
Check valve	Screwed	±30%
	Flanged	{ + 200% −80%
Sleeve check valve		Multiply flanged values by 0.2–0.5
Tilting check valve		Multiply flanged values by 0.13–0.19
Drainage gate check		Multiply flanged values by 0.03–0.07
Angle valve	Screwed	±20%
	Flanged	±50%
Basket strainer		±50%
Foot valve		±50%
Couplings		±50%
Unions		±50%
Reducers		±50%

Source: Reproduced with permission from *Pipe Friction Manual,* 3rd ed., Hydraulic Institute, New York, 1961.

Table 6.4 Losses in Partially Open Valves

	Ratio of K to K fully open	
	Gate Valve	Globe Valve
Fully open	1	1
1/4 closed	3–5	1.5–2
1/2 closed	12–22	2–3
3/4 closed	70–120	6–8

Data From: White, F. M., *Fluid Mechanics,* McGraw-Hill Book Co, New York, 1979.

In addition to valves and fittings, most pipelines have abrupt changes at entrances, exits, reducers, increasers, diffusers, and bends. One of these losses, the case of loss at a sudden enlargement, can be calculated, with the results agreeing reasonably well with experiment. When this is done, the head loss is found to be

$$h = \frac{(V_1 - V_2)^2}{2g}$$
(6.17)

where V_1 is the upstream velocity and V_2 the downstream velocity. Such a condition exists when a pipe exits into a large tank or reservoir. For this case equation (6.17) gives a loss equal to one velocity head in the pipe. Alternatively, we can argue that the kinetic energy in the pipe is converted to internal energy and does not appear again as a pressure head. In this sense one velocity head is "lost" at the exit of the pipe. For a sudden contraction, Figure 6.14 gives the loss coefficient. It will be noted that if this curve is applied to the sharp entrance of a pipe from a large reservoir, the loss coefficient is half a velocity head. As will also be seen from Figure 6.11, this loss can be greatly reduced by rounding off the entrance to the pipe.

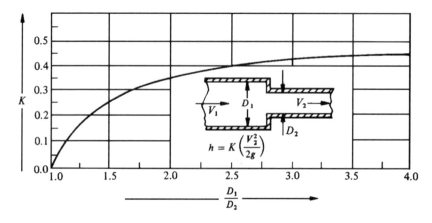

Figure 6.14   Resistance coefficients for reducers. (Reproduced with permission from *Pipe Friction Manual,* 3rd ed., Hydraulic Institute, New York, 1961.)

In addition to the foregoing, Figures 6.15 and 6.16 can be used to obtain the resistance coefficient, K, for increasers, diffusers, and bends. It should be noted that many of these losses can be decreased dramatically by good design techniques. In large air-handling systems, vanes are frequently used to decrease the turbulence that occurs in elbows and bends. Figure 6.17 on page 240 shows the effect of using vanes in such a system. These vanes serve to decrease the turbulence and concurrently they also reduce the noise caused by the air flowing in the ducts. The same principle of good design also applies to fluid systems.

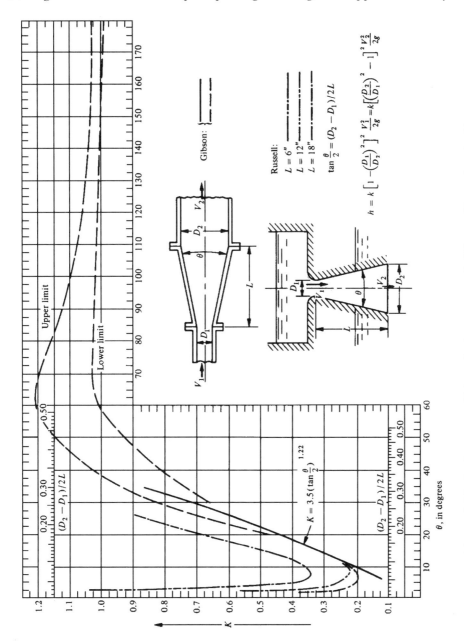

Figure 6.15   Resistance coefficients for increasers and diffusers. (Reproduced with permission from *Pipe Friction Manual*, 3rd ed., Hydraulic Institute, New York, 1961.)

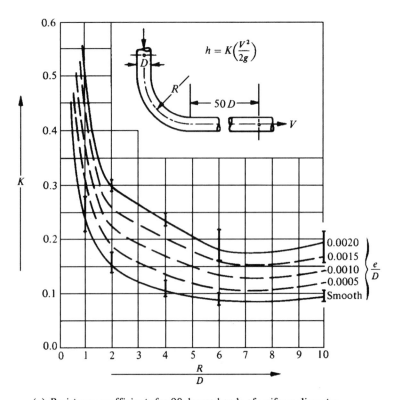

(a) Resistance coefficients for 90 degree bends of uniform diameter.

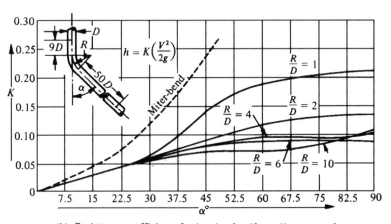

(b) Resistance coefficients for bends of uniform diameter and
smooth surface at Reynolds No. ~ 2.25 × 10³

Figure 6.16   (Reproduced with permission from *Pipe Friction Manual,* 3rd ed., Hydraulic
Institute, New York, 1961.)

TABLE 6.5   Equivalent Length of New Straight Pipe for Valves and Fittings for Turbulent Flow Only

Fittings	Connection	Material	¼	⅜	½	¾	1	1¼	1½	2	2½	3	4	5	6	8	10	12	14	16	18	20	21
Regular 90° ell	Screwed	Steel	2.3	3.1	3.6	4.4	5.2	6.6	7.4	8.5	9.3	11	13										
	Flanged	Steel			0.92	1.2	1.6	2.1	2.4	3.1	3.6	4.4	5.9	7.3	8.9	12	14	17	18	21	23	25	30
	Flanged	Cast iron											4.8		7.2	9.8	12	15	17	19	22	24	28
Long radius 90° ell	Screwed	Steel	1.5	2.0	2.2	2.3	2.7	3.2	3.4	3.6		4.0	4.6										
	Flanged	Steel			1.1	1.3	1.6	2.0	2.3	2.7		3.3	3.7	5.0	5.7	7.0	8.0	9.0	9.4	10	11	12	14
	Flanged	Cast iron													4.7	5.7	6.8	7.8	8.6	9.6	11	11	13
Regular 45° ell	Screwed	Steel	0.34	0.52	0.71	0.92	1.3	1.7	2.1	2.7	3.2	4.0	5.5										
	Flanged	Steel			0.45	0.59	0.81	1.1	1.3	1.7	2.0	2.6	3.5	4.5	5.6	7.7	9.0	11	13	15	16	18	22
	Flanged	Cast iron											2.9		4.5	6.3	8.1	9.7	12	13	15	17	20
Tee line flow	Screwed	Steel	0.79	1.2	1.7	2.4	3.2	4.6	5.6	7.7	9.3	12	17										
	Flanged	Steel			0.69	0.82	1.0	1.3	1.5	1.8	1.9	2.2	2.8	3.3	3.8	4.7	5.2	6.0	6.4	7.2	7.6	8.2	9.6
	Flanged	Cast iron										1.9	2.2		3.1	3.9	4.6	5.2	5.9	6.5	7.2	7.7	8.8
Tee branch flow	Screwed	Steel	2.4	3.5	4.2	5.3	6.6	8.7	9.9	12	13	14	21										
	Flanged	Steel			2.0	2.6	3.3	4.4	5.2	6.6	7.5	9.4	12	15	18	24	30	34	37	43	47	52	62
	Flanged	Cast iron										7.7	10		15	20	25	30	35	39	44	49	57
180° return bend	Regular flanged	Steel	2.3	3.1	3.6	4.4	5.2	6.6	7.4	8.5	9.3	11	13										
		Cast iron			0.92	1.3	1.6	2.1	2.4	3.1	3.6	4.4	5.9	5.0	5.7	7.0	8.0	9.0	9.4	10	11	12	14
	Long radius flanged	Steel			1.1	1.3	1.6	2.1	2.4	2.9	3.4	3.4	4.2		4.7	5.7	6.8	7.8	8.6	9.6	11	11	13
		Cast iron										2.8											
Globe valve	Screwed	Steel	21	22	22	24	29	37	42	54	62	79	110	150	190	260	310	390					
	Flanged	Steel			38	40	45	54	59	70	77	94	120		150	210	270	330					
		Cast iron										77	99										
Gate valve	Screwed	Steel	0.32	0.45	0.56	0.67	0.84	1.1	1.2	1.5	1.7	1.9	2.5		3.2	3.2	3.2	3.2	3.2	3.2	3.2	3.2	3.2
	Flanged	Steel								2.6	2.7	2.8	2.9	3.1	2.6	2.7	2.8	2.9	2.9	3.0	3.0	3.0	3.0
		Cast iron										2.3	2.4										
Angle valve	Screwed	Steel	12.8	15	15	15	17	18	18	18	18	18	15	50	63	90	120	140	160	190	210	240	300
	Flanged	Steel			15	15	17	18	21	22	22	28	31	50	52	74	98	120	150	170	200	230	280
		Cast iron										23	31										
Swing check valve	Screwed	Steel	7.2	7.3	8.0	8.8	11	13	15	19	22	27	38		63	90	120	140		190	210	240	300
	Flanged	Steel			3.8	5.3	7.2	10	12	17	21	22	38	50	63	90	120	140		170	200	230	280
		Cast iron										27	31		52	74	98	120					
Coupling or union	Screwed	Steel	0.14	0.18	0.21	0.24	0.29	0.36	0.39	0.45	0.47	0.53	0.65		1.6	2.3	2.9	3.5	4.0	4.7	5.3	6.1	7.6
		Cast iron	0.04	0.07	0.10	0.13	0.18	0.26	0.31	0.43	0.52	0.44	0.52	1.3	1.3	1.9	2.4	3.0	3.6	4.3	5.0	5.7	7.0
Bell mouth inlet			0.44	0.68	0.96	1.3	1.8	2.6	3.1	4.3	5.2	0.67	0.95		16	23	29	35	40	47	53	61	76
												0.55	0.77		13	19	24	30	36	43	50	57	70
Square mouth inlet			0.88	1.4	1.9	2.6	3.6	5.1	6.2	8.5	10	6.7	9.5	26	32	45	58	70	80	95	110	120	150
Reentrant pipe												5.5	7.7			37	49	61	73	86	100	110	140
Sudden enlargement												13	15										

$$h = \frac{(V_1 - V_2)^2}{2g} \text{ feet of liquid; if } V_2 = 0, \; h = \frac{V_1^2}{2g} \text{ feet of liquid}$$

Source: Reproduced with permission from *Pipe Friction Manual*, 3rd ed., Hydraulic Institute, New York, 1961.

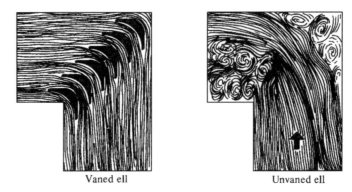

Vaned ell   Unvaned ell

Figure 6.17   Effects of vanes on turbulence.

ILLUSTRATION 6.16   OTHER LOSSES

A large open reservoir is connected to a steel tube whose inside diameter is 2 in. and whose length is 50 ft. A fully screwed gate valve is placed in the tube near its outlet. If the level of water in the tank is 50 ft above the outlet of the tube, determine the rate at which the water flows when the valve is fully opened. Assume that the loss coefficients for bends and elbows add up to 3.0 and that the tube discharges to the atmosphere.

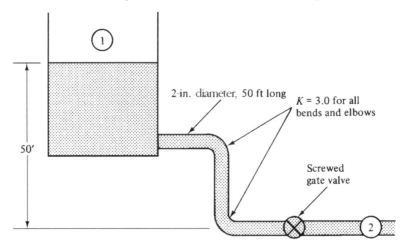

2-in. diameter, 50 ft long

$K = 3.0$ for all bends and elbows

Screwed gate valve

50′

Figure 6.18   Illustrative Problem 6.16

ILLUSTRATIVE PROBLEM 6.16

Given: $D = 2$ in., $L = 50$ ft, gate valve fully open, $Z_1 = 50$ ft, $K = 3.0$, $Z_2 = 0$, $p_1 = 0$, $p_2 = 0$, $V_1 = 0$.

Find: V_2

Assumptions: Steady, incompressible flow; losses

Basic Equations: Energy: $\dfrac{p_1}{\gamma} + \dfrac{V_1^2}{2g} + Z_1 = \dfrac{p_2}{\gamma} + \dfrac{V_2^2}{2g} + Z_2 + h_L$

Continuity: $Q = A_1V_1 = A_2V_2$

Darcy: $h_f = f\dfrac{L}{D}\dfrac{V^2}{2g}$

Solution: Writing an energy equation between sections ① and ② in Figure 6.18,

$$\dfrac{p_1}{\gamma} + \dfrac{V_1^2}{2g} + Z_1 = \dfrac{p_2}{\gamma} + \dfrac{V_2^2}{2g} + Z_2 + h_L$$

Since we have atmospheric pressure at sections ① and ② and V_1 can be taken to be zero,

$$\dfrac{V_2^2}{2g} + h_L = Z_1 - Z_2$$

where we have already termed $V_2^2/2g$ to be the exit loss. Inserting an entrance loss, a loss term for the valve, and a friction loss term as well as the other losses, we have

$$\dfrac{V_2^2}{2g}\left(1 + f\dfrac{L}{D} + 3.0 + 0.5 + 0.17\right) = Z_1 - Z_2 = 50$$

In general, f is a function of the velocity, V, and an explicit solution cannot be obtained for this problem. Let us assume, however, that the flow is fully turbulent:

$$\dfrac{e}{D} = \dfrac{0.00015}{2/12} = 0.0009$$

and f for fully developed turbulent flow is 0.019. Therefore,

$$\dfrac{V_2^2}{2g}\left(1 + \dfrac{0.019 \times 50}{2/12} + 3.0 + 0.5 + 0.17\right) = 50$$

and

$$V_2 = 17.6 \text{ ft/s}$$

At this point it is necessary to check the Reynolds number to verify the assumption of fully turbulent flow. Since $(VD) = 17.6 \times 2 = 35.2$, Re $\simeq 2.0 \times 10^5$ from Figure 6.8. From Figure 6.8 we have $f = 0.021$. As a second trial assume that $f = 0.021$, and solve to obtain $V_2 = 17.1$ ft/s. For this velocity Re $\simeq (17.1/17.6) \times 2 \times 10^5 = 1.9 \times 10^5$, and $f = 0.021$. Therefore, the solution is that water flows at the rate of 17.1 ft/s in the pipe.

ILLUSTRATION 6.17   TOTAL LOSSES

Air flows in a $\frac{1}{2}$-in. schedule 80 steel pipe (whose inside diameter is 0.546 in.) from a large storage tank to a second tank. The piping system will contain twelve 90° regular screwed elbows, two screwed gate valves, one screwed globe valve, three screwed tees with line flow, and 200 ft of straight pipe. If 0.5 lb/min flows, and the air has a specific weight of 0.21 lb/ft³, determine the pressure drop in the system. Assume that the tank pressure is 50 psia, that the air temperature is 100°F, and that the pressure drop in the pipe is less than 10% of the tank pressure. If the pressure drop exceeds 10% of the tank pressure, it is incorrect to apply the equations for incompressible flow to this flow situation.

ILLUSTRATIVE PROBLEM 6.17

Given: D = 0.546 in., p_1 = 100 psia, twelve 90° regular screwed elbows, two screwed gate valves, one screwed globe valve, three screwed tees with line flow; L = 200 ft, $\dot{w}$ = 0.5 lb/min, γ = 0.21 lb/ft³, t = 100°F, V_1 = 0, $Z_1 = Z_2$

Find: Δp due to flow

Assumptions: $\Delta p \leq 0.1p_1$, sharp entrance, no velocity head recovery at outlet

Basic Equations: Energy: $\dfrac{p_1}{\gamma} + \dfrac{V_1^2}{2g} + Z_1 = \dfrac{p_2}{\gamma} + \dfrac{V_2^2}{2g} + Z_2 + h_L$

Darcy: $h_f = f\dfrac{L}{D}\dfrac{V^2}{2g}$,   Continuity: $\dot{w} = \gamma AV$;   Total $K = \Sigma K$

Solution: $\dot{w} = \gamma AV$, or $V = \dfrac{\dot{w}}{\gamma A}$

For this problem, $V = \dfrac{(0.5\ \text{lb/min})/(60\ \text{s/min})}{0.21\ \text{lb/ft}^3 \times \pi/4\ (0.546\ \text{in./}(12\ \text{in./ft}))^2} = 24.4$ ft/s

$$(VD) = 24.4 \times 0.546 = 13.3$$

From Figure 6.10, Re $\simeq 10^4$

For commercial steel pipe,

$$e = 0.00015\ \text{ft}$$

Therefore,

$$\frac{e}{D} = \frac{0.00015}{0.546/12} = 0.0033$$

From Figure 6.8, $f = 0.035$

Expressing the friction loss in terms of velocity heads yields

$$K_f = f\frac{L}{D}$$

or

$$K_f = 0.035\,\frac{200}{0.546/12} = 153.8$$

For twelve 90° regular screwed elbows in a $\frac{1}{2}$-in pipe, $K = 2$ for each elbow.

For two screwed gate valves, $K = 0.32$ per valve.

For one screwed glove valve, $K = 14$.

For three screwed tees with line flow, $K = 0.9$ per tee.

For entrance, $K = 0.5$. (assumed sharp entrance)

For exit, $K = 1$. (assumed no recovery of $V_2^2/2g$ as pressure)

Adding all these terms, we have

$$K = 153.8 + 12(2) + 2(0.32) + 14 + 3(0.9) + 1 + 0.5 = 196.6$$

Therefore,

$$h_L = 196.6\left[\frac{(24.4)^2}{2g}\right]\frac{0.21}{144} = 2.65\ \text{psi}$$

Since this figure is within the 10% limit the assumption of incompressible flow is valid for this problem.

Before continuing, two cautions must be noted at this time. The tables and charts in this chapter for the calculation of friction and minor losses are based on nominal IPS (iron pipe size). In all the illustrative problems, we have used the actual inside diameter, thereby introducing a small error. In most instances the error introduced is well within the limits of accuracy of the data. For certain cases (particularly small pipe sizes) this may not be true, and the data given in Appendix B for pipe sizes should be used.

The second caution concerns the use of the Bernoulli equation. Because of the nonuniform velocity distribution across any section of a pipe, a correction should be made in the velocity head terms. Again, in most cases of concern to engineers this correction can be neglected. In laminar flow, however, the correction factor is 2. For turbulent flow it varies from 1.01 to about 1.10 and is usually neglected, except for very precise work.

6.8   NONCIRCULAR PIPE SECTIONS

The problem of determining the pressure drop in pipes having noncircular cross sections would at first appear to be quite difficult. However, if the shear stress at the wall of the noncircular pipe is equated to the shear stress at the wall of an equivalent circular pipe, we can obtain an "equivalent" diameter for the noncircular pipe. The equivalent diameter, D_{eq}, is given by the equation

$$D_{eq} = \frac{4A}{P}$$

(6.18)

where A is the cross-sectional flow area and P is the wetted perimeter (i.e., the perimeter in contact with the fluid). The calculation of pressure drop in a noncircular pipe involves the calculation of the equivalent diameter and its use to obtain the friction factor. Care should be exercised when using equation (6.18) for shapes that are widely different from being nearly circular. Also, equation (6.18) gives better results when applied to turbulent flow situations—but if used for laminar flow situations, large errors can be introduced.

ILLUSTRATION 6.18   EQUIVALENT DIAMETER

Three pipes, one in the shape of a square of side L, another an equilateral triangle of side L, and the third a rectangle whose width is half of its depth, are to be used in a flow system. Determine the equivalent diameter of each of these pipes.

ILLUSTRATIVE PROBLEM 6.18

Given: A square, an equilateral triangle, a rectangle with $w = \frac{1}{2}L$

Find: D_{eq} for each shape

Assumptions: Equation (6.18) applicable

Basic Equations: $D_{eq} = \dfrac{4A}{P}$

Solution:

1. *Square:*

$$A = L^2 \qquad P = 4L$$

$$D_{eq} = \frac{4A}{P} = \frac{4(L^2)}{4L} = L$$

The equivalent diameter of a square is simply the dimension of any side.

2. *Equilateral triangle:*

$$A = \frac{\sqrt{3}}{4}L^2 \qquad P = 3L$$

$$D_{eq} = \frac{4A}{P} = \frac{4\sqrt{3}L^2}{4(3L)} = \frac{L}{\sqrt{3}}$$

3. *Rectangle:*

$$A = \frac{L^2}{2} \qquad P = \frac{L}{2} + \frac{L}{2} + L + L = 3L$$

$$D_{eq} = \frac{4A}{P} = \frac{4(L^2/2)}{3L} = \frac{2}{3}L$$

6.9   INTERSECTING PIPES—PARALLEL FLOW

When pipes are connected in parallel, as shown in Figure 6.19, we can state that the sum of the flows in the three branches must equal the total flow in the main, and loss in pressure between sections ① and ② must be the same regardless of the path taken. It is interesting to note the analogy that this problem has with a direct-current network in which the circuit elements are resistors connected in parallel. We can compare the current in the electrical circuit with the quantity of fluid flowing in the pipes; the resistance (in ohms) of the electrical circuits has its counterpart in the flow resistance in the pipes, and the potential drop (voltage) in the electrical circuit is analogous to the pressure drop in the pipe. This type of electrical analogy has provided the basis for the solution of many fluid flow problems on analog computers.

If we consider any of the branches, such as branch A between points ① and ② in Figure 6.19, we can write the following:

$$Q_A = A_A V_A \tag{6.19}$$

and

$$V = K'_A \sqrt{h} \tag{6.20}$$

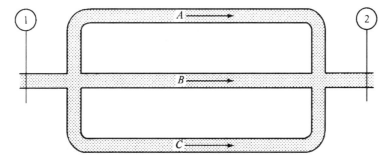

Figure 6.19   Pipes in parallel.

where h is the loss of head between sections ① and ②, K' is $\sqrt{2g/K_{eq}}$, and K_{eq} is the sum of all the loss coefficients in branch A. For each of the other branches, we can write similar relations. Thus we can write for the flow in the main, Q,

$$Q = Q_A + Q_B + Q_C$$

or

$$Q = (A_A K'_A + A_B K'_B + A_C K'_C)\sqrt{h} \tag{6.21}$$

Equation (6.21) can be used to obtain either the flow in the main or the flow in any of the branches. The following example will illustrate this application.

ILLUSTRATION 6.19   INTERSECTING PARALLEL PIPES

A steel horizontal branch piping system is as shown in Figure 6.20. Neglect minor losses and determine the pressure drop from ① to ② if water at 68°F is flowing into the system at section ① at the rate of 750 gal/min. All diameters are inside diameters.

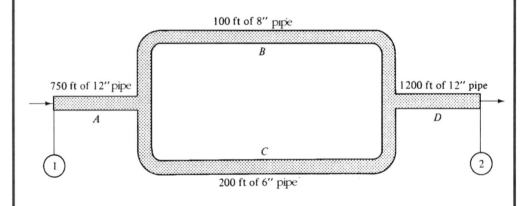

100 ft of 8″ pipe

B

750 ft of 12″ pipe

A

1200 ft of 12″ pipe

D

① ②

C

200 ft of 6″ pipe

Figure 6.20   Illustrative Problem 6.19.

ILLUSTRATIVE PROBLEM 6.19

Given: L_A = 750 ft, D_A = 12 in., L_D = 1200 ft, D_D = 12 in., L_B = 100 ft, D_B = 8 in., L_C = 200 ft, D_C = 6 in.

Find: $p_2 - p_1 = \Delta p$

Assumptions: Neglect minor losses, turbulent flow

Basic Equations: Darcy : $\Delta p_f = f \dfrac{L}{D} \dfrac{V^2}{2g} \dfrac{\gamma}{144}$

Continuity: $Q_A = Q_B + Q_C = Q_D$; also $Q = AV$

Solution: The pressure drop in branches A and D can be determined explicitly as follows. For branch A,

$$750 \text{ gal/min} \times \frac{231}{1728} = 100.5 \text{ ft}^3/\text{min}$$

$$\text{flow area} = \frac{\pi}{4} (1)^2 = 0.7854 \text{ ft}^2$$

$$\text{velocity} = \frac{Q}{A} = \frac{100.5}{60 \times 0.7854} = 2.13 \text{ ft/s}$$

$$(VD) = 2.13 \times 12 = 25.5$$

$$e = 0.00015 \qquad \frac{e}{D} = 0.00015$$

From Figure 6.8,

$$\text{Re} = 1.7 \times 10^5 \qquad f = 0.017$$

$$\Delta p_f = f \frac{L}{D} \frac{V^2}{2g} \frac{\gamma}{144} = 0.017 \times \frac{750}{1} \times \frac{2.13^2}{2g} \times \frac{62.4}{144} = 0.39 \text{ psi}$$

For branch D, all calculations are the same as for branch A with only the length changed. Therefore,

$$\Delta p_f = 0.39 \times \frac{1200}{750} = 0.62 \text{ psi}$$

Branches B and C cannot be solved explicitly, since the quantity flowing in each branch is unknown. However,

$$Q_B + Q_C = 750 \text{ gal/min}$$

Thus,

$$Q_B = 750 - Q_C \text{ gal/min}$$

$$Q_B = (750 - Q_C) \frac{231}{1728} \left(\frac{1}{60} \right) \text{ ft}^3/\text{s}$$

and

$$Q_B = 1.67 - 0.00223 Q_C \text{ ft}^3/\text{s}$$

But,

$$V_B = \frac{Q_B}{A_p} = \frac{Q_B}{(\pi/4)[(8/12)^2]} = 2.87 Q_B$$

Then

$$V_B = 4.79 - 0.0064Q_C$$

We also have the condition that

$$(\Delta p_f)_B = (\Delta p_f)_C$$

As a first approximation, let $f_B = f_C$. Then

$$\left(\frac{L}{D}\right)_B \frac{V_B^2}{2g} = \left(\frac{L}{D}\right)_C \frac{V_C^2}{2g}$$

Substituting for V_B and rearranging, we have the following quadratic equation for Q_C:

$$Q_C^2 + 204Q_C - 76{,}000 = 0$$

and

$$Q_C \approx 190 \text{ gal/min}$$

Therefore,

$$Q_B \approx 560 \text{ gal/min}$$

Now check branch C:

$$V = \frac{190 \times 231}{1728 \times 60} \frac{1}{0.196} = 2.16 \text{ ft/s}$$

$$(VD) = 6 \times 2.16 = 13$$

$$\frac{e}{D} = \frac{0.00015}{12} = 0.0003$$

$$\text{Re} = 9 \times 10^4$$

$$f = 0.02$$

$$(\Delta p_f)_C = 0.02 \times \frac{200}{1/2} \times \frac{(2.16)^2}{2g} = 0.58 \text{ ft}$$

Now check branch B:

$$V_B = \frac{560}{60} \times \frac{231}{1728} \frac{1}{(\pi/4)[(8/12)^2]} = 3.59 \text{ ft/s}$$

$$(VD) = 3.59 \times 8 = 28.7$$

$$\frac{e}{D} = \frac{0.00015}{8/12} = 0.000225$$

$$Re = 2 \times 10^5$$

$$f = 0.0178$$

$$(\Delta p_f)_B = 0.0178 \times \frac{(3.59)^2}{2g} \times \frac{100}{8/12} = 0.53 \text{ ft}$$

At this point we note that the two pressure drops are not equal (as they must be), and, if the desired accuracy warrants the effort, another iteration can be made using the results of the foregoing solution. The quantity of water flowing in branch C will be less than 190 gal/min, and in branch B it will be greater than 560 gal/min. The total pressure drop between ① and ② will be close to

$$0.39 + 0.62 + \frac{0.55(62.4)}{144} = 1.25 \text{ psi}$$

The preceding problem is relatively simple yet it can be seen that a considerable expenditure of time and effort is required before a solution can be obtained. For more complex problems the time required to obtain a solution in the manner indicated is prohibitive. Numerical procedures, such as the Hardy-Cross method, have been devised to decrease the effort in problems of this type, but these procedures can become quite lengthy, too. Fortunately, these problems can be programmed relatively easily for solution on both analog and digital computers, and at present almost all complex problems of this type are solved using computers. It is important that the reader understand the basis of this class of problems so that the necessary inputs for the computerized solution can be obtained.

6.10 REVIEW

In this chapter we have placed the emphasis on the practical aspects of calculating the pressure drop in pipes and ducts when the fluid flowing is incompressible. Starting with an examination of the character of the flow, we were able to state that the flow regimes could be determined by knowing the Reynolds number. In laminar flow we found that the pressure drop was independent of the roughness of the pipe. In turbulent flow the pressure drop is a complex function of the Reynolds number and the relative roughness. Based on an extensive analysis of existing data, Moody devised a logarithmic plot that enables us to determine the friction factor to use in the Darcy-Weisbach formula for evaluating the pressure drop in turbulent flow. Although our approach was based on both analytic and empirical data, we noted that even with the best available data, it is possible to find deviations of more than 30% between calculated and measured results for new or clean commercial pipes. For old or severely corroded pipes, it is almost impossible to obtain a satisfactory comparison between calculated and measured pressure drops.

Pressure changes occur at changes in section, valves, bends, branches, fittings, inlets, and outlets, and must be included with the friction pressure drop in any calculation of head requirements or pump size. For these losses, termed other or minor losses, the available data leave much to be desired. In the worst instance a variation of $+80$ to -200% can be found for a flanged check valve. It is evident that the designer of a piping system must exercise a great deal of judgment in evaluating the pressure drop and must allow for this inaccuracy when sizing pumps or establishing the estimate of the head on a given system. It is important to keep this in mind, to avoid blind dependence on calculated flow rates or pressure drops. The calculations need to be tempered with experience and judgment.

KEY TERMS

Terms of importance in this chapter:

Absolute roughness: arbitrary values of the roughness of pipes used by Moody in formulating his correlation of pressure drop in pipes.

Boundary layer: the region adjacent to the pipe wall in which a large pressure gradient occurs.

Critical velocity: the velocity at which laminar flow breaks up and starts to become turbulent in character.

Darcy-Weisbach equation: a general relation that yields the head loss as a function of L/D and the velocity head, $V^2/2g$.

Equivalent diameter: for noncircular sections, a diameter that can be used in the Darcy-Weisbach equation; equal to $4A/P$.

Equivalent length: a fictitious length that is assumed to be added to the pipe length to account for the minor losses.

Friction factor: a variable in the Darcy-Weisbach equation; that is, $h_f = f(L/D)(V^2/2g)$.

Hagen-Poiseuille equation: an analytic expression for the calculation of the pressure drop in laminar flow.

Laminar flow: flow characterized as consisting of concentric parallel cylindrical layers flowing past each other; also, flow in which the Reynolds number ≤ 2000.

Minor losses: losses due to entrance, valves, fittings, and so on—usually losses are a small percentage of the loss due to friction.

Moody diagram: a logarithmic diagram enabling one to obtain the friction factor in turbulent flow when the relative roughness and the Reynolds number are known—devised by L. F. Moody in 1944 based on a review of the existing literature.

Other losses: *see* minor losses.

Relative roughness: the ratio of the absolute roughness to the diameter of the pipe, e/D.

Reynolds number: the ratio of the inertia forces to viscous forces; also given as $DV\rho/\mu$.

Streamline flow: *see* laminar flow.

Transition region: the flow regime between laminar and turbulent flow; $2000 < Re < 4000$.

Turbulent flow: flow in which the motion of individual particles is random transverse to the main direction of flow, causing the particles to intermingle randomly; $Re \geq 4000$.

Viscous flow: *see* laminar flow.

KEY EQUATIONS:

Reynolds number	$\text{Re} = \dfrac{DV\rho}{\mu}$	(6.1)
Reynolds number	$\text{Re} = \dfrac{DV}{\nu}$	(6.2)
Pressure drop in laminar flow	$\Delta p = 128\mu\,\dfrac{LQ}{\pi D^{4}}$	(6.3)
Head loss in laminar flow	$h_f = \dfrac{32\mu L\bar{V}}{\gamma D^{2}}$	(6.5)
Darcy-Weisbach equation	$h_f = f\dfrac{L}{D}\dfrac{V^{2}}{2g}$	(6.7)
Friction factor—laminar flow	$f = \dfrac{64}{\text{Re}}$	(6.9)
Other losses	$h = K\dfrac{V^{2}}{2g}$	(6.14)
Equivalent length	$L_{eq} = \dfrac{KD}{f}$	(6.16)
Sudden enlargement	$h = \dfrac{(V_1 - V_2)^{2}}{2g}$	(6.17)
Equivalent diameter	$D_{eq} = \dfrac{4A}{P}$	(6.18)

QUESTIONS

1. Distinguish between laminar and turbulent flow.
2. If the Reynolds number is calculated using English units, what is the value of the conversion factor to obtain the Reynolds number in SI units?
3. What does the Reynolds number represent?
4. Why is the Reynolds number so important when we consider the steady flow of incompressible fluids in pipes?
5. Is it possible to have laminar flow if the Reynolds number exceeds 2000?
6. Can you give a logical explanation of why the roughness of the pipe does not enter into the determination of the friction factor in laminar flow?

7. Would you expect the Hagen-Poiseuille equation to give accurate results for corroded pipes?

8. For most cases would you expect that the friction factor will be greater or less in the laminar flow regime than in the turbulent flow regime?

9. What does "complete turbulence, rough pipes," shown on the Moody diagram, mean?

10. Is the Moody diagram applicable to old pipes?

11. The zone between Re 2000 and 4000 is shown on the Moody diagram as the "critical zone." What does this signify?

12. Is the Darcy-Weisbach equation applicable to both the laminar and turbulent regions?

13. Why does the roughness of the pipe enter into the determination of the friction factor in the turbulent region when it does not enter into the determination of the friction factor in the laminar region?

14. Explain why screwed fittings always have higher losses than flanged fittings.

15. Losses in fittings are not indicated to be a function of the velocity or of the Reynolds number. Do you think that this is correct? Why are the data shown as being only a function of the diameter?

16. There is a basic problem with the concept of equivalent length. Can you explain it?

17. Note that the resistance coefficients for bends are given as functions of both the Reynolds number and the relative roughness of the pipe. Do you feel that this is appropriate?

18. A rectangular duct is made 48 in. wide and 12 in. high. Do you think that the concept of the equivalent diameter applies reasonably to this duct? Explain your reasoning.

19. In what way do pipes in parallel resemble electrical circuits that are connected in parallel?

20. Based on your answer to Question 19, can you suggest a way in which an electrical circuit can be made analogous to an hydraulic circuit?

PROBLEMS

Use Tables 2.4 and 2.5 for the properties of water. Also use Figures 2.13, 2.15, and 6.10 as necessary for viscosity data and flow.

Reynolds Number

6.1. Water flows in a pipe whose inside diameter is 2 in. at an average velocity of 10 ft/s. Determine the Reynolds number if the temperature of the water is 100°F. Use Table 2.4.

6.2. Water at 20°C flows in a pipe whose inside diameter is 50 mm. If the average velocity is 5 m/s, what is the Reynolds number? Use Table 2.5.

6.3. Oil flows in a pipe that has an inside diameter of 1.5 in. If the oil has a specific weight of 45 lb/ft³ and a viscosity of 3 cP, determine its Reynolds number if the average velocity is 14 ft/s.

6.4. What is the least value of friction factor that you would expect in laminar flow?

6.5. Oil flows in an 8-in. diameter pipe at a volume flow of 22 ft³/s. If the oil has a specific gravity of 0.85 and a viscosity of 325×10^{-5} lb·s/ft², determine whether the flow is laminar or turbulent.

6.6. Oil has a kinematic viscosity of 1.5×10^{-4} ft²/s. Determine the Reynolds number if the pipe has a 2 in. diameter and the flow is 80 gallons per minute.

6.7. Determine whether the flow of gasoline is at 10°C ($\mu = 7 \times 10^{-6}$ lb·s/ft²) at the rate of 1.5 liters per second in a 3 in. diameter pipe will be laminar or turbulent if the specific gravity of the gasoline is 0.68.

6.8. Oil flows through a tube having a $\frac{1}{4}$-in. inside diameter. It is desired that the flow shall always be laminar. If the viscosity of the oil is 1 P and its specific weight is 52 lb/ft³, determine the maximum velocity in the tube. Also determine the average velocity.

6.9. Water flows in a 5-in. diameter pipe in laminar flow. If the Reynolds number is 1200 and the average velocity is 0.1 ft/s, what is the pressure drop per 100 ft of pipe?

6.10. Show that for a given volumetric flow the pressure drop in a pipe is inversely proportional to the fourth power of the diameter in laminar flow.

6.11. Crude oil having a specific gravity of 0.86 at 100°F flows through a 2-in. diameter pipe at the rate of 5 gallons per minute. Is the flow laminar or turbulent?

6.12. Water at 10°C flows in a 100 mm diameter pipe at a velocity of 1 m/s. Is the flow laminar or turbulent?

6.13. SAE 10 oil at 100°F has a kinematic viscosity of 0.0004 ft²/s. If this oil flows in a pipeline that has a 6 in. diameter with an average velocity of 1 ft/s, determine whether the flow is laminar or turbulent and the maximum velocity in the pipe.

6.14. The Reynolds number for an oil flowing past section ① in Figure P6.14 is 800. Calculate the Reynolds number at section ② of the pipe.

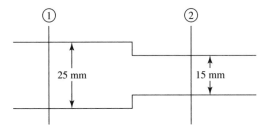

Figure P6.14

6.15. Calculate the Reynolds number for water at 70°F flowing in a 2-in. diameter tube at the rate of 300 liters per minute.

6.16. Kerosene at 100°F flows in a 3-in. diameter pipe at the rate of 10 liters per minute. Will the flow be laminar or turbulent?

6.17. Crude oil at 150°F and having a specific gravity of 0.86 flows in a 4-in. diameter pipe at the rate of 200 gallons per minute. Calculate the Reynolds number.

6.18. Water at 70°C flows through a heating coil at the rate of 2 gallons per minute. If the coil is made of 1-in. diameter pipe, determine the Reynolds number.

6.19. A fluid flows in a horizontal pipe at the rate of $\frac{1}{4}$ ft^3/s, and the friction head loss is measured to be 25 ft. If the pipe is 1000 ft long and has an inside diameter of 2 in., determine whether the flow is laminar or turbulent.

6.20. Determine the pressure drop in a 2-in. inside diameter pipe that is 1000 ft long, in which oil flows at an average velocity of 3 ft/s, if the specific weight of the oil is 51 lb/ft^3 and its viscosity is 56 cP.

6.21. Oil flows in a 3-in. inside diameter steel pipe. If the absolute viscosity of the oil is 0.92 cP and its specific weight is 50 lb/ft^3, determine the flow rate, the Reynolds number, and the pressure drop per foot length if the maximum velocity is 4 ft/s. Note that for laminar flow the average velocity is one-half of the maximum velocity.

6.22. If the oil in Problem 6.11 is at 50°F, is the flow laminar or turbulent?

6.23. A fluid has a kinematic viscosity of 1.5 centistokes. If this fluid flows at the rate of 150 liters per minute through a 1 1/2-in. diameter pipe, determine if the flow is laminar or turbulent.

6.24. Oil has a kinematic viscosity of 1.9 centistokes. A flow of 50 gallons per minute is maintained in a pipe having a diameter of 25 mm. Calculate the Reynolds number.

6.25. Glycerin at 100°F has a viscosity of 4×10^{-3} lb·s/ft^2 and a specific gravity of 1.26. Determine if the flow of 50 gallons per minute of glycerin in a 10-cm pipe is laminar or turbulent.

6.26. A volumetric flow of water of 8×10^{-3} liters per second flows in a 6-mm tube at 10°C. Is this flow laminar or turbulent?

6.27. Gasoline at 50°F flows in a pipe at the rate of 0.5 liters per second. Determine whether the flow is laminar or turbulent if the pipe has a diameter of 40 mm.

6.28. Crude oil at 100°F (sg = 0.56) flows in a tube having a diameter of 1-in. What is the maximum possible average velocity in the tube if the flow is laminar?

6.29. Water at 50°F flows in a 25-mm tube. If the flow is laminar, what is the maximum volumetric flow rate in gallons per minute?

6.30. An oil has a kinematic viscosity of 20 centistokes. Determine the Reynolds number if the pipe diameter is 1 in. and the flow rate is 50 gallons per minute.

6.31. An oil has a specific gravity of 0.9 and a kinematic viscosity of 0.0003 m^2/s, and flows in a 10 mm diameter tube. Determine the Reynolds number if the flow rate in the tube is 0.01 m^3/s.

6.32. An oil flows in a 2-in. diameter pipe with a Reynolds number of 1000. Calculate the average and maximum velocity in the pipe if the oil has a specific gravity of 0.9 and a kinematic viscosity of 0.0001 ft^2/s.

6.33. Determine the velocity at a distance of 1/8 in. from the wall of the pipe in Problem 6.32.

6.34. Oil at 60°F has a specific gravity of 0.92 and a kinematic viscosity of 0.002 ft^2/s. Calculate the Reynolds number if the oil flows in an 8-in. diameter pipe with a flow rate of 2000 ft^3/h.

6.35. A fluid flows in a round pipe with a Reynolds number less than 2000. The velocity

   a) is maximum at the wall.
   b) is constant over the cross section.
   c) is zero at the wall and varies parabolically across the section.
   d) is none of the above.

6.36. A fluid flows through a round pipe with a Reynolds number less than 2000. The discharge varies

   a) directly as the square of the diameter.
   b) linearly with the viscosity.
   c) directly with the pressure drop.
   d) none of the above.
   e) all of the above.

6.37. Air having a kinematic viscosity of 15×10^{-5} ft^2/s flows through a pipe at 3 ft/s. What is the maximum diameter for laminar flow?

6.38. Determine the average velocity and the maximum velocity when a fluid flows in a 2-in. diameter pipe at the maximum Reynolds number for laminar flow. Assume a kinematic viscosity of 5×10^{-5} ft^2/s.

6.39. An oil pipeline has a 12 in. diameter, and the oil in it flows at an average velocity of 4.0 ft/s. If the kinematic viscosity of the oil is 2.2×10^{-4} ft^2/s, determine whether the flow is laminar or turbulent.

6.40. Calculate the pipe size required to deliver 100 gpm of a fuel oil having a kinematic viscosity of 6×10^{-5} ft^2/s. Assume that the flow is laminar and the Reynolds number is 2000.

6.41. Calculate the Reynolds number if 0.05 m^3/s of oil having a specific gravity of 0.8 and a viscosity of 0.03 Pa·s flows in a 6-in. diameter pipe.

6.42. An oil has a kinematic viscosity of 0.0002 ft^2/s. This oil flows in a 5-in. diameter pipe. Calculate the maximum average velocity for the flow to be laminar.

6.43. Determine whether the flow of an oil having a kinematic viscosity of 0.004 ft^2/s and flowing in a 4-in. diameter pipe with a velocity of 8 ft/s is either laminar or turbulent.

6.44. An oil has a specific gravity of 0.8 and a kinematic viscosity of 2×10^{-5} m^2/s. Is the flow either laminar or turbulent when it flows in a 5-in. diameter pipe at the rate of 59 gpm?

Pipe Flow

6.45. Water at 20°C flows in a commercial steel pipe that is 25 cm in diameter and 5000 m long. If the average velocity is 4 m/s, determine the head loss due to friction.

6.46. A commercial steel pipe whose inside diameter is 6 in. is used to deliver water at the rate of 700 gal/min to a chemical plant. Determine the friction loss per mile of pipe. Assume water at 68°F.

6.47. An oil at 68°F is pumped through a commercial steel pipeline that has a 15 cm diameter and is 10,000 m long. If the flow rate is 80 liters per second, determine the horsepower required to overcome friction. The oil has a specific gravity of 0.9 and a kinematic viscosity of 1.6×10^{-5} m²/s.

6.48. Oil having a viscosity of 0.5 cP and a specific weight of 52 lb/ft³ flows upward in a 1-in. inside diameter pipe with a velocity of 0.1 ft/s. Determine the difference in pressure between two static taps located 100 ft apart.

6.49. If water at 68°F flows in a 2-in. inside diameter pipe that is 20 ft long, determine the pressure drop if the flow is 250 gal/min. Neglect minor losses.

6.50. Calculate the head loss if 100 liters per second of a fluid flows through a 5-in. diameter cast iron pipe that is 400 feet long. The fluid has a specific gravity of 1.25 and a viscosity of 0.86 Pa·s.

6.51. Water at 50°F flows through a 5-in. diameter commercial steel pipe at the average velocity of 0.2 m/s. The head loss in 100 m of pipe is found to be 0.5 m. Determine the friction factor for this flow.

*6.52. Determine the flow in m³/s for the discharge of water at 20°C through a cast iron pipe that is 20 cm in diameter and 200 m long if the head loss due to friction is 6 m.

6.53. Calculate the horsepower required to pump water at 60°F through a 6-in. diameter commercial steel pipe at the rate of 7 ft³/s if the pipe is 1 mile long.

*6.54. Water at 70°F is being pumped through a 6-in. diameter cast iron pipe. If a head loss of 20 ft was observed in a 500 ft length of pipe, determine the discharge in ft³/s through the pipe.

6.55. Gasoline at 70°F (sg = 0.68) is pumped at the rate of 250 liters/s through a 20-cm diameter cast iron pipe that is 10 km long. Determine the theoretical power required to pump the gasoline.

6.56. Oil having a specific gravity of 0.86 and a kinematic viscosity of 0.00003 ft²/s flows at the rate of 2 ft³/s through a horizontal 8-in. diameter cast iron pipe that is 2000 ft long. Calculate the head loss due to friction.

6.57. Water at 60°F flows through a 4-in. diameter wrought iron pipe with a velocity of 10 m/s. Compute the head loss due to friction per mile of pipe.

6.58. Calculate the pressure drop in 400 ft of 6-in. diameter commercial steel pipe that is carrying water at 70°F with an average velocity of 6 ft/s.

*6.59. A test is conducted to determine the wall roughness of an older pipe installation. Water at 68°F is pumped through the pipe at the rate of 50 gallons per minute. The pipe has a 1 in. diameter. Pressure gages located 100 ft apart horizontally read 170 psi and 100 psi respectively. Calculate the roughness of the pipe wall.

6.60. Air flows at the rate of 900 ft³/min in a 6-in. diameter commercial steel pipe. Determine the pressure drop per 100 ft of pipe if the air has a density of 0.002

slugs/ft^3 and the density remains essentially constant. The viscosity of the air can be taken to be 3.7×10^{-7} lb · s/ft^2.

6.61. Oil having a specific gravity of 0.9 and a kinematic viscosity of 1×10^{-4} ft^2/s flows at the rate of 5 ft^3/s through 2000 ft of 8 in. cast iron pipe. Determine the pressure difference between the ends of the pipe if the pipe slopes down from the horizontal at an angle of 8° in the direction of flow.

6.62. A fluid has a specific gravity of 0.8 and a kinematic viscosity of 2.2×10^{-4} ft^2/s. If it flows in a smooth brass pipe that has a 3 in. diameter and is 1000 ft long with a flow rate of 10 gallons/minute, determine the frictional head loss per foot of pipe.

6.63. An oil has a specific gravity of 0.85 and is pumped through a horizontal 3-in. diameter pipe that is 0.2 miles long at the rate of 0.05 ft^3/s. Determine the viscosity of the oil if the frictional pressure drop is 20 psi and the flow is laminar.

*6.64. A fuel oil flows in a horizontal commercial steel pipeline that is 5/8 of a mile long. The diameter of the pipeline is 8 in. and the pressure drop is 125 psi due to friction. Calculate the flow rate if the specific gravity of the oil is 0.9 and the kinematic viscosity is 0.0004 ft^2/s.

*6.65. Prove that for a constant value of the friction factor f, the frictional head loss in a pipe is proportional to Q^2/D^5.

*6.66. Two pipes have the same value of f and connect two reservoirs whose surfaces are at different elevations. Using the results of problem 6.65, determine the ratio of the flow rates in these pipes if one has a 4 in. diameter and the other has a 6 in. diameter.

6.67. Calculate the pressure drop per foot in a 2-in. diameter smooth pipe when oil having a specific gravity of 0.86 and a viscosity of 2×10^{-4} lb · s/ft^2 flows at the rate of 0.2 ft^3/s.

6.68. Determine h required to deliver 0.01 ft^3/s through the 1-in. diameter commercial steel pipe shown in Figure P6.68. Neglect minor losses.

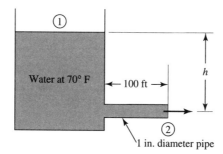

Water at 70° F   ← 100 ft → h

①

② 1 in. diameter pipe   **Figure P6.68**

*6.69. Oil is being pumped from a truck to a tank 10 ft higher than the truck through a 2-in. inside diameter steel pipe 100 ft long (Figure P6.69 on page 258). Assume that the pump outlet pressure is 15 psig, and determine the flow rate in gallons per minute. The oil has an absolute viscosity of 105 cP and a specific weight of 56 lb/ft^3.

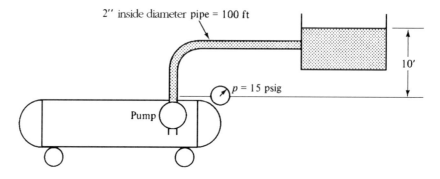

Figure P6.69

Noncircular Pipe Sections

6.70. A rectangular sheet metal duct is 7 in. × 16 in. Determine its equivalent diameter.

6.71. Calculate the equivalent diameter of the shape shown in Figure P6.71 if a fluid flows in the shaded portion.

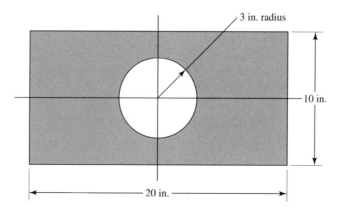

Figure P6.71

6.72. Calculate the equivalent diameter and Reynolds number if water at 70°F flows in the shaded portion of the unit shown in Figure P6.72 at the rate of 15 liters per second.

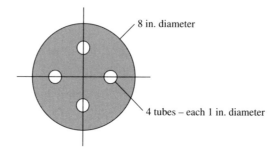

Figure P6.72

6.73. Calculate the equivalent diameter and Reynolds number if water at 10°C flows in the shaded portion of the unit shown in Figure P6.73 at the rate of 200 gpm.

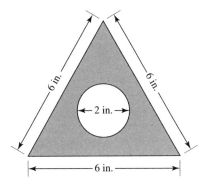

6 in.   6 in.

2 in.

6 in.

Figure P6.73

6.74. A duct has a 10 in. diameter. Another duct is made in the shape of a square with sides equal to 10 in. Compare the equivalent diameter of these two ducts.

6.75. A circular duct and a square duct each have the same cross-sectional area. Which duct has the larger equivalent diameter?

6.76. A square duct and an equilateral triangular duct have the same cross-sectional area. Which duct has the larger equivalent diameter?

6.77. A heat exchanger has a 60-in. diameter shell and contains 100 1-in. diameter tubes. Determine the equivalent diameter for fluid flowing along the length of the tubes.

6.78. Determine the equivalent diameter of the duct shown in Figure P6.78.

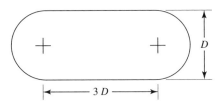

$+$ $+$ D

$3\,D$

Figure P6.78

6.79. If oil has a specific gravity of 0.86 and a viscosity of 2.03×10^{-4} lb·s/ft^2 and flows in the shaded portion shown in Figure P6.79, calculate the Reynolds number for a flow of 100 gpm.

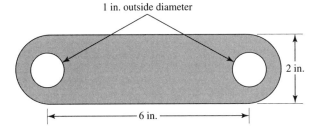

1 in. outside diameter

2 in.

6 in.

Figure P6.79

6.80. Determine the equivalent diameter of the duct shown in Figure P6.80. Assume the flow is in the shaded area.

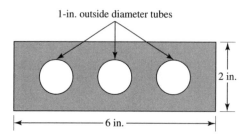

1-in. outside diameter tubes

2 in.

6 in.

Figure P6.80

6.81. Oil flows in a square pipe that is 100 mm on a side. If the Reynolds number is 1000 and the velocity of the oil is 0.5 m/s, determine the pressure drop per meter of pipe.

*6.82.** Assuming that the concept of an equivalent diameter can be applied to an annulus between two concentric tubes having diameters D_2 and D_1, determine the equivalent diameter of such an annulus in terms of D_2 and D_1, assuming that the fluid wets both surfaces.

6.83. Based upon the results of Problem 6.82 ($D_{eq} = D_2 - D_1$), calculate the pressure drop per foot of an annulus having D_2 equal to 1 in. and D_1 equal to $\frac{3}{4}$in. if the fluid flowing is an oil with a viscosity of 90 cP and a specific weight of 50 lb/ft³. The average flow velocity is 10 ft/s.

*6.84.** Show that the equivalent diameter is $2h$ when a fluid flows between two parallel plates that are h apart.

6.85. Using the results of Problem 6.84, calculate the head loss per foot when water at 70°F flows with an average velocity of 5 ft/s between two parallel smooth plates that are 2 in. apart.

6.86. A wind tunnel is made of smooth sheet metal and has a cross section that is 20 in. × 40 in. If the tunnel is 200 ft long, calculate the head loss if the air flowing in it has a kinematic viscosity of 1.52×10^{-4} ft²/s and a density of 0.00242 slugs/ft³. The velocity of the air is 50 ft/s.

6.87. Calculate the power required by an ideal fan for Problem 6.86.

6.88. Air has a density of 0.0024 slugs/ft³ and a kinematic viscosity of 0.00016 ft²/s. The air flows in a smooth horizontal sheet metal duct that is 14 in. × 10 in. and 80 ft long. If the flow rate is 1800 ft³/min, estimate the pressure drop in psf.

6.89. Determine the horsepower required for a perfect blower to sustain the flow in Problem 6.88.

6.90. Water at 68°F flows in a vertical square steel pipe whose sides are 4 in. (Figure P6.90). Determine the difference in pressure between two static pressure taps located 100 ft apart if the average water velocity is 15 ft/s and the flow is up.

6.91. The flow in Problem 6.90 is reversed. Determine the pressure difference between the taps.

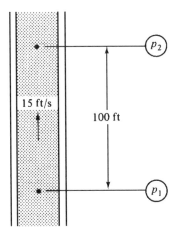

Figure P6.90

Pipe Flow with Minor Losses

6.92. A company wishes to pump crude oil from a storage tank to its plant 35 mi away (Figure P6.92). Assume that an 18-in. inside diameter steel pipeline is to be used and that the plant is at an elevation of 50 ft above the tank. The oil has a viscosity of 1.50 cP and a specific weight of 52 lb/ft³. Determine the required pressure at the pump outlet if the average velocity in the pipe is to be limited to 1.5 ft/s. Include entrance and exit losses.

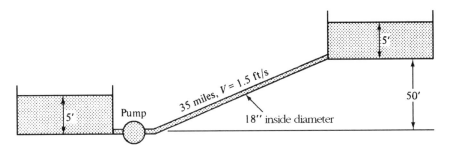

Figure P6.92

6.93. In Problem 6.46, determine the friction loss per mile if there are two regular flanged elbows, three wide-open flanged gate valves, a flanged check valve, and four 90° flanged long-radius elbows in each mile of pipe. Include entrance and exit losses.

6.94. An 8-in. inside diameter pipe is 2500 ft long and carries water from a pump to a reservoir whose water surface is 200 ft above the pump. If 4 ft³/s is being pumped, determine the gage pressure at the discharge of the pump. Assume that the pipe is rough cast iron and neglect minor losses. If entrance and exit losses are included, determine the gage pressure at the discharge of the pump.

6.95. Water at 68°F flows in a horizontal steel pipe at the rate of 15 gal/min. If the pipe has an inside diameter of 3 in., is 500 ft long, has three 90° regular elbows, four fully open gate valves, three regular 180° radius bends, and a check valve, determine the pressure drop in the line. All fittings are flanged. Include entrance and exit losses.

6.96. A diffuser is placed in a pipeline to connect a 6-in. inside diameter pipe to a 12-in. inside diameter pipe. If the total included angle of the diffuser is 30° and the flow is from the 6 in. to the 12 in. pipe at 10 ft/s, determine the pressure drop in the diffuser.

6.97. If the diffuser included angle is 7° in Problem 6.96, determine the pressure drop in the diffuser.

6.98. Air at 50 psia and 100°F flows in a horizontal pipe that has a 6-in. inside diameter. It is desired to limit the pressure drop due to the pressure losses to 10% of the initial absolute air pressure. What is the maximum velocity in the pipe if minor losses are neglected and the pipe is 500 ft long? (Use $\bar{\gamma} = 0.229$ lb/ft³.)

6.99. An abrupt contraction is placed in a horizontal pipeline that has an inside diameter of 4 in. If the contracted flow diameter is 2 in., determine the pressure loss for 68°F water flowing at 10 ft/s in the 4 in. pipe.

6.100. In Problem 6.99 the flow is reversed. Determine the pressure drop if the velocity in the 4 in. line is kept at 10 ft/s.

6.101. Water at 20°C flows in an abrupt contraction in a pipe whose diameter goes from 100 mm to 50 mm. If the velocity before the contraction is 3 m/s, what is the pressure loss?

6.102. If the flow is reversed in Problem 6.101, what will the pressure drop be? Both velocities are to be taken to remain the same.

6.103. Water is supplied from a reservoir to a 500-ft-long horizontal round concrete pipe 36 in. in diameter. What head is required to cause a flow of 20,000 gal/min? Neglect entrance and exit losses.

6.104. Compare the results of Problem 6.103 with the results obtained if the entrance and exit losses had not been neglected. What do you conclude from the comparison?

*6.105. Water at 68°F flows in a 2-in. inside diameter steel pipe at the rate of 250 gal/min. If the pipe has a globe valve and a check valve and it delivers water to a tank that is 60 ft above the pipe inlet, determine the pressure difference between the pipe inlet and outlet. All fittings are screwed; neglect inlet and outlet losses. The total pipe length is 250 ft.

6.106. A 4 in. diameter commercial steel pipeline is 380 ft long. Assume that K for the bends, elbows, entrance, and exit equals 5.4 and that a fully open flanged globe valve is in the system. If the fluid is water at 60°F and the head available for flow is 50 ft, determine the flow rate.

6.107. A 48 in. diameter concrete pipe is one mile long and connects two reservoirs. If the water is at 10°C and flows at the rate of 50 ft³/s, determine the difference in elevation between the two reservoirs. Take into account the entrance and exit loss, and the loss due to 3 90° regular flanged elbows and one fully open flanged gate valve.

6.108. A reservoir containing water at 60°F is to be drained to a lower one. Assuming 6-in. diameter commercial steel pipe is to be used and that the difference in level between the surfaces of the reservoirs is 50 ft, determine the flow rate. The connecting pipe is 200 ft long and has 3 90° screwed regular elbows, a screwed gate valve fully open, and a screw type swing check valve that is fully open.

6.109. A horizontal commercial steel pipe discharges water to the atmosphere from an open reservoir as shown in Figure P6.109. If the inside diameter of the pipe is 6 in., and the pipe is 400 ft long, calculate the flow rate. Assume that the outlet of the pipe is located 30 ft below the water level in the top of the reservoir.

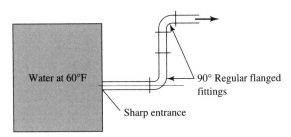

Plan view   **Figure P6.109**

6.110. Determine the pressure required to make 1.0 ft³/s of 60°F water flow through 1000 ft of a 4-in. diameter commercial steel pipe as shown in Figure P6.110. Assume that the sum of all the minor losses equals 5 velocity heads.

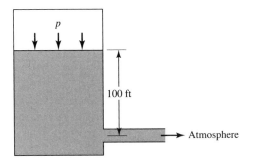

Figure P6.110

DYNAMIC FORCES

LEARNING GOALS

After reading and studying this chapter you should be able to:

1. Define *impulse* and *momentum*.
2. Apply the concepts of impulse and momentum to a control volume.
3. Calculate the forces (and their components) that are required to change the direction of a stream.
4. Calculate the forces that are exerted on fixed vanes that are used as deflectors of streams.
5. Calculate the forces that a stream exerts on a moving vane.
6. Apply the concepts of impulse and momentum to a fan or propeller.
7. Calculate the efficiency of fans or propellers using the impulse–momentum concept.

7.1 INTRODUCTION

Human beings have been familiar since ancient times with the tremendous forces that fluid streams can exert. The awesome power of the wind in a hurricane, the destruction caused by tidal waves, and the cresting and flooding of rivers have all been described in great detail. It is also known that these forces have been utilized to humanity's benefit. Windmills, water wheels, jets for excavation, and so on, are just some of the recorded efforts made to utilize the dynamic forces exerted by fluid streams. In this chapter our attention is directed primarily to the computation of these forces using both the energy (Bernoulli) equation and Newton's second law in its appropriate form.

7.2   FORCE, IMPULSE, AND MOMENTUM

Newton's second law of motion states that for a constant mass (m) the resultant force acting on a system is simply the product of its mass and its acceleration. Mathematically,

$$F = ma \tag{7.1}$$

If we now recall that acceleration (a) is the rate at which velocity changes with respect to time,

$$a = \frac{\Delta V}{\Delta t} \tag{7.2}$$

where $\Delta V = V_2 - V_1$ is the change in velocity, and Δt is the change in time during which the velocity was changing. Using equation (7.2) in conjunction with equation (7.1) gives us

$$F = \frac{m\Delta V}{\Delta t} \tag{7.3}$$

Equation (7.3) can be used in its present form or recast into two other useful forms. The first is

$$F\Delta t = m\Delta V = m(V_2 - V_1) \tag{7.3a}$$

The left side of equation (7.3a) is called the *impulse* imparted to the body and the right side of this equation is termed the change of *momentum* of the body.

  The other form of equation (7.3) that will be useful to us is

$$F = \frac{m}{\Delta t}\Delta V = \dot{m}\Delta V = \dot{m}(V_2 - V_1) \tag{7.3b}$$

where we have used the symbol $\dot{m}$ for the term $m/\Delta t$, which is the time rate of flow. In SI units $\dot{m}$ is given in kg/s and in English units $\dot{m}$ is given in slugs/s. Equation (4.1c) permits us to express $\dot{m}$ in terms of fluid properties,

$$\dot{m} = \rho AV \tag{4.1c}$$

Therefore, we can write equation (7.3b) as

$$F = \rho AV(V_2 - V_1) \tag{7.4}$$

  But $AV = Q$; therefore,

$$F = \rho Q(V_2 - V_1) = \frac{\gamma}{g}Q(V_2 - V_1) \tag{7.5}$$

The units of Q are m^3/s or ft^3/s, ρ is in kg/m^3 or slugs/ft^3, V is in m/s or ft/s, and γ is in N/m^3 or lb/ft^3.

It must be noted at this time that *force, impulse,* and *momentum* are vectors. Therefore, it is necessary to write equations (7.3), (7.4), and (7.5) in each of the three coordinates, x, y, and z. If the force in the x direction is desired, the x velocities are used. Similarly, for the y and z directions, the appropriate velocities must be used.

ILLUSTRATION 7.1   FORCE, IMPULSE, AND MOMENTUM

In a turbojet engine, air enters and is compressed, fuel is added and burned, energy is removed by a turbine, and finally the combustion gases are exhausted. For such an engine let V_1 be the inlet air velocity relative to the engine, $\dot{m}_a$ the mass rate of airflow, $\dot{m}_f$ the mass rate of fuel flow, and V_2 the exhaust velocity relative to the engine. Assume that the inlet pressure, atmospheric pressure, and the exhaust pressure are all equal and that the inlet and exit areas are equal and determine the thrust of the engine. Figure 7.1 shows an aircraft turbine.

Figure 7.1   Aircraft gas turbine. (Courtesy of Pratt & Whitney of United Technologies Corp.)

ILLUSTRATIVE PROBLEM 7.1

Given: $V_1, \dot{m}_a, \dot{m}_f$, and V_2

Find: Engine thrust

Assumptions: One-dimensional flow, $p_{inlet} = p_{outlet} = p_a$; also, $A_1 = A_2$

Basic Equations: Equation (7.3b): $F = \dot{m}(V_2 - V_1)$

Solution: Consider the control volume shown in Figure 7.2. Since V_1 and V_2 are specified relative to the engine, we may consider the engine fixed and the inlet and exhaust stream velocities as absolute. The momentum in is $\dot{m}_a(V_1)$ and the momentum out is

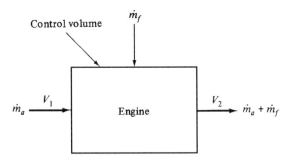

Figure 7.2   Illustrative Problem 7.1.

$(\dot{m}_a + \dot{m}_f)V_2$. Since the pressures and areas are assumed to be equal, they will cancel out on both sides of the control volume and we do not have to consider them. Therefore,

$$F_x = (\dot{m}_a + \dot{m}_f)V_2 - \dot{m}_a(V_1)$$

In words, the thrust in the direction of the gas flow is simply the rate of change of momentum into and out of the control volume.

ILLUSTRATION 7.2   PIPE FLOW

A 75-mm inside diameter pipe carries water at the rate of 0.2 m³/s. If there is a 180° bend in the horizontal plane, determine the thrust exerted by the water on the pipe. Use $\rho = 1000$ kg/m³. Neglect any pressure drop in the bend.

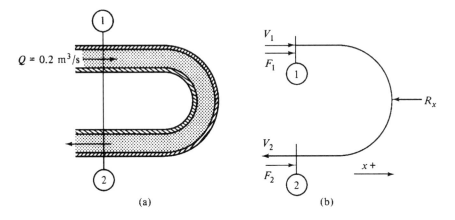

(a)   (b)

Figure 7.3   Illustrative Problem 7.2.

ILLUSTRATIVE PROBLEM 7.2

Given: Horizontal, 180° pipe bend; $D = 75$ mm, $Q = 0.2$ m³/s, $\rho = 1000$ kg/m³

Find: Thrust on bend

Assumptions: $\Delta p = 0$, steady flow, pipe bend in one plane

Basic Equations: Continuity: $Q = AV$

$$\text{Impulse–momentum: } F = \rho Q(V_2 - V_1)$$

Solution: Figure 7.3a shows a sketch of the bend, and Figure 7.3b shows a free-body diagram of the bend. Consider x to be positive to the right. Since Q is specified as 0.2 m³/s,

$$V = \frac{Q}{A} = \frac{0.2 \text{ m}^3/\text{s}}{[\pi(0.075)^2/4] \text{ m}^2} = 45.3 \text{ m/s}$$

The reaction R_x due to the dynamic motion of the water is given by equation (7.5) as

$$R_x = \rho Q(V_2 - V_1)$$

Since $V_1 = 45.3$ m/s, V_2 must equal -45.3 m/s (since the stream leaves the 180° bend facing in the negative direction). It is most important to keep in mind at all times that *velocity is a vector!* Therefore,

$$R_x = 1000 \text{ kg/m}^3 \times 0.2 \text{ m}^3/\text{s} \times (-45.3 - 45.3) \text{ m/s}$$

$$R_x = -18\ 120 \text{ kg·m/s}^2 = -18\ 120 \text{ N}$$

The negative sign indicates that the reaction force is to the left on the water. Note also that this solution does not take into account pressure differences between the inlet and the outlet of the bend. The 18 120-N force is the force required to alter the direction of flow of the water, and is sometimes called *dynamic thrust.*

CALCULUS ENRICHMENT

An interesting application of both the momentum and continuity equations is the determination of the speed of a small disturbance in a channel. Consider the channel shown in Figure A.

	p	
V		$V + dV$
p		$p + dp$
ρ		$\rho + d\rho$
A		A

Figure A   Velocity of a small disturbance.

The change at section p gives rise to a change in velocity, pressure, and density. Writing the continuity equation for mass conservation,

$$\rho VA = (\rho + d\rho)(V + dV)A \qquad \text{(a)}$$

Simplifying,

$$\rho V + V d\rho = 0 \qquad \text{(b)}$$

Applying the momentum equation yields

$$pA - (p + dp)A = \rho VA(V + dV - V) \tag{c}$$

or

$$dp = -\rho V dV \tag{d}$$

Eliminating ρdV between equations (b) and (d),

$$V^2 = \frac{dp}{d\rho} \tag{e}$$

From Equation (e) we conclude that a small disturbance can only occur when a velocity equal to $\sqrt{dp/d\rho}$ exists in the channel. If we assume that the velocity in the channel to the left of section P is $\sqrt{dp/d\rho}$, the analysis is as above and the disturbance is propagated through the fluid at rest. The velocity given by Equation (e) is called the speed of sound in the fluid.

7.3   DEFLECTION OF STREAMS BY STATIONARY BODIES

As was seen in Illustrative Problem 7.2, the alteration of the direction of flow of a fluid or of its velocity means that a net force must have been exerted on the fluid stream. Concurrently, an equal and opposite force must have been exerted by the fluid stream on the body, causing the velocity change. If the deflector is fixed, it is necessary for the structure to be capable of resisting the resultant force acting on it, while the same force exerted on a moving deflector (vane) is capable of doing useful work. The ability of a fluid stream to do work on a moving vane is the basis upon which the theory of turbo-machinery is based.

Let us consider the situation of a fixed vane, shown in Figure 7.4, and apply the impulse–momentum relations developed earlier. The fluid stream of cross-sectional area A_0 with an initial velocity V_0 is deflected through the angle θ. Assume the frictional resistance of the stream on the vane to be negligible and the velocity of the stream to be uniform throughout. For a free jet the pressure at both the inlet to the vane and the outlet of the vane must be equal, since they must equal the surrounding (atmospheric) pressure. Therefore, the speed (not velocity) of the jet must be the same at inlet as at outlet; the vane only changes the direction of the stream. The force exerted by the vane on the stream in the horizontal direction is

$$F_x = \dot{m}(V_x - V_0) = \dot{m}(V_0 \cos\theta - V_0) \tag{7.6}$$

and

$$F_x = -\rho AV_0^2(1 - \cos\theta) = -\rho QV_0(1 - \cos\theta) \tag{7.7}$$

In the vertical direction,

$$F_y = \dot{m}(V_y - 0) = \dot{m}(V_0 \sin\theta) \tag{7.8}$$

and

$$F_y = \rho AV_0^2 \sin\theta = \rho QV_0 \sin\theta \tag{7.9}$$

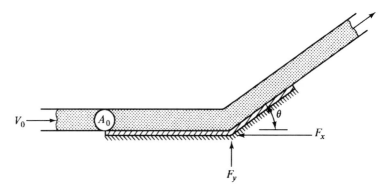

Figure 7.4   Fixed deflector vane.

The resultant force R is given by

$$R = \sqrt{F_x^2 + F_y^2} = \rho A V_0^2 \sqrt{2(1 - \cos \theta)} \tag{7.10}$$

or

$$R = \rho Q V_0 \sqrt{2(1 - \cos \theta)} \tag{7.10a}$$

The derivation of equations (7.6) through (7.10) has assumed that the difference in elevation between the ends of the vane is negligible. Note, too, that the forces given by these equations are the forces exerted on the fluid stream. The forces on the fixed vane will be equal and opposite to these forces.

ILLUSTRATION 7.3   IMPULSE–MOMENTUM

Water is deflected by a 90° bend that is in a horizontal plane (both ends at the same elevation); see Figure 7.5. If the stream is 3 in. in diameter, γ is 62.4 lb/ft³, and the stream velocity is 4 ft/s, determine the forces exerted on the stream by the vane due to the dynamic action of the stream.

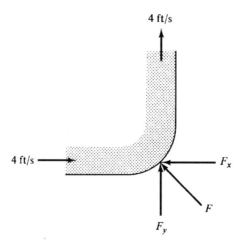

Figure 7.5   Illustrative Problem 7.3.

ILLUSTRATIVE PROBLEM 7.3.

Given: 90° bend, 3-in. diameter water stream, $\gamma = 62.4$ lb/ft³, $V = 4$ ft/s

Find: Forces on stream

Assumptions: Steady, incompressible flow; horizontal plane

Basic Equations: Continuity: $Q = AV$

Impulse–momentum: $F_x = \rho AV_0^2(1 - \cos\theta)$, $F_y = \rho AV_0^2 \sin\theta$

Solution:

$$A = \frac{\pi}{4}\left(\frac{3\text{ in.}}{12\text{ in./ft}}\right)^2 = 0.0491\text{ ft}^2$$

Using equation (7.7), we obtain

$$F_x = -\rho AV_0^2(1 - \cos\theta) = \frac{-62.4\text{ lb/ft}^3}{32.17\text{ ft/s}^2}(0.0491\text{ ft}^2)(4\text{ ft/s})^2(1 - \cos 90)$$

$$= 1.52\text{ lb}\left(\frac{\gamma}{g} = \rho,\ \cos 90 = 0\right)$$

From equation (7.9),

$$F_y = \rho AV_0^2 \sin\theta = \frac{62.4\text{ lb/ft}^3}{32.17\text{ ft/s}^2}(0.0491\text{ ft}^2)(4\text{ ft/s})^2 = 1.52\text{ lb}\ (\sin 90 = 1)$$

and the resultant force is found from equation (7.10) as

$$R = \sqrt{F_x^2 + F_y^2} = \sqrt{(-1.52)^2 + (1.52)^2} = 2.15\text{ lb}$$

ILLUSTRATION 7.4   IMPULSE–MOMENTUM

A jet directs a stream of water against a fixed vane as indicated in Figure 7.4. If the angle θ is 45° and the jet has a velocity of 10 m/s, determine the magnitude of the forces on the vane. Also determine the magnitude and direction of the resultant of these forces. The density of water can be taken as 1000 kg/m³, and Q is 0.1 m³/s.

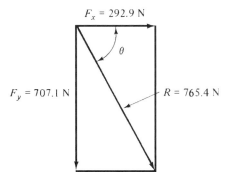

$F_x = 292.9$ N

θ

$F_y = 707.1$ N

$R = 765.4$ N

Figure 7.6   Illustrative Problem 7.4.

ILLUSTRATIVE PROBLEM 7.4

Given: $\theta = 45°$, $\rho = 1000$ kg/m^3, $Q = 0.1$ m^3/s, $V = 10$ m/s

Find: Forces on vane

Assumptions: Horizontal vane, steady, incompressible flow

Basic Equations: Continuity: $Q = AV$

Impulse–momentum: $F_x = -\rho Q V_0 (1 - \cos \theta)$, $F_y = \rho Q V_0 \sin \theta$

Solution: Applying equations (7.7) and (7.9), we have

$$F_x = -\rho Q V_0 (1 - \cos \theta) = -1000 \text{ kg/m}^3 \times 0.1 \text{ m}^3/\text{s} \times (10 \text{ m/s})(1 - 0.7071)$$

$$F_x = -292.9 \text{ N}$$

and

$$F_y = \rho Q V_0 \sin \theta = 1000 \text{ kg/m}^3 \times 0.1 \text{ m}^3/\text{s} \times 10 \text{ m/s} \times 0.7071$$

$$F_y = 707.1 \text{ N}$$

Notice that these forces are on the fluid. For the vane,

$$F_x = 292.9 \text{ N}$$

$$F_y = -707.1 \text{ N}$$

Therefore,

$$R = \sqrt{F_x^2 + F_y^2} = \sqrt{(292.9)^2 + (-707.1)^2} = 765.4 \text{ N}$$

We can determine the line of action of this resultant from Figure 7.6. Therefore,

$$\theta = \arccos \frac{292.9}{765.4}$$

$$= 67.5°$$

ILLUSTRATION 7.5   IMPULSE–MOMENTUM

If the 180° bend of Illustrative Problem 7.2 is also subjected to a difference in pressure at inlet and outlet due to the pressure drop in the bend, what is the total force on the pipe? Assume that the pressure at the inlet is 250 kPa and the pressure at the outlet is 240 kPa.

ILLUSTRATIVE PROBLEM 7.5

Given: Inputs same as Illustrative Problem 7.2; also, $p_1 = 250$ kPa, $p_2 = 240$ kPa

Find: Force on pipe

Assumptions: Same as Illustrative Problem 7.2

Basic Equations: Same as Illustrative Problem 7.2 with pressure forces at inlet and outlet equal to pA.

Solution: The dynamic forces remain the same. In addition, there is a force at the inlet acting to the right and equal to pA. Therefore, at the inlet,

$$F_{x1} = pA = 250\ 000 \times \frac{\pi}{4} (0.075)^2 = 1104.5 \text{ N}$$

At the outlet there is also a pressure force to the right equal to pA. Therefore,

$$F_{x2} = pA = 240\ 000 \times \frac{\pi}{4} (0.075)^2 = 1060.3 \text{ N}$$

The *total* force on the pipe due to both pressure differences and change in velocity is

$$F_{x1} + F_{x2} - R_x = 1104.5 + 1060.3 - (-18\ 120) = 20\ 284.8 \text{ N}$$

where the value of R_x is obtained from Illustrative Problem 7.2

ILLUSTRATION 7.6   BASIC SOLUTION OF THE IMPULSE–MOMENTUM EQUATION

Write a BASIC program to determine the force and its components that act on a fixed deflector vane. Use as inputs ρ, Q, V_0, and θ. Use Illustrative Problem 7.4 to provide a numerical check on your solution.

ILLUSTRATIVE PROBLEM 7.6

Given: ρ, Q, V_0, θ

Find: Force on vane—a BASIC solution

Assumptions: Incompressible, steady flow

Basic Equations: Equation (7.7): $F_x = -\rho Q V_0 (1 - \cos \theta)$

Equation (7.9): $F_y = \rho Q V_0 \sin \theta$

Equation (7.10a): $R = \rho Q V_0 \sqrt{2(1 - \cos \theta)}$

Solution: The necessary equations that we need are provided by equations (7.7), (7.9), and (7.10a). In addition, the angle of the line of action of the resultant force will be given by $\theta = \arctan (F_y/F_x)$. One word of caution must be noted. The angles and their inverses are given in radians, which requires a conversion from degrees to radians and from radians to degrees within the program. The program is listed below and the numerical values for Illustrative Problem 7.4 agree exactly with our previous solution.

```
10  INPUT "density = "; R
20  INPUT "volumetric flow rate = "; Q
30  INPUT "initial velocity = "; V
40  INPUT "vane angle = "; T
45  D = T*(3.1416/180)
```

```
50   X = -(R*Q*V)*(1 - (COS(D)))
60   Y = (R*Q*V)*(SIN(D))
70   F = SQR( (X*X) + (Y*Y))
80   P = ATN(Y/X)
85   E = P*(180/3.1416)
90   LPRINT "density = "; R
100  LPRINT "volumetric flow rate = "; Q
110  LPRINT "initial velocity = "; V
120  LPRINT "vane angle = "; T
130  LPRINT "x component of vane force ="; X
140  LPRINT "y component of vane force = "; Y
150  LPRINT "resultant vane force = "; F
160  LPRINT "angle of line of action of force = "; E
170  END
```

density = 1000
volumetric flow rate = .1
initial velocity = 10
vane angle = 45
x component of vane force = −292.89451748876
y component of vane force = 707.10807985945
resultant vane force = 765.36856152935
angle of line of action of force = −67.499789541343

ILLUSTRATION 7.7   EXCEL SPREADSHEET SOLUTION OF THE IMPULSE–MOMENTUM RELATION

Solve Illustrative Problem 7.6 using the Excel spreadsheet.

ILLUSTRATIVE PROBLEM 7.7

Given: Same inputs as Illustrative Problem 7.6

Find: Excel spreadsheet solution

Assumptions: Same as Illustrative Problem 7.6

Basic Equations: Same as Illustrative Problem 7.6

Solution: The inputs are placed in boxes A3, B3, C3, and D3. The angle is entered in radians, which = 45*PI( )/180, where Excel uses PI( ) for π. Box E3 is the solution for F_x, box F3 is the solution for F_y, and box G3 is the resultant. The angle of the resultant with the horizontal is given by box H3. Note that ATAN is the inverse tangent, which has been multiplied by 180/PI( ) to obtain the answer in degrees.

	A	B	C	D	E	F	G	H
1			Illustrative Problem 7.7					
2	rho	Q	V₀	angle	Fx	Fy	R	angle of force
3	1000	0.1	10	0.78539816	−292.89322	707.106781	765.366865	−67.5
	D3 = 45 * PI()/180							
	E3 = −A3 * B3 * C3 * (1−COS(D3))							
	F3 = A3 * B3 * C3 * SIN(D3)							
	G3 = (E3 * E3 + F3 * F3)^0.5							
	H3 = (ATAN(F3/E3)) * 180/PI( )							

In Chapter 8 we will study nozzles in some detail. Let us at this time look at a simple nozzle, as shown in Figure 7.7. The basic purpose of this device is to increase the velocity at its outlet by decreasing the area. Consider the impulse–momentum relation (repeated here for convenience),

$$F_x = \rho Q (V_2 - V_1) \tag{7.5}$$

where F_x is the net force acting in the x direction. At the inlet there is a pressure force equal to $p_1 A_1$, and at the outlet the pressure force is $p_2 A_2$. These pressure forces act opposite to each other. If we denote the force exerted *by the nozzle* on the fluid as F_R, equation (7.5) applied to this device becomes

$$p_1 A_1 - p_2 A_2 - F_R = \rho Q(V_2 - V_1) \tag{7.11}$$

Solving for F_R, we obtain

$$F_R = p_1 A_1 - p_2 A_2 - \rho Q(V_2 - V_1) \tag{7.12}$$

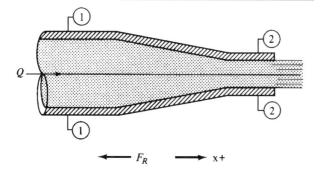

Figure 7.7   Nozzle.

ILLUSTRATION 7.8   NOZZLE REACTION

A pump discharges 500 gpm through a 4-in. diameter pipe. At the end of the pipe there is a nozzle with an outlet diameter of 1 in. What is the force exerted by the fluid on the hose? Take $p_1 = 100$ psig, p_2 to be atmospheric pressure, and the specific weight of water to be 62.4 lb/ft³.

ILLUSTRATIVE PROBLEM 7.8

Given: $Q = 500$ gpm, $D_1 = 4$ in., $D_2 = 1$ in., $p_1 = 100$ psig, $p_2 = 0$, $\gamma = 62.4$ lb/ft³

Find: Force on nozzle

Assumptions: Steady, incompressible flow $p_2 = 0$

Basic Equations: Continuity: $Q = A_1 V_1 = A_2 V_2$

Equation (7.12): $F_R = p_1 A_1 - p_2 A_2 - \rho Q(V_2 - V_1)$

Solution: At the inlet,

$$p_1 A_1 = 100 \frac{\text{lb}}{\text{in.}^2} \times \frac{\pi}{4} (4 \text{ in.})^2 = 1256.6 \text{ lb}$$

At the outlet,

$$p_2 A_2 = 0$$

To obtain Q in cfs $\left(\dfrac{ft^3}{s}\right)$,

$$Q = 500 \text{ gpm} \times \frac{231 \text{ in.}^3/\text{gal}}{1728 \text{ in.}^3/\text{ft}^3} \times \frac{1}{60 \text{ s/min}} = 1.11 \text{ ft}^3/\text{s}$$

$$V_1 = \frac{Q}{A_1} = \frac{1.11 \text{ ft}^3/\text{s}}{\pi/4(4 \text{ in.}/(12 \text{ in.}/\text{ft}))^2} = 12.7 \text{ ft/s}$$

$$V_2 = \frac{Q}{A_2} = \frac{1.11 \text{ ft}^3/\text{s}}{\pi/4(1 \text{ in.}/(12 \text{ in.}/\text{ft}))^2} = 203.5 \text{ ft/s}$$

Applying equation (7.12), we obtain

$$F_R = p_1 A_1 - p_2 A_2 - \rho Q (V_2 - V_1)$$

$$F_R = 1256.6 \text{ lb} - 0 \text{ lb} - \frac{62.4 \text{ lb/ft}^3}{32.17 \text{ lb/slug}} \times (1.11 \text{ ft}^3/\text{s})(203.5 \text{ ft/s} - 12.7 \text{ ft/s})$$

$$\therefore F_R = 845.8 \text{ lb}$$

Since this is the force on the fluid, the force on the nozzle will be equal to the force on the fluid and directed opposite to the direction of flow.

7.4   MOVING VANES

If the vane is moving with the velocity u in the same direction as the initial velocity of the stream, we have the situation shown schematically in Figure 7.8. At the inlet the absolute velocity of the stream is V_0 and its velocity relative to the vane is $V_0 - u$. The relative velocity is turned by the vane through the angle θ without a change in its magnitude. This relative velocity vector $V_0 - u$ is now added vectorially to u to give the final absolute velocity leaving the vane. Figure 7.9 shows the vector diagram at the outlet of the vane, where V_1 is the resultant absolute velocity at the outlet. The component of the absolute velocity in the horizontal direction at the outlet is $(V_0 - u) \cos (\theta + u)$, and at the inlet it is V_0. Therefore,

$$F_x = \dot{m} [(V_0 - u) \cos (\theta + u) - V_0] = \dot{m}(V_0 - u)(1 - \cos \theta) \tag{7.13}$$

and

$$F_y = \dot{m}[(V_0 - u) \sin \theta] \tag{7.14}$$

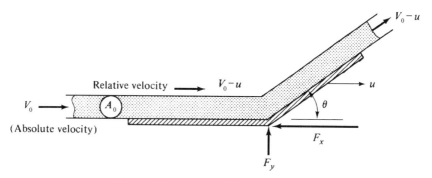

Figure 7.8 Moving deflector vane.

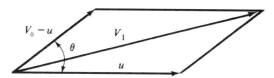

Figure 7.9 Velocity diagram.

Notice from equations (7.13) and (7.14) that if $V_0 - u$ is replaced by the relative velocity, the problem of the moving vane is reduced to the problem of the fixed vane with a velocity equal to the relative velocity.

ILLUSTRATION 7.9 MOVING VANE

If the vane in Illustrative Problem 7.3 has a velocity of 2 ft/s to the right, determine the forces acting on the stream.

ILLUSTRATIVE PROBLEM 7.9

Given: Same inputs as Illustrative Problem 7.3 with the vane velocity $u = 2$ ft/s to the right.

Find: Forces acting on stream

Assumptions: See Illustrative Problem 7.3

Basic Equations: See Illustrative Problem 7.3. Also, Equations (7.13) and (7.14) with $V_0 - u$ replaced by the relative velocity.

Solution: Refer to the sketch of Illustrative Problem 7.3. Using the data of Illustrative Problem 7.3 and equations (7.13) and (7.14) gives us

$$V_0 - u = 4 - 2 = 2 \text{ ft/s}$$

Therefore, using the data from Illustrative Problem 7.3,

$$F_x = -\frac{62.4 \times 0.0491 \times 4}{32.17}(2 - 0) = -0.76 \text{ lb}$$

$$F_y = \frac{62.4 \times 0.0491 \times 4}{32.17} (2) = 0.76 \text{ lb}$$

and

$$R = \sqrt{(-0.76)^2 + (0.76)^2} = 1.07 \text{ lb}$$

ILLUSTRATION 7.10   MOVING VANE

If the stream in Illustrative Problem 7.9 is turned through an angle of 180° (a U bend), determine the forces acting on the stream.

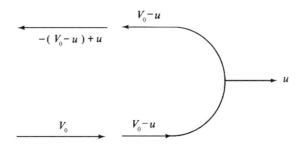

Figure 7.10   Illustrative Problem 7.10.

ILLUSTRATIVE PROBLEM 7.10

Given: Illustrative Problem 7.9 with an angle of 180°

Find: Force acting on stream

Assumptions: See Illustrative Problem 7.9

Basic Equations: See Illustrative Problem (7.9).

Solution: Refer to Figure 7.10. Using the absolute velocities from the diagram we obtain

$$F_x = \dot{m}[-(V_0 - u) + u - V_0] = \dot{m}[-2(V_0 - u)]$$

$$= -2\dot{m}(V_0 - u)$$

$$= -2\frac{62.4 \times 0.0491 \times 4}{32.17} (2) = -1.56 \text{ lb}$$

and

$$F_y = 0$$

The resultant force is therefore 1.56 lb acting to the left.

The power transferred from a stream to a moving blade is the product of $F_x(u)$. If the machine is frictionless, this power is given by

$$P = F_x(u) = \dot{m}(V_0 - u)(1 - \cos\theta)u \tag{7.15}$$

Examination of equation (7.15) shows that no power output will be obtained if the blade is either stationary or operating at $u = V_0$. If equation (7.15) is plotted as a function of u, it is found that P is a maximum when $u = V_0/2$. If u is kept constant and P is plotted as a function of θ, P is found to be maximum when $\theta = 180°$. Using both of these conditions, we find that the maximum power transferred, P, is exactly equal to the power of the free jet. A blade operated at half the stream speed with the stream turned through 180° will have an efficiency of energy transfer theoretically equal to 100%.

7.5   FAN OR PROPELLER

The function of a fan or propeller is to transform efficiently its rotary motion into forward linear motion of an airstream to provide an increased air velocity. The blades of such a device can be considered to be a number of airfoils of very short span set end to end that must absorb the power furnished to the propeller by its driving motor. The theory of the action of these small airfoils is known as the Drzewiecki theory or blade element theory, and its development will be found in standard texts on aerodynamics or propellers. For the present we shall consider a somewhat simplified treatment in which the fan or propeller will be considered as a disk that exerts a uniform pressure or thrust on the airstream without causing any rotation of the fluid. The only result of the thrust is to increase the momentum of the air. The additional assumptions will be made that the pressure at stations far enough in front of and in back of the propeller is atmospheric and that the flow is streamline from entrance to exit of the device.

Before proceeding with the analysis, let us first recast the Bernoulli equation into a more convenient form. Repeating equation (5.2), we have

$$\frac{p_1}{\gamma} + \frac{V_1^2}{2g} + Z_1 = \text{constant} \tag{5.2}$$

Multiplying equation (5.2) by γ and noting that $\rho = \gamma/g$,

$$p_1 + \frac{V_1^2\gamma}{2g} + \gamma Z_1 = p_1 + \frac{\rho V_1^2}{2} + \gamma Z_1 = \text{constant} \tag{7.16}$$

For problems in gas flow and gas dynamics the elevation term is usually negligible when compared to the other terms, which leads to a form of the Bernoulli equation sometimes called the aerodynamic form of the Bernoulli equation: namely,

$$p + \frac{\rho V^2}{2} = \text{constant} \tag{7.17}$$

If a body is immersed in a fluid such as is shown in Figure 7.11, there will be a point A on the body where the velocity of the fluid is zero. Since the fluid is stagnant at this point, it is known as the *stagnation point,* and the pressure at this point will exceed the pressure of the surrounding stream by $\rho V^2/2$ as determined by equation (7.17).

We are now ready to consider an idealized fan or propeller as shown in Figure 7.12. As was stated earlier, the basic action of the fan or propeller is to impart a momentum to the fluid, thereby developing a propulsive thrust. For our analysis we will assume the density of the air to remain essentially constant and we will also assume that there are no flow losses in the device.

If we consider sections ② and ③ and write the momentum relation between these sections, we obtain

$$F = (p_3 - p_2)A = \rho Q(V_4 - V_1) \tag{7.18}$$

where A is the plane area swept out by the propeller. Since $Q = AV$,

$$(p_3 - p_2) = \rho V(V_4 - V_1) \tag{7.19}$$

The assumption that there are no energy losses in this device permits us to write the energy equation in aerodynamic form between sections ① and ②, and then between sections ③ and ④.

$$p_1 + \frac{\rho V_1^2}{2} = p_2 + \frac{\rho V_2^2}{2} \tag{7.20}$$

and

$$p_3 + \frac{\rho V_3^2}{2} = p_4 + \frac{\rho V_4^2}{2} \tag{7.21}$$

Since p_1 and p_4 are the pressures of the undisturbed stream, $p_1 = p_4$. We can therefore solve for $p_3 - p_2$ from equations (7.20) and (7.21) as

$$p_3 - p_2 = \tfrac{1}{2}\rho(V_4^2 - V_1^2) \tag{7.22}$$

But equation (7.19) also gives us $p_3 - p_2$. Therefore,

$$\tfrac{1}{2}\rho(V_4^2 - V_1^2) = \rho V(V_4 - V_1)$$

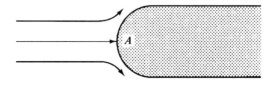

Figure 7.11   Streamlines around a submerged body.

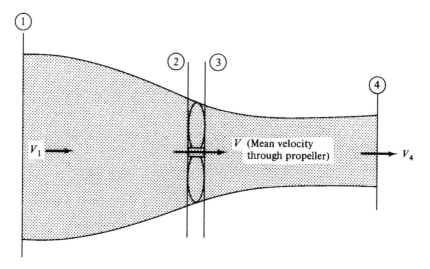

Figure 7.12   Ideal propeller.

which yields V as

$$V = \frac{V_1 + V_4}{2}$$

(7.23)

From equation (7.23) we conclude that the velocity through the propeller area is the arithmetic mean of the initial and final velocities, with half of the total change in velocity occurring *before* the air passes through the propeller and half being given to the air *after* it passes through the propeller.

The useful work done on the airstream is the product of the thrust and velocity. We can obtain the power input to the propeller as

$$P = Q\gamma E_p$$

(7.24)

E_p is obtained from the Bernoulli equation as

$$E_p = \frac{V_4^2 - V_1^2}{2g}$$

(7.25)

Using equations (7.24) and (7.25) as well as equation (7.23) yields the following:

$$P = Q\gamma\left(\frac{V_4^2 - V_1^2}{2g}\right)$$

but $\gamma = \rho g$. Therefore,

$$P = Q\rho\left(\frac{V_4^2 - V_1^2}{2}\right) = Q\rho(V_4 - V_1)\left(\frac{V_4 + V_1}{2}\right)$$

(7.26)

and

$$P_{\text{input}} = Q\rho(V_4 - V_1)V \tag{7.27}$$

The power in the airstream is the force at the inlet multiplied by the velocity of the stream at the inlet, giving us

$$P_{\text{airstream}} = FV_1$$

But F is given by equation (7.18) as $\rho Q(V_4 - V_1)$. Thus, the power in the airstream is

$$P_{\text{airstream}} = \rho Q(V_4 - V_1)V_1 \tag{7.28}$$

If we now define the efficiency of the propeller as the ratio of the power in the airstream to the power input by the propeller, we obtain

$$\eta = \left(\frac{V_1}{V}\right)100\% \tag{7.29}$$

At this point let us define $V_4 - V_1$ as ΔV. Using this definition and equations (7.23) and (7.29) gives us the propeller efficiency as

$$\eta = \left(\frac{V_1}{V_1 + \Delta V/2}\right)100\% \tag{7.30}$$

Equation (7.30) expresses the efficiency of an ideal fan or propeller in which the only energy loss considered is the loss in kinetic energy due to the change in velocity through the device. An actual propeller will have an efficiency less than that given by equation (7.30) because of the fact that the flow is not streamline and because there are friction losses in the blades, pulsations due to the finite number of blades, tip losses, and stream rotation after the stream leaves the blades. Propellers have actual efficiencies of 80% to 90%. The high efficiency of fans and propellers has made their application to wind chargers for power generation very attractive for alternative power sources. Recent propeller developments have also shown great economies of operation for turboprop aircraft applications.

ILLUSTRATION 7.11 PROPELLER

What is the maximum efficiency of a propeller if the air speed of a plane is 150 mi/h and the airstream velocity leaving the propeller is 200 mi/h?

ILLUSTRATIVE PROBLEM 7.11

Given: Plane speed = 150 mph, airstream leaving = 200 mph

Find: Maximum efficiency

Assumptions: Steady, incompressible flow of air through an ideal propeller

Basic Equations: Equation (7.30), $\eta = \left(\dfrac{V_1}{V_1 + \Delta V/2}\right)100\%$

Solution: The increment in velocity across the entire unit is $200 - 150 = 50$ mi/h. Since half of this is added before the propeller, the velocity at the propeller is 175 mi/h. Therefore, $\Delta V/2 = 25$ mi/h and

$$\eta = \left(\frac{V_1}{V_1 + \Delta V/2}\right)100 = \left(\frac{150}{150 + 25}\right)100$$

$$= 85.7\%$$

ILLUSTRATION 7.12   PROPELLER

An airplane is traveling at 300 mi/h and has a propeller that is 6 ft in diameter. If the specific weight of air is 0.0755 lb/ft³, determine the thrust provided by the propeller if it is 90% efficient.

ILLUSTRATIVE PROBLEM 7.12

Given: $D = 6$ ft, $V_1 = 300$ mph, $\gamma = 0.0755$ lb/ft³, $\eta = 90\%$

Find: Thrust

Assumptions: Ideal propeller

Basic Equations: Equation (7.23): $V = \dfrac{V_1 + V_4}{2}$

Equation (7.18): $F = \rho_1 A_1 V_1 (V_4 - V_1) = \rho_1 A_1 V_1 (\Delta V)$

Solution:

$$\eta = \frac{V_1}{V_1 + \Delta V/2}$$

and V_1 is given as 300 mi/h or 440 ft/s

$$0.9 = \frac{440}{440 + \Delta V/2}$$

$$0.9\left(440 + \frac{\Delta V}{2}\right) = 440$$

$$0.9\frac{(\Delta V)}{2} = 440 - 396 = 44$$

and

$$\Delta V = 97.8 \text{ ft/s} \qquad \text{and} \qquad \frac{\Delta V}{2} = 48.9 \text{ ft/s}$$

Therefore, the velocity at the propeller is $440 + 48.9 = 488.9$ ft/s. At the exit, $V_4 = 488.9 + 48.9 = 537.8$ ft/s. From equation (7.18),

$$F = \rho_1 A_1 V_1 (V_4 - V_1) = \rho_1 A_1 V_1 (\Delta V)$$

Since the propeller is 6 ft in diameter and γ is 0.0755 lb/ft^3,

$$F = \frac{0.075}{32.17} \times \frac{\pi}{4}\, (6)^2 \times 488.9(97.8)$$

$$= 3152 \text{ lb}$$

7.6   REVIEW

In Chapters 4 and 5 we considered certain fluid dynamics applications that could be solved using the energy or the Bernoulli equations. There are many instances, however, where it is necessary to invoke other considerations to solve problems in fluid mechanics. In this chapter our work started with Newton's second law as a basis. We were able to rearrange it to yield the impulse–momentum relation, which we subsequently applied to the solution of certain problems that are of importance in fluid dynamics. These problems have all dealt with forces exerted on or by fluids that are in motion.

The first problem that we considered was the force exerted by a flowing stream that changes its direction of flow due to a fixed deflector vane. We considered the effects on the stream using the impulse–momentum relation. As noted in our derivation of this relation, both impulse and momentum are vectors and must be treated accordingly. The nozzle, a device that is used to increase the velocity of a stream, was also investigated. We then considered the effects that occur when a stream is deflected by a moving vane. Our final application was to fans or propellers. We took the fan or propeller to consist of a disk that exerted a uniform pressure or thrust on the airstream without causing any rotation of the fluid. Thus the only result of the thrust is to increase the momentum of the fluid. Using a modified form of the Bernoulli equation and the impulse–momentum relation, we arrived at a solution to the efficiency and power of fans or propellers.

KEY TERMS

Terms of importance in this chapter:

Aerodynamic Bernoulli equation: $p + \rho V^2/2 = $ constant.

Drzewiecki (blade element theory):   considers a fan or propeller to consist of a series of small airfoils.

Efficiency: the ratio of the power in the airstream to the power input by the propeller.

Fan: a disk that exerts a uniform pressure or thrust on an airstream without causing rotation of the air.

Impulse: for a constant force F, equal to $F\Delta t$, where Δt is the time that the force acts on a body.

Momentum or change of momentum: $m\Delta V$.

Nozzle: a device that is used principally to increase the velocity of a stream.

Propeller: *see* fan.

Stagnation point: a point on a body where the velocity of a fluid is zero.

KEY EQUATIONS

Impulse–momentum relation	$F\Delta t = m(V_2 - V_1)$	(7.3a)
Impulse–momentum relation	$F = \dot{m}(V_2 - V_1)$	(7.3b)
Impulse–momentum relation	$F = \rho Q(V_2 - V_1) = \dfrac{\gamma}{g} Q(V_2 - V_1)$	(7.5)
Fixed vane	$F_x = {}^-\rho Q V_0(1 - \cos\theta)$	(7.7)
Fixed vane	$F_y = \rho Q V_0 \sin\theta$	(7.9)
Fixed vane	$R = \rho Q V_0\sqrt{2(1 - \cos\theta)}$	(7.10a)
Nozzle	$F_R = p_1 A_1 - p_2 A_2 - \rho Q(V_2 - V_1)$	(7.12)
Moving vane	$F_x = \dot{m}(V_0 - u)(1 - \cos\theta)$	(7.13)
Moving vane	$F_y = \dot{m}[(V_0 - u)\sin\theta]$	(7.14)
Moving blade	$p = \dot{m}(V_0 - u)(1 - \cos\theta)u$	(7.15)
Aerodynamic Bernoulli equation	$p + \dfrac{\rho V^2}{2} = \text{constant}$	(7.17)
Fan or propeller	$V = \dfrac{V_1 + V_4}{2}$	(7.23)
Fan or propeller	$P_{input} = Q\rho(V_4 - V_1)V$	(7.27)
Fan or propeller	$P_{airstream} = \rho Q(V_4 - V_1)V_1$	(7.28)
Fan or propeller	$\eta = \left(\dfrac{V_1}{V}\right)100\%$	(7.29)
Fan or propeller	$\eta = \left(\dfrac{V_1}{V_1 + \Delta V/2}\right)100\%$	(7.30)

QUESTIONS

1. Cite several instances in your life where you have experienced or seen the force exerted by a fluid stream.

2. Show that the units of impulse and momentum are the same.

3. If a force is exerted on a car that does not move, is there an impulse? Is there a change of momentum?

4. Will a horizontally directed impulse cause a vertical change in momentum? Explain your answer.

5. *True or false:* When a fluid stream is deflected by a fixed vane, the force exerted by the vane equals the force exerted by the stream on the vane.

6. What angle of a fixed vane will cause it to exert the greatest force on a moving stream?

7. Explain why firefighters must use great strength to control the hoses that are used to put out fires.

8. Under what two conditions will we obtain the maximum power transfer to a moving blade?

9. What is the maximum efficiency of power transfer to a moving blade?

10. What percentage of the total velocity change in an airstream occurs before and after a propeller blade?

11. What factors limit the speed of propellers?

12. Why are propeller blades made with a twisting form from their roots to their tips?

13. Most fan blades are not made twisted. Why not?

PROBLEMS

Impulse–Momentum

7.1. A ball weighing 1 lb is thrown horizontally with a velocity of 60 ft/s at a wall. It rebounds elastically from the wall so that its velocity is 60 ft/s. What is the change in its momentum?

7.2. A baseball weighing 0.4 lb is struck by a bat and is deflected 135° from its original horizontal direction as shown in Figure P7.2. If the initial velocity of the ball is 135 ft/s and the final velocity is 115 ft/s, calculate the horizontal and vertical components of the impulse of the bat on the ball.

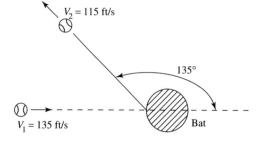

Figure P7.2

7.3. A body of 10 kg experiences a force of 5 N for a period of 20 s. What will be its change in velocity?

*7.4. Show that the thrust developed by a turbojet engine can be approximately written as

$$F = (1 + R)V_2\dot{m}_a$$

where R is the fuel-air ratio, $\dot{m}_f/\dot{m}_a$. State all your assumptions.

7.5. A turbojet engine is tested on a fixed test stand. Air having a specific weight of 0.075 lb/ft^3 enters the engine at the rate of 7500 ft^3/s. Fuel is added to the engine at the rate of 1 lb of fuel for each 150 lb of air. Determine the thrust produced by the engine if the exhaust velocity is found to be 820 ft/s. Neglect pressure differences at the inlet and outlet of the engine and take the inlet area to be 40 ft^2.

7.6. If the engine described in Problem 7.5 operates with 250 m^3/s of air at entry having a specific weight of 11.78 N/m^3, and uses 1 kg of fuel for each 150 kg of air, what thrust will it produce? The exit velocity is found to be 300 m/s and the inlet and outlet pressures are taken to be equal. The inlet area of the engine is 4 m^2.

7.7. Use the results of Problems 7.4 to solve Problem 7.5. Compare your answers.

7.8. Use the results of Problem 7.4 to solve Problem 7.6. Compare your answers.

Deflection of Streams by Fixed Vanes

7.9. A jet of water 3 in. in diameter strikes a flat plate placed normal to the direction of flow. If the plate is stationary, calculate the force on it if the initial velocity of the jet is 40 ft/s. Use 62.4 lb/ft^3 for the specific weight of water.

7.10. A jet of water (density equal to 1000 kg/m^3) issues from a nozzle having a diameter of 50 mm. What force will this water jet exert on a plate placed normal to the direction of flow if the velocity of the jet is 15 m/s?

7.11. A stream of water impinges on a flat plate. If the stream exerts a force of 1000 N on the plate, what is the velocity of the stream? Assume the fluid density to be 1000 kg/m^3 and the stream to have a diameter of 75 mm.

7.12. A stream of air has a horizontal velocity of 60 ft/s. What force is required to deflect this stream of air 180° if it has a cross-sectional area of 20 ft^2? The specific weight of air equals 0.0765 lb/ft^3.

7.13. If air has a specific weight of 12.02 N/m^3 and an airstream has a velocity of 30 m/s, what force is required to deflect this stream 180°? Assume that the cross-sectional area of the stream is 3 m^2.

7.14. A 90° pipe bend in a horizontal plane carries water at a velocity of 20 ft/s. If the pipe has an inside diameter of 5 in., what force is exerted on the bend by the dynamic action of the stream? The specific weight of water is 62.4 lb/ft^3.

7.15. A stream of water issues from a hose that has a 20-mm inside diameter. If 400 liters/min is delivered, what is the reaction force on the hose? Use $\gamma = 9810$ N/m^3 and consider that the jet is brought to rest.

7.16. What is the reaction of the water on the vane shown in Figure P7.16 on page 288 if the jet has a velocity of 20 m/s? Use $\rho = 1000$ kg/m^3 and $Q = 0.2$ m^3/s.

7.17. Solve Problem 7.16 if the angle of the vane is 120°.

7.18. Solve Problem 7.12 if friction reduces the velocity to 50 ft/s.

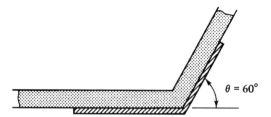

Figure P7.16

*7.19. A stream of water strikes a plate as shown in Figure P7.19. What force does it exert on the plate? Assume that the water ($\gamma = 62.4$ lb/ft^3) issues from a 3-in. diameter nozzle at the rate of 60 ft/s.

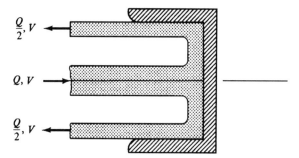

Figure P7.19

7.20. A disk weighing 150 N is supported by a vertical jet of water issuing from a nozzle. If the jet velocity is 7 m/s, what is the diameter of the jet?

*7.21. A horizontal jet of fluid strikes a large flat plate that tilts at an angle θ to the horizontal as shown in Figure P7.21. Assume no friction between the fluid and the plate. If the volume flow of the jet is Q, show that the volume flow upward, Q_u, and the volume flow downward, Q_d, are given by

$$Q_u = \frac{Q}{2}(1 + \cos \theta)$$

$$Q_d = \frac{Q}{2}(1 - \cos \theta)$$

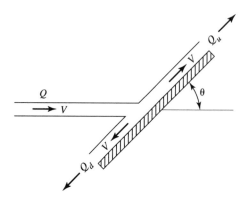

Figure P7.21

7.22. A pump discharges 200 gal/min through a 3-in. diameter pipe (Figure P7.22). At the end of the pipe there is a nozzle having an outlet diameter of 1 in. If $\gamma = 62.4$ lb/ft³, $p_1 = 50$ psig, and p_2 is atmospheric pressure, what force does the fluid exert on the hose?

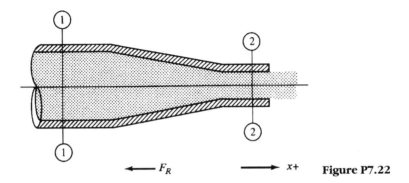

F_R $x+$ **Figure P7.22**

Forces Exerted by Moving Vanes

7.23. A fluid stream exerts a force of 500 N on a vane that is moving in the direction of the stream with a velocity of 10 m/s. What power does the stream deliver to the vane?

7.24. A boat is propelled steadily at 20 mi/h by a 25-hp motor. What thrust does the propeller exert on the boat?

7.25. Solve Problem 7.13 if the vane has a velocity of 5 m/s to the right.

7.26. Solve Problem 7.14 if the bend has a velocity of 5 ft/s to the right.

7.27. Solve Problem 7.16 if the vane is moving to the right with a velocity of 5 m/s.

Fan or Propeller

7.28. An airplane travels at 300 mi/h. If atmospheric pressure is 14.6 psia and the specific weight of air is 0.075 lb/ft³, determine the stagnation pressure exerted on the airplane.

7.29. An airplane travels at 480 km/h. If atmospheric pressure is 100 kPa and the specific weight of air is 11.78 N/m³, determine the stagnation pressure on the airplane.

7.30. If a propeller on an airplane going at 200 mi/h causes the air to leave with an absolute velocity of 280 mi/h, determine its efficiency.

7.31. A propeller 5 ft in diameter is mounted on a plane traveling at 350 mi/h. If the airstream leaves the propeller with an absolute velocity of 410 mi/h, determine the efficiency of the propeller and the thrust imparted to the aircraft. Assume that $\gamma = 0.075$ lb/ft³.

7.32. A propeller is 2 m in diameter and is mounted on an aircraft going at 500 km/h. The airstream leaves the propeller with a velocity of 650 km/h. Determine the efficiency of the propeller and the thrust. Assume that the γ of air equals 11.8 N/m^3.

7.33. If a propeller has an efficiency of 90%, determine its thrust if it is 6 ft in diameter and is mounted on a plane traveling at 275 mi/h. Use $\gamma = 0.075$ lb/ft^3.

7.34. Determine the horsepower absorbed by a propeller if it is on a plane traveling at 240 mi/h, is 7 ft in diameter, is 80% efficient, and $\gamma = 0.075$ lb/ft^3.

7.35. A propeller is 2.5 m in diameter and can exert a thrust of 7 kN on an airplane traveling at 360 km/h. What is its efficiency? Use $\gamma = 11.8$ N/m^3.

7.36. A propeller 7 ft in diameter is capable of exerting a 1500-lb thrust on an airplane traveling at 280 mi/h. Determine its efficiency. Use $\gamma = 0.075$ lb/ft^3.

MEASUREMENTS

LEARNING GOALS

After reading and studying this chapter you should be able to:

1. State why there is a need to define a property clearly and then to reduce this definition to a measurable quantity.
2. Define the various units in which pressure is measured and show how to convert from one set of units to another.
3. Show the corrections that are needed for extremely accurate pressure measurements.
4. Indicate some applications of manometers and micromanometers.
5. State the principle of operation of the Fortin barometer.
6. Explain the principle of operation of the McLeod gage and how it is used to measure low pressures.
7. Describe the operation of the Bourdon gage and some of its modifications.
8. Indicate the application of various pressure transducers for the measurement of pressure.
9. Use the data from an orifice to calculate velocity and flow.
10. Apply the trajectory method to calculate the velocity coefficient of an orifice.
11. Describe the operation of the piezometer, Pitot, and Pitot-static devices.
12. Calculate the velocity of flow using the Pitot-static tube.
13. Determine the coefficients of the venturi meter and calculate the flow from venturi meter measurements.

8.1   INTRODUCTION

Measurements, be they of time, length, area, or volume, are obvious to most of us. The measurement of properties represents a large part of the application and needs of technology. Therefore, in this chapter we spend some time on the fundamental measurement of pressure, pressure in static fluids, pressure in moving fluids, velocity, and the volumetric quantity of flow. Because of the basic nature of much of this material, a significant amount of it will be descriptive, and wherever needed we will supplement the descriptive portion with quantitative data. In some of the earlier chapters we considered some of the items that are covered here. In particular, the manometer and barometer are again discussed even though some of the discussion is a repeat of earlier discussions. This has been done for completeness, and it would be entirely proper to review some of this earlier material prior to or while studying it in this chapter.

8.2   PRESSURE MEASUREMENTS IN STATIC FLUIDS[1]

Pressure is a property of a system, and, as a fundamental parameter, it is of the utmost importance that it be measured accurately. In the text of Chapter 2 pressure was defined to be the normal force exerted by a fluid on a surface. It would be more correct to qualify this definition and for the present restrict it to fluids at rest. No definition is really useful, however, unless and until it can be converted into measurable characteristics. Figure 8.1 illustrates the basic definition of pressure and the elementary static concepts upon which the entire field of pressure measurement is based.

There is unfortunately a confusing number of units used to express pressure. While all of these units are based on the definition of a normal force per unit area, their usage has entered the technical literature based on particular applications. Table 8.1 on page 294 gives conversion factors between some of the commonly used units of pressure. For convenience, there are two approximations that can be used to give the student a "feel" for some of the units:

$$1 \text{ mm} = 1000 \text{ microns} \simeq 0.04 \text{ in. Hg}$$

and

$$1 \text{ micron} = 0.001 \simeq 0.00004 \text{ in. Hg}$$

In performing any measurement it is necessary to have a standard of comparison in order to calibrate the measuring instrument. In Section 8.3 and following sections five pressure standards currently used as the basis for all pressure-measurement work will be discussed. Table 8.2 on page 294 summarizes these standards, the pressure range in which they are used, and their accuracy.

[1] Portions of the material in this chapter have been taken with permission from "Pressure and Its Measurement" by R. P. Benedict, *Electro-Technology* (New York), **80,** October 1967, p. 69ff.

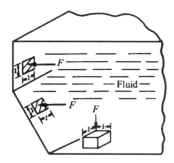

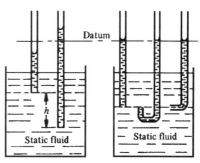

Pressure is the normal force F (lb/in.^{2}, lb/ft^2, dynes/cm^2) exerted on a unit area of a surface bounding a fluid.

Fluid pressure varies with depth, but it is the same in all directions at a given depth.

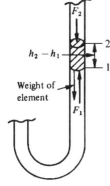

Variation of fluid pressure with elevation is found by balancing the forces on a static-fluid element (F_1 is equal to F_2 plus the weight of the element). For a constant-density fluid, the pressure difference $p_2 - p_1$ is equal to the specific weight γ times $(h_2 - h_1)$.

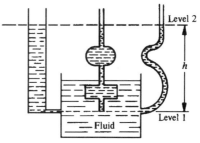

Pressure is independent of the shape and size of the vessel. The pressure difference between level 1 and level 2 is always $p_1 - p_2 = \gamma h$, where γ is the specific weight of the constant-density fluid in the vessel.

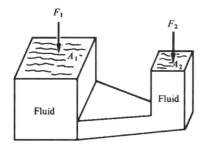

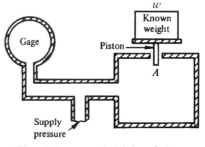

Constant pressure transmission in a confined fluid can be used to multiply force by the relation $p = F_1/A_1 = F_2/A_2$.

When the known weight is balanced, the gage pressure is $p = w/A$; this is the basic principle of dead-weight testing.

Figure 8.1   Basic pressure-measurement concepts.

Table 8.1   Pressure-Unit Conversion Factors

Pressure unit	psi	in. H_2O	in. Hg	Atmospheres	Microbars	mm Hg	Microns
1 psi	1.000	27.730	2.0360	6.8046 $\times 10^{-2}$	68,947.6	51.715	51,715.0
1 in. H_2O (68°F)	0.036063	1.00	0.073424	2.4539 $\times 10^{-3}$	2,486.4	1.8650	1,865.0
1 in Hg (°32F)	0.49115	13.619	1.000	3.3421 $\times 10^{-2}$	33,864.0	25.400	25,400.0
1 atm	14.69595	407.513	29.9213	1.000	1.01325 $\times 10^6$	760.000	7.6000 $\times 10^5$
1 microbar (dyn/cm²)	1.4504 $\times 10^{-5}$	4.0218 $\times 10^{-4}$	2.9530 $\times 10^{-5}$	9.8692 $\times 10^{-7}$	1.000	7.5006 $\times 10^{-4}$	0.75006
1 mm Hg (32°F)	0.019337	0.53620	0.03937	1.3158 $\times 10^{-3}$	1,333.2	1.000	1,000.0
1 micron (32°F)	1.9337 $\times 10^{-5}$	5.3620 $\times 10^{-4}$	3.9370 $\times 10^{-5}$	1.3158 $\times 10^{-6}$	1.3332	0.0010	1.000

Table 8.2   Characteristics of Pressure Standards

Type	Range	Accuracy
Dead-weight piston gage	0.01 to 10,000 psig	0.01 to 0.05% of reading
Manometer	0.1 to 100 psig	0.002 to 0.02% of reading
Micromanometer	0.0002 to 20 in. H_2O	1% of reading to 0.001 in. H_2O
Barometer	27 to 31 in. Hg	0.001 to 0.03% of reading
McLeod gage	0.01 micron to 1 mm Hg	0.5 to 3% of reading

8.3   DEAD-WEIGHT PISTON GAGE

The dead-weight, free-piston gage consists of an accurately machined piston inserted into a close-fitting cylinder. Masses of known weight are loaded on one end of the free piston, and pressure is applied to the other end until enough force is developed to lift the piston–weight combination. When the piston is floating freely between the cylinder limit stops, the gage is in equilibrium with the unknown system pressure. The dead-weight pressure can then be defined as

$$P_{dw} = \frac{F_e}{A_e} \tag{8.1}$$

where F_e, the equivalent force of the piston–weight combination, depends on such factors as local gravity and air buoyancy, and A_e, the equivalent area of the piston–cylinder combination, depends on such factors as piston–cylinder clearance, pressure level, and temperature.

A fluid film provides the necessary lubrication between the piston and cylinder. In addition, the piston—or, less frequently, the cylinder—may be rotated or oscillated to reduce friction even further. Because of fluid leakage, system pressure must be continuously trimmed upward to keep the piston–weight combination floating. This is often achieved by decreasing system volume using a pressure–volume apparatus (as shown in Figure 8.2). As long as the piston is freely balanced, system pressure is defined by equation (8.1).

Corrections must be applied to the indication of the dead-weight piston gage, p_i, to obtain the accurate system pressure, P_{dw}. The two most important corrections concern air buoyancy and local gravity. According to Archimedes' principle, the air displaced by the weights and the piston exerts a buoyant force that causes the gage to indicate too high a pressure. The correction term for this effect, C_b, is the ratio of two specific weights,

$$C_b = - \frac{\gamma_{air}}{\gamma_{weights}}$$

Whenever gravity at the measurement site varies from the standard sea level value of 32.1740 ft/s² because of latitude or altitude variations, a gravity correction term, C_g, must be applied:

$$C_g = \frac{g_{local}}{g_{std}} - 1$$

$$= -2.637 \times 10^{-3} \cos 2\phi - 9.6 \times 10^{-8}h - 5 \times 10^{-5} \qquad (8.2)$$

where ϕ is latitude in degrees and h is altitude above sea level in feet.

The corrected dead-weight piston gage pressure is therefore

$$P_{dw} = p_i(1 + C_b + C_g) \qquad (8.3)$$

The effective area of the dead-weight piston gage is usually taken as the mean of the cylinder and piston areas, but temperature affects this dimension. The effective area increases between 13 and 18 ppm/°F for commonly used materials, and a suitable correction for this effect may also be applied.

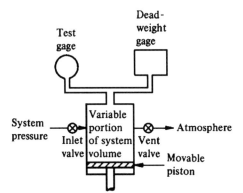

Figure 8.2   Pressure–volume regulator to compensate for fluid leakage in a dead-weight gauge.

8.4   MANOMETER

The manometer has been discussed earlier in this book as a means of measuring pressure. For accurate measurements, however, there are certain factors that must be taken into account. Therefore, for the sake of continuity and completeness a brief discussion of the manometer will be included in this section, and those corrections most necessary for accurate use of this instrument will also be discussed.

A manometer can simply consist of a transparent tube in the form of an elongated U, partially filled with a suitable liquid. Mercury and water are the most commonly used manometric fluids because detailed information is available on their specific weights.

The fluid whose pressure is unknown is applied to the top of one of the tubes of the manometer, while a reference fluid pressure is applied to the other tube, as shown in Figure 8.3. In the steady state, the difference between the unknown pressure and the reference pressure is balanced by the weight/unit area of the displaced manometer liquid, so that

$$\Delta p_{mano} = \gamma_m \Delta h_e \tag{8.4}$$

where γ_m, the corrected specific weight of the manometer fluid, depends on such factors as temperature and local gravity, and h_e, the equivalent manometer fluid height, depends on such factors as scale variations with temperature, relative specific weights, heights of the fluids involved, and capillary effects. This equation holds as long as manometer fluid displacement is constant.

The value of specific weight γ_m corrected for local gravity is

$$\gamma_c = \gamma_{st}(1 + C_g) \tag{8.5}$$

where C_g is the correction term given in equation (8.3) and γ_c is the corrected value of specific weight for any fluid whose specific weight is γ_{st} at standard gravity (32.1740 ft/s²) and a given Fahrenheit temperature t.

The variation of specific weight with temperature is described by empirical relationships; for mercury,

$$\gamma_{st} = \frac{0.491154}{1 + 1.01(t - 32) \times 10^{-4}} \tag{8.6}$$

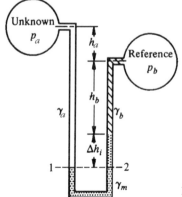

Figure 8.3   U-tube manometer.

and for water,

$$\gamma_{st} = \frac{62.2523 + 0.978476 \times 10^{-2}t - 0.145 \times 10^{-3}t^2 + 0.217 \times 10^{-6}t^3}{1728} \tag{8.7}$$

where γ_{st} is in pounds per cubic inch. These relationships have proved satisfactory for accurate manometer measurements; some typically useful values are listed in Table 8.3.

One important correction needed to find the equivalent manometer fluid height h_e is associated with the relative weights and heights of the fluids involved. The hydraulic correction, C_h, is

$$C_h = 1 + \frac{\gamma_h}{\gamma_m}\left(\frac{h_h}{\Delta h_i}\right) - \frac{\gamma_a}{\gamma_m}\left(\frac{h_a + h_h}{\Delta h_i} + 1\right) \tag{8.8}$$

The quantities in this equation are depicted in Figure 8.3.

The second main correction needed to determine h_e concerns capillary effects. The shape of the interface between two fluids at rest depends on the relative gravity and cohesion and adhesion forces between the fluids and their containing walls. In water–air–glass combinations, the crescent shape of the liquid surface (called the meniscus) is concave upward, and the water is said to wet the glass. In this situation, adhesive forces dominate, and water in a tube will be elevated by capillary action. For mercury–air–glass combinations, cohesive forces dominate, the meniscus is concave downward, and the mercury level in a tube will be depressed by capillary action, as shown in Figure 8.4.

Based upon the discussion of surface tension in Chapter 2, the capillary correction factor for manometers, C_c, can be written as

$$C_c = \frac{2\cos\theta_m}{\gamma_m}\left(\frac{\sigma_{a-m}}{r_a} - \frac{\sigma_{b-m}}{r_b}\right) \tag{8.9}$$

Table 8.3 Specific Weights of Mercury and Water (at Standard Gravity Value of 32.1740 Ft/S²)

Temperature (°F)	Specific weight γ_{st} (lb/in.³) Hg	H₂O	Temperature (°F)	Specific weight γ_{st} (lb/in.³) Hg	H₂O
32	0.491154	0.036122	68	0.489375	0.036062
36	0.490956	0.036126	72	0.489178	0.036045
40	0.490757	0.036126	76	0.488981	0.036026
44	0.490559	0.036124	80	0.488784	0.036005
48	0.490362	0.036120	84	0.488588	0.035983
52	0.490164	0.036113	88	0.488392	0.035958
56	0.489966	0.036104	92	0.488196	0.035932
60	0.489769	0.036092	96	0.488000	0.035905
64	0.489572	0.036078	100	0.487804	0.035877

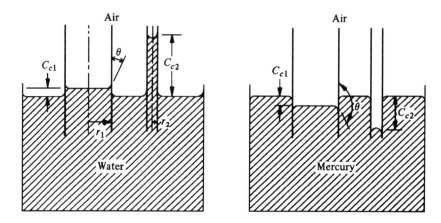

Figure 8.4   Meniscus effects in water and mercury.

where θ_m is the angle of contact between the manometer fluid and the glass, σ_{a-m} and σ_{b-m} are the surface-tension coefficients of the manometer fluid (m) with respect to the fluids (a and b) above it, and r_a and r_b are the radii of the tubes containing fluids a and b. Typical values for these capillary variables are given in Table 8.4. For mercury manometers, C_c is positive when the larger capillary effect occurs in the tube in which the manometer fluid is highest. For water manometers under the same conditions, C_c is negative.

When the same fluid is applied to both legs of a standard U-tube manometer, the capillary effect is often neglected. This can be done because the tube bores are approximately equal, and capillarity in one tube counterbalances that in the other. Capillary effect can be extremely important, however, and must also be considered in manometer-type instruments.

To minimize the effect of a variable meniscus, which can be caused by dirt, the method of approaching equilibrium, tube bore, and so on, the tubes are always tapped before reading and the measured liquid height is always based on readings taken at the center of the meniscus in each leg of the manometer. To reduce the capillary effect itself, the use of large-bore tubes (over $\frac{3}{8}$-in. diameter) is most effective.

When both hydraulic and capillary corrections are taken into account, the equivalent manometer fluid height is

$$\Delta h_e = C_b \Delta h_i \pm C_c \tag{8.10}$$

where Δh_i is the indicated height shown in Figure 8.3.

Table 8.4   Capillary Effects

| Condition | Surface tension σ | | Contact angle, θ (deg) |
	dyn/cm	lb/in.	
Mercury, vacuum, glass	480	2.74×10^{-3}	140
Mercury, air, glass	470	2.68×10^{-3}	140
Mercury, water, glass	380	2.17×10^{-3}	140
Water, air, glass	73	0.416×10^{-3}	0

8.5   MICROMANOMETER

While the manometer is a useful and convenient tool for pressure measurements, it is unfortunately limited when making low-pressure measurements. To extend its usefulness in the low-pressure range, micromanometers have been developed that have extended the useful range of low-pressure manometer measurements to pressures as low as 0.0002 in. H_2O.

One type of micromanometer is the so-called Prandtl-type micromanometer, in which capillary and meniscus errors are minimized by returning the meniscus of the manometer liquid to a null position before measuring the applied pressure difference. As shown in Figure 8.5, a reservoir, which forms one side of the manometer, is moved vertically to locate the null position. This position is reached when the meniscus falls within two closely scribed marks on the near-horizontal portion of the micromanometer tube. Either the reservoir or the inclined tube is then moved by a precision leadscrew arrangement to determine the micromanometer liquid displacement (Δh), which corresponds to the applied pressure difference. The Prandtl-type micromanometer is generally accepted as a pressure standard within a calibration uncertainty of 0.001 in. H_2O.

Another method for minimizing capillary and meniscus effects in manometry is to measure liquid displacements with micrometer heads fitted with adjustable, sharp, index points. Figure 8.6 shows a manometer of this type; the micrometers are located in two connected transparent containers. In some commercial micromanometers, contact with the surface of the manometric liquid may be sensed visually by dimpling the surface with the index point, or even by electrical contact. Micrometer-type micromanometers also serve as pressure standards within a calibration uncertainty of 0.001 in. H_2O.

An extremely sensitive high-response micromanometer uses air as the working fluid and thus avoids all the capillary and meniscus effects usually encountered in liquid manometry. In this device, shown in Figure 8.7, the reference pressure is mechanically amplified by centrifugal action in a rotating disk. The disk speed is adjusted until the amplified reference pressure just balances the unknown pressure. This null position is recognized by observing

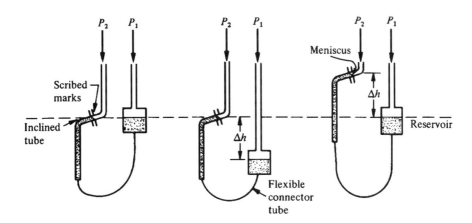

Figure 8.5   Two variations of the Prandtl-type micromanometer.

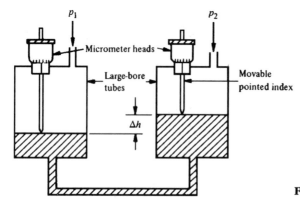

Figure 8.6   Micrometer-type manometer.

the lack of movement of minute oil droplets sprayed into a glass indicator tube. At balance, the air micromanometer yields the applied pressure difference

$$\Delta p_{micro} = Kpn^2 \qquad (8.11)$$

where p is the reference air density, n the rotational speed of the disk, and K a constant that depends on disk radius and annular clearance between the disk and the housing. Measurements of pressure differences as small as 0.0002 in. H_2O can be made with this type of micromanometer within an uncertainty of 1%.

A word on the reference pressures employed in manometry is pertinent at this point in the discussion. If atmospheric pressure is used as a reference, the manometer yields gage pressures. Because of the variability of air pressure, gage pressures vary with time, altitude, latitude, and temperature. If, however, a vacuum is used as reference, the manometer yields absolute pressures directly, and it may serve as a barometer. In any case, the absolute pressure is always equal to the sum of the gage and ambient pressures; by ambient pressure we mean the pressure surrounding the gage, which is usually atmospheric pressure.

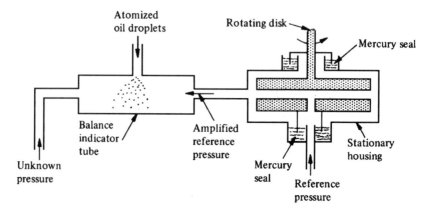

Figure 8.7   Air-type centrifugal micromanometer.

8.6   BAROMETERS

The reservoir or cistern barometer consists of a vacuum-reference mercury column immersed in a large-diameter ambient-vented mercury column that serves as a reservoir. The most common cistern barometer in general use is the Fortin type, in which the height of the mercury surface in the cistern can be adjusted. The operation of this instrument can best be explained with reference to Figure 8.8.

The datum-adjusting screw is turned until the mercury in the cistern makes contact with the ivory index, at which point the mercury surface is aligned with zero on the instrument scale. Next, the indicated height of the mercury column in the glass tube is determined. The lower edge of a sighting ring is lined up with the top of the meniscus in the tube. A scale reading and a vernier reading are taken and combined to yield the indicated mercury height at the barometer temperature.

Since atmospheric pressure on the mercury in the cistern is exactly balanced by the weight per unit area of the mercury column in the glass tube,

$$p_{baro} = \gamma_{Hg} h_{t0} \qquad (8.12)$$

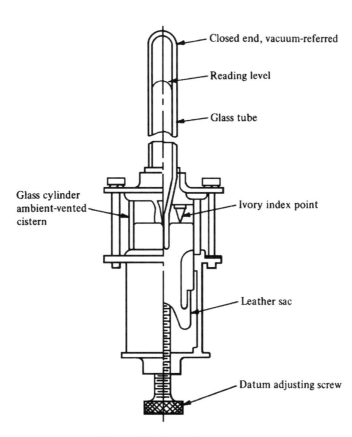

Figure 8.8   Fortin barometer

The referenced specific weight of mercury, γ_{Hg}, depends on such factors as temperature and local gravity; the referenced height of mercury, h_{t0}, depends on such factors as thermal expansion of the scale and the mercury.

When the scale zero of the Fortin barometer is adjusted to agree with the mercury level in the cistern, the correct height of mercury at temperature t will be

$$h_t = h_{ti}[1 + S(t - t_s)] \tag{8.13}$$

where S is the linear coefficient of thermal expansion of the scale, h_{ti} the indicated height of mercury, and t_s the temperature at which the scale was calibrated. Whenever t is greater than reference temperature t_0, the correct height of mercury at t will be greater than the height of mercury at t_0:

$$h_t = h_{t0}[1 + m(t - t_0)] \tag{8.14}$$

where m is the cubical coefficient of thermal expansion of mercury and h_{t0} the referenced height of the mercury column.

A temperature correction factor can be defined by

$$C_t = h_{t0} - h_{ti} \tag{8.15}$$

From equations (8.13) and (8.14),

$$C_t = \left[\frac{S(t - t_s) - m(t - t_0)}{1 + m(t - t_0)}\right]h_{ti} \tag{8.16}$$

When standard values of $S = 10.2 \times 10^{-6}/°F$, $m = 101 \times 10^{-6}/°F$, $t_s = 62°F$, and $t_0 = 32°F$ are substituted in equation (8.16),

$$C_t = -\left[\frac{9.08(t - 28.63)10^{-5}}{1 + 1.01(t - 32)10^{-4}}\right]h_{ti} \tag{8.17}$$

This temperature correction is zero at $t = 28.63°F$ for all values of h_{ti}. It can be approximated with very little loss of accuracy as

$$C_t \approx -9(t - 28.6)10^{-5}h_{ti} \tag{8.18}$$

The uncertainty introduced in h_{t0} by equation (8.18) is always less than 0.001 in. Hg for values of h_{ti} from 28.5 to 31.5 in. Hg and values of t from 60 to 100°F.

The specific weight γ_{Hg} in equation (8.12) must be based upon the local value of gravity and the reference temperature t_0; $\gamma_{Hg} = \gamma_{t0}(1 + C_g)$, where C_g is the gravity correction term given by equation (8.2) and γ_{t0} is 0.491154 lb/in.³ for Hg when $t_0 = 32°F$.

When the barometer is read at an elevation other than that of the test site, an altitude correction factor must be applied to the local absolute barometric pressure. The altitude correction factor, C_z, may be obtained from the pressure–height relationship. When this is done, we obtain

$$C_z = p_{baro}[e^{(Z_{baro} - Z_{site})/RT} - 1] \tag{8.19}$$

where Z_{baro} and Z_{site} are altitudes in feet, R is the gas constant (53.35 ft/°R), and T is the absolute temperature in degrees Rankine.

Other factors may also contribute to the uncertainty of h_{ti}. Proper illumination is essential to define the location of the crown of the meniscus. Precision meniscus sighting under optimum viewing conditions can approach ± 0.001 in. With proper lighting, contact between the ivory index and the mercury surface in the cistern can be detected to much better than ± 0.001 in.

To keep the uncertainty in h_{ti} within 0.01% ($\approx$0.003 in. Hg), the mercury temperature must be known within ± 1°F. Scale temperature need not be known to better than ± 10°F for comparable accuracy. Uncertainties caused by nonequilibrium temperature conditions can be avoided by installing the barometer in a uniform temperature room.

The barometer tube must be vertically aligned for accurate pressure determination. This is accomplished by a separately supported ring encircling the cistern; adjustment screws control the horizontal position.

Depression of the mercury column in commercial barometers is accounted for in the initial calibration setting at the factory. The quality of the barometer is largely determined by the bore of the glass tube. Barometers with a bore of $\frac{1}{4}$ in. are suitable for readings of about 0.01 in., whereas barometers with a bore of $\frac{1}{2}$ in. are suitable for readings down to 0.002 in.

8.7 McLEOD GAGE

The *McLeod Gage* is used in making low-pressure measurements. This instrument is shown in Figure 8.9 to consist of glass tubing arranged so that a sample of gas at unknown pressure can be trapped and then isothermally compressed by a rising mercury column. This amplifies the unknown pressure and allows measurement by conventional manometric means. All the mercury is initially contained in the area below the cutoff level. The gage is first exposed to the unknown gas pressure, p_1; the mercury is then raised in tube A beyond the cutoff, trapping a gas sample of initial volume $\overline{V}_1 = \overline{V} + ah_c$. The mercury is continuously forced upward until it reaches the zero level in the reference capillary B. At this time the mercury in the measuring capillary C reaches a level h, where the gas sample is at its final volume, $\overline{V}_2 = ah$, and at the amplified pressure, $p_2 = p_1 + h$. Then

$$p_1 \overline{V}_1 = p_2 \overline{V}_2 \tag{8.20}$$

$$p_1 = \frac{ah^2}{\overline{V}_1 - ah} \tag{8.21}$$

If $ah \ll \overline{V}_1$, as is usually the case,

$$p_1 = \frac{ah^2}{\overline{V}_1} \tag{8.22}$$

The larger the volume ratio $(\overline{V}_1/\overline{V}_2)$, the greater will be the amplified pressure p_2 and the manometer reading h. Therefore, it is desirable that measuring tube C have a small bore. Unfortunately, for tube bores under 1 mm, the compression gain is offset by reading uncertainty caused by capillary effects.

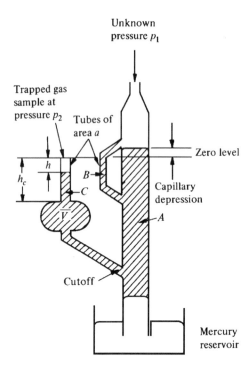

Figure 8.9   McLeod gage.

Reference tube B is introduced to provide a meaningful zero for the measuring tube. If the zero is fixed, equation (8.22) indicates that manometer indication h varies nonlinearly with initial pressure p_1. A McLeod gage with an expanded scale at the lower pressures exhibits a higher sensitivity in this region. The McLeod pressure scale, once established, serves equally well for all the permanent gases (those whose critical pressure is appreciably below room temperature).

There are no corrections to be applied to the McLeod gage reading, but certain precautions should be taken. Moisture traps must be provided to avoid taking any condensable vapors into the gage. Such vapors occupy a larger volume at the initial low pressure than they occupy in the liquid phase at the high reading pressures. Thus, the presence of condensable vapors always causes pressure readings to be too low. Capillary effects, although partially counterbalanced by using a reference capillary, can still introduce significant uncertainties, since the angle of contact between mercury and glass can vary $\pm$ 30°. Finally, since the McLeod gage does not give continuous readings, steady-state conditions must prevail for the measurements to be useful.

The mercury piston of the McLeod gage can be actuated in a number of ways. A mechanical plunger may force the mercury up tube A, while a partial vacuum over the mercury reservoir holds the mercury below the cutoff until the gage is charged. The mercury can then be raised by bleeding dry gas into the reservoir. There are also several types of swivel gages where the mercury reservoir is located above the gage zero during charging. A 90° rotation of the gage causes the mercury to rise in tube A by the action of gravity. In a variation of the McLeod principle, the gas sample may be compressed between two mercury columns, thus avoiding the need for a reference capillary and a sealed-off measuring capillary.

8.8   PRESSURE TRANSDUCERS

In the earlier sections of this chapter we discussed five pressure standards that can be used for either calibration or the measurement of pressure in static systems. In this section we discuss some commonly used devices that are used for making measurements using an elastic element to convert fluid energy to mechanical energy. Such a device is known as a *pressure transducer*. Examples of mechanical pressure transducers having elastic elements are deadweight free-piston gages, manometers, Bourdon gages, bellows, and diaphragm gages.

Electrical transducers have an element that converts this displacement to an electrical signal. Active electrical transducers generate their own voltage or current output as a function of displacement. Passive transducers require an external signal. The piezo-electric pickup is an example of an active electrical pressure transducer. Electric elements employed in passive electrical pressure transducers include strain gages, slide-wire potentiometers, capacitance pickups, linear differential transformers, and variable-reluctance units.

The Bourdon gage was described briefly earlier in Chapter 2. In this transducer the elastic element is a small-volume tube that is fixed at one end but free at the other end to allow displacement under the deforming action of the pressure difference across the tube walls. In the most common model, shown in Figure 8.10, a tube with an oval cross section is bent in a circular arc. Under pressure the tube tends to become circular, with a subsequent increase in the radius of the arc. By an almost frictionless linkage, the free end of the tube rotates a pointer over a calibrated scale to give a mechanical indication of pressure.

The reference pressure in the case containing the Bourdon tube is usually atmospheric, so that the pointer indicates gage pressures. Absolute pressure can be measured directly without evacuating the complete gage casing by biasing a sensing Bourdon tube against a reference Bourdon tube, which is evacuated and sealed as shown in Figure 8.11 on page 306. Bourdon gages are available for wide ranges of absolute, gage, and differential pressure measurements within a calibration uncertainty of 0.1% of the reading.

Another common elastic element used in pressure transducers is the bellows, shown as a gage element in Figure 8.12 on page 306. In one arrangement, pressure is applied to one side of a bellows and the resulting deflection is partially counterbalanced by a spring. In a differential arrangement, one pressure is applied to the inside of one sealed bellows, the other pressure is led to the inside of another sealed bellows, and the pressure difference is indicated by a pointer.

A final elastic element to be mentioned because of its widespread use in pressure transducers is the diaphragm. One such arrangement is shown in Figure 8.13 on page 308. Such elements may be flat, corrugated, or dished plates; the choice depends on the strength and amount of deflection desired. In high-precision instruments, a pair of diaphragms is used back to back to form an elastic capsule. One pressure is applied to the inside of the capsule; the other pressure is external. The calibration of this differential transducer is relatively independent of pressure magnitude.

Thus far we have discussed a few mechanical pressure transducers. In many applications it is more convenient to use transducer elements that depend on the change in the electrical parameters of the elements as a function of the applied pressure. The only active electrical pressure transducer in common use is the piezoelectric transducer. Sound-pressure instrumentation makes extensive use of piezoelectric pickups in such forms as hollow cylinders and disks. Piezoelectric pressure transducers are also used in measuring rapidly fluctuating or transient pressures. These transducers are difficult to calibrate by static

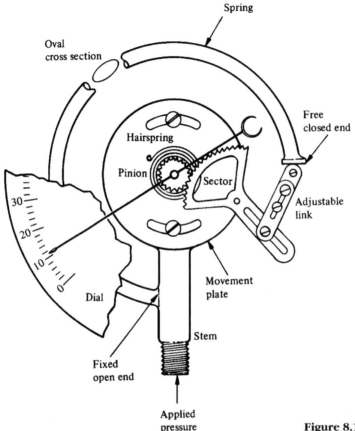

Figure 8.10   Bourdon gage.

procedures. In a recently introduced technique called electrocalibration, the transducer is calibrated by electric field excitation rather than by physical pressure. The most common passive electrical pressure transducers are the variable-resistance types.

The strain gage is probably the most used pressure transducer element. Strain gages operate on the principle that the electrical resistance of a wire varies with its length under load. In unbonded strain gages, four wires run between electrically insulated pins located on a fixed frame and other pins located on a movable armature, as shown in Figure 8.14 on page 308. The wires are installed under tension and form the legs of a bridge circuit. Under pressure the elastic element (usually a diaphragm) displaces the armature, causing two of the wires to elongate while reducing the tension in the remaining two wires. The resistance change causes a bridge imbalance proportional to the applied pressure.

The bonded strain gage takes the form of a fine wire filament, set in cloth, paper, or plastic and fastened by a suitable cement to a flexible plate, which takes the load of the elastic element. This is show in Figure 8.15 on page 309. Two similar strain gage elements are often connected in a bridge circuit to balance unavoidable temperature effects. The nominal bridge output impedance of most strain gage pressure transducers is 350 Ω, nominal excitation voltage is 10 V (ac or dc), and natural frequency can be as high as 50 Hz. Transducer resolution is infinite, and the usual calibration uncertainty of such gages is within 1% of full scale.

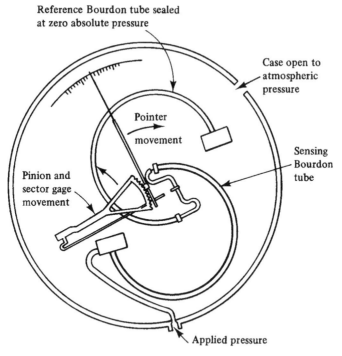

Figure 8.11   Bourdon gage for absolute pressure measurement.

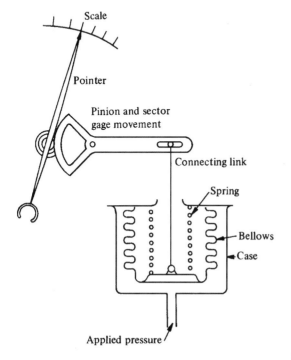

Figure 8.12   Bellows gage.

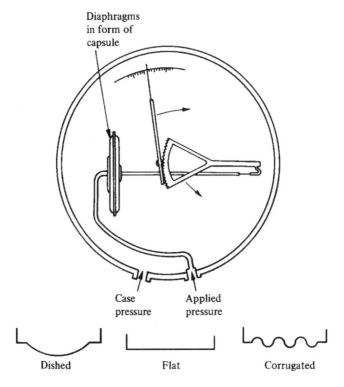

Figure 8.13   Diaphragm-based pressure transducer.

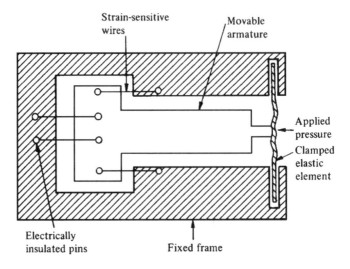

Figure 8.14   Typical unbonded strain gage.

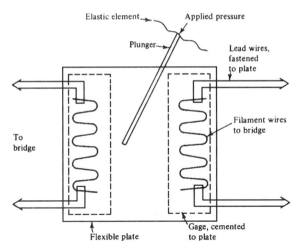

Figure 8.15 Typical bonded strain gage.

Many other forms of electrical pressure transducers are in use in industry, and only a few of these will be discussed briefly. These elements fall under the categories of potentiometer, variable-capacitance, linear variable differential transformer (LVDT), and variable-reluctance transducers. The potentiometer types are those that operate as variable resistance pressure transducers. In one arrangement, the elastic element is a helical Bourdon tube, while a precision wire-wound potentiometer serves as the electric element. As pressure is applied to the open end of the Bourdon tube, it unwinds, causing the wiper (connected directly to the closed end of the tube) to move over the potentiometer.

In the variable-capacitance pressure transducer, the elastic element is usually a metal diaphragm that serves as one plate of a capacitor. Under an applied pressure the diaphragm moves with respect to a fixed plate. By means of a suitable bridge circuit, the variation in capacitance can be measured and related to pressure by calibration.

The electric element in an LVDT is made up of three coils mounted in a common frame to form the device shown in Figure 8.16. A magnetic core centered in the coils is free to be displaced by a bellows, Bourdon, or diaphragm elastic element. The center coil, the primary winding of the transformer, has an ac excitation voltage impressed across it. The two outside coils form the secondaries of the transformer. When the core is centered, the induced voltages in the two outer coils are equal and 180° out of phase; this represents the zero-

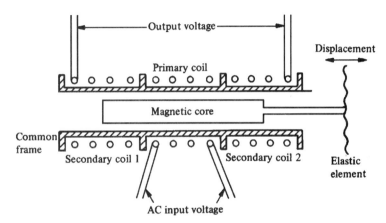

Figure 8.16 Linear variable differential transformer.

pressure position. When the core is displaced by the action of an applied pressure, however, the voltage induced in one secondary increases, while that in the other decreases. The output-voltage difference varies essentially linearly with pressure for the small core displacements allowed in these transducers. The voltage difference is measured and related to the applied pressure by calibration.

The last electric element that will be discussed is the variable-reluctance pressure transducer. The basic element of the variable-reluctance pressure transducer is a movable magnetic vane in a magnetic field. In one type the elastic element is a flat magnetic diaphragm located between two magnetic output coils. Displacement of the diaphragm changes the inductance ratio between the output coils and results in an output voltage proportional to pressure.

In this section the pressure elements that have been discussed were primarily for the measurement of pressures in systems that are at rest and in which the fluid is in equilibrium. The measurement of pressure in systems in which fluid is moving is discussed in the following sections of this chapter. The usual elements for making these measurements are the static or wall tap, static tubes, aerodynamic probes, Pitot tubes, Pitot-static tubes, and so on.

8.9   MEASUREMENTS IN A MOVING STREAM

The measurement of any property of a system requires that the method used should not change the magnitude of the property being measured, thereby yielding an erroneous result. There are many methods available to determine the pressure and velocity at a given location in a flow system, and the student is referred to the technical literature for detailed discussions on the subject. In particular, the most recent publication of the American Society of Mechanical Engineers, *Flow Meters: Their Theory and Application,* and the *Transactions of the ASME* are recommended for this purpose. At present we shall be concerned solely with those devices to which we can apply the Bernoulli equation to evaluate the desired results.

8.9.1 Piezometer

The measurement of the pressure of a fluid in motion is particularly difficult, since the presence of the measuring device will almost invariably alter the flow or change the magnitude of the pressure. An opening in the wall of a pipe is quite often used to measure the pressure of a flowing fluid. This static pressure is presumed to be the pressure of the undisturbed fluid, and the opening is known as a *piezometer opening*. The reading obtained with a single opening of this type will not be correct if the opening is not normal to the pipe surface or if the edges of the opening have burrs, and this method of pressure measurement is subject to large errors due to the presence of small perturbations in the installation of the piezometer opening (Figure 8.17). To obtain a more correct reading several holes are often placed around the periphery of a pipe in a given plane and connected together. This device is known as a *piezometer ring,* and the pressure is read directly on a manometer attached to the common pressure connection.

When a body is submerged in a fluid, such as in the case of an airfoil, it is often desirable to obtain the static pressure distribution along its boundaries. This can be determined by making piezometer openings along the surface and taking the corresponding static pressure readings by attaching tubes to these openings. In the case of an airfoil we can use these pressure readings to obtain the velocity distribution by applying the Bernoulli equation.

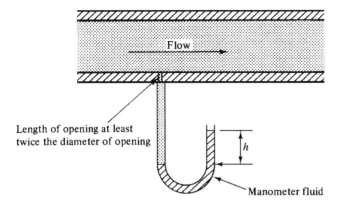

Length of opening at least
twice the diameter of opening

Manometer fluid   **Figure 8.17**   Piezometer opening.

8.9.2 The Pitot Tube and Pitot-Static Tube

Henri Pitot, a Frenchman, inserted an open-ended tube into the Seine River near Paris early in the eighteenth century and found that the rise of water in the vertical leg of the tube was proportional to the square of the velocity of the stream at the location at which he made his measurements. Tubes of this type are used to measure the local velocity of a flowing fluid and are called Pitot tubes in honor of Henri Pitot. Figure 8.18 shows a simple Pitot tube inserted into a stream with its open end facing into the stream flow. Both the tube and its opening are made as small as is practical to keep flow disturbances due to the presence of the tube to a minimum. The assumption is made that the streamline at the center of the Pitot tube is not disturbed by the tube. At the inlet to the tube the stream velocity is reduced to zero, and the level of liquid rises in the tube until it reaches the level h above the centerline of the tube. It is assumed that the pressure at section ② is the stagnation pressure of the streamline in question. By applying the Bernoulli equation to the streamline at the center of the tube successively at sections ① and ② in the flow, we obtain

$$\frac{V_1^2}{2g} + \frac{p_1}{\gamma} = \frac{p_2}{\gamma}$$

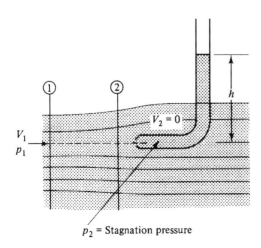

p_2 = Stagnation pressure

Figure 8.18   Pitot tube.

and

$$V = \sqrt{\frac{2g(p_2 - p_1)}{\gamma}}$$ (8.23)

The term $p_2 - p_1$ is sometimes known as the impact, dynamic, or velocity pressure of a stream and is equal to $\frac{1}{2}\rho V^2$. In equation (8.23) the specific weight, γ, is that of the flowing stream and not that which exists in the manometer leg. If the Pitot tube is installed in an open channel, then $(p_2 - p_1)/\gamma$ is simply h, as shown in Figure 8.19. For a closed conduit it is necessary to evaluate $(p_2 - p_1)/\gamma$ by making a second measurement. This can be done by combining a piezometer tube with the Pitot tube or using a piezometer opening in the pipe wall. Let us assume that a piezometer opening is used in conjunction with a Pitot tube, as shown in Figure 8.20, to measure the velocity of a fluid in a pipe. Further assume that a suitable manometer fluid whose specific weight is γ_M will be used to evaluate the pressure differential between the Pitot tube and the piezometer (as shown in Figure 8.20), and let us write the Bernoulli equation between points ① and ②:

$$\frac{V_1^2}{2g} + \frac{p_1}{\gamma} = \frac{p_2}{\gamma}$$ (8.24)

By evaluation of the manometer, we have

$$p_1 + \gamma h_1 + \gamma_M h_2 - \gamma(h_1 + h_2) = p_2$$

Simplifying yields

$$\frac{p_2 - p_1}{\gamma} = h_2\left(\frac{\gamma_M}{\gamma} - 1\right)$$ (8.25a)

and

$$V_1 = \sqrt{2gh_2\left(\frac{\gamma_M}{\gamma} - 1\right)}$$ (8.25b)

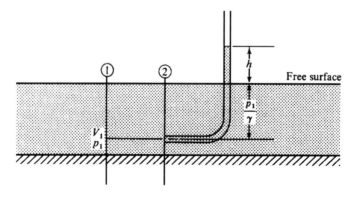

Figure 8.19   Pitot tube in an open channel.

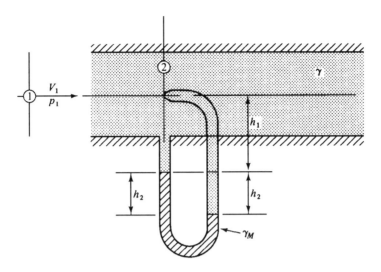

Figure 8.20   Pitot tube-piezometer for velocity measurement.

In the system shown in Figure 8.20 it is necessary to exercise care in installing the piezometer tap to avoid errors in the measurement that have been previously noted.

ILLUSTRATION 8.1   THE PITOT TUBE–PIEZOMETER

Water flows in a pipe. A Pitot-tube piezometer uses mercury as the metering fluid, and a deflection of 50 mm (h_2) is noted on the manometer. Calculate the flow velocity if $\gamma_w = 9810$ N/m^3 and the specific gravity of mercury is 13.6.

ILLUSTRATIVE PROBLEM 8.1

Given: Water flowing with $\gamma = 9810$ N/m^3; manometer fluid is mercury, sg = 13.6; $h_2 = 50$ mm

Find: V_1

Assumptions: Steady, incompressible flow; Bernoulli equation applicable

Basic Equations: Equation (8.25b): $V_1 = \sqrt{2gh_2\left(\dfrac{\gamma_M}{\gamma} - 1\right)}$

Solution: Apply Equation (8.25b) directly:

$$V_1 = \sqrt{2gh_2\left(\frac{\gamma_M}{\gamma} - 1\right)}$$

$$V_1 = \sqrt{2 \times 9.81\,\frac{\text{m}}{\text{s}^2} \times 0.050\text{ m}\left(\frac{13.6\gamma}{\gamma} - 1\right)}$$

$$V_1 = 3.52 \text{ m/s}$$

The arrangement shown in Figure 8.20 has the disadvantage of requiring two openings in the pipe wall, requires care in making the piezometer opening, and is usually rigid so that only a single measurement point can be obtained. To overcome these objections, the Pitot tube is combined with a static tube in a single unit called the Pitot-static tube. It usually consists of two tubes installed one inside of the other and connected to a manometer, as shown in Figure 8.21. The inner tube is open at the end and measures the total pressure of the stream, while the outer tube has holes drilled in it normal to the direction of the flow and measures the static pressure of the stream. With the manometer connections as shown in Figure 8.21, the manometer reads the dynamic pressure. Using an analysis similar to that used for the system shown in Figure 8.20, we obtain equation (8.26) for the system shown in Figure 8.21:

$$V_1 = \sqrt{2gh_2\left(\frac{\gamma_M}{\gamma} - 1\right)} \tag{8.26}$$

The accuracy of the Pitot-static tube is dependent on its construction and proper installation in the pipe. In addition, it is quite sensitive to misalignment with the flow. Although there are many Pitot-static tubes of standard size, shape, and construction that are intended to minimize these problems, it is always advisable to calibrate a particular installation. When this is done, equation (8.26) is modified to account for the nonideal action of the tube by introducing an empirical coefficient C:

$$V_1 = C\sqrt{2gh_2\frac{\gamma_M}{\gamma} - 1)} \tag{8.27}$$

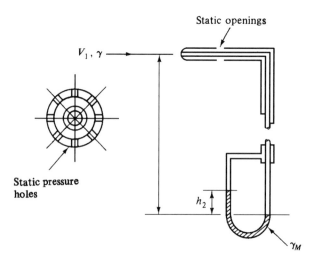

Figure 8.21   Combined Pitot-static tube.

ILLUSTRATION 8.2   PITOT-STATIC TUBE

A stream of water flows at 3 m/s, and the manometer (Figure 8.21) indicates a differential of 50 mm Hg. Determine C for this Pitot-static tube if the specific gravity of mercury is 13.6.

ILLUSTRATIVE PROBLEM 8.2

Given: $V_1 = 3$ m/s, $h_2 = 50$ mm Hg, sg $= 13.6$ for Hg, water flowing

Find: C, the coefficient of the Pitot-static tube

Assumptions: Steady, incompressible flow, Bernoulli equation applicable

Basic Equations: Equation (8.27): $V_1 = C \sqrt{2gh_2\left(\dfrac{\gamma_M}{\gamma} - 1\right)}$

Solution: Using equation (8.27) and solving for C,

$$C^2 = \frac{V_1^2}{2gh_2(\gamma_M/\gamma - 1)}$$

$$C^2 = \frac{(3 \text{ m/s})^2}{2 \times 9.81 \text{ m/s}^2 \times 0.05 \text{ m} (13.6 - 1)}$$

and $C^2 = 0.728$

yielding $C = 0.85$

The Pitot tube or the Pitot-static tube gives us the velocity at a point. As was noted in Chapter 6, the velocity at a given section of a pipe varies across the section. In order to obtain the average velocity in a pipe at a given section, a traverse across the diameter at 10 points is taken using annular circles of equal areas. The ten velocities are calculated and their arithmetic average is taken as the average velocity across the section, which is then used to calculate Q. Figure 8.22 on page 316 shows the location for the probe points.

8.9.3 Venturi Meter

As previously noted, the Pitot tube is used to measure the velocity at a point in a flow stream. For a permanent installation in a pipe that can be used to monitor the rate of fluid continuously with a minimum of flow losses, a venturi meter is frequently utilized. The venturi meter consists of a converging-diverging conical section of pipe arranged to give an increase in velocity and kinetic energy of the stream, thereby causing a measurable drop in pressure in the converging section. The diverging section is used to reconvert the increased kinetic energy of the stream back to pressure energy at the outlet of the meter with a minimum of turbulence and friction loss. The cone angle at the entrance to the throat is usually 20 to 30°, while the cone of the exit section is usually from 5 to 14°.

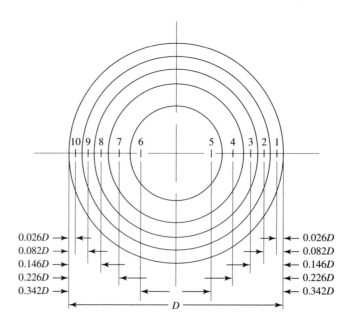

Figure 8.22   Location of velocity measurement points.

In a simplified installation, as shown in Figure 8.23, a manometer is connected to the inlet and throat of the venturi meter. Let us assume that there are no flow losses in the meter and write the Bernoulli equation for sections ① and ②; we obtain

$$\frac{p_1}{\gamma} + \frac{V_1^2}{2g} = \frac{p_1}{\gamma} + \frac{V_2^2}{2g}$$

and

$$\frac{V_2^2}{2g} - \frac{V_1^2}{2g} = \frac{p_1}{\gamma} - \frac{p_2}{\gamma} \tag{8.28}$$

The continuity equation requires that for the steady flow of an incompressible fluid $A_1V_1 = A_2V_2$. Therefore, equation (8.28) becomes

$$\frac{p_1}{\gamma} - \frac{p_2}{\gamma} = \frac{V_2^2}{2g}\left[1 - \left(\frac{A_2}{A_1}\right)^2\right] \tag{8.29}$$

and the volume rate of fluid flow (AV) is

$$AV = A_2V_2 = \left[\frac{A_2}{\sqrt{1 - (A_2/A_1)^2}}\right]\sqrt{\frac{2g(p_1 - p_2)}{\gamma}} \tag{8.30}$$

where the specific weight γ is that of the flowing fluid and not that of the manometer fluid shown in Figure 8.23. Using the manometer and a recording instrument, it is possible to

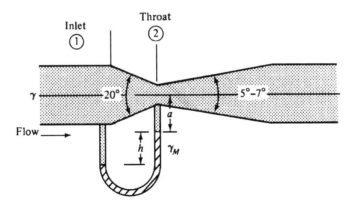

Figure 8.23   Simple venturi meter.

obtain a continuous record of the flow in a pipe. For the case of a real fluid with friction, we introduce a coefficient C_V into equation (8.30). Therefore,

$$(AV)_{actual} = \frac{C_V A_2}{\sqrt{1 - (A_2/A_1)^2}} \sqrt{\frac{2g(p_1 - p_2)}{\gamma}} \qquad (8.31)$$

In terms of the manometer shown in Figure 8.23

$$(AV)_{actual} = \frac{C_V A_2}{\sqrt{1 - (A_2/A_1)^2}} \sqrt{2g\left(\frac{\gamma_M}{\gamma} - 1\right)h} \qquad (8.32)$$

The velocity coefficient of a venturi meter is not constant and is found to be a function of the Reynolds number, the area ratio A_2/A_1, and the shape of the venturi. For greatest accuracy an in-place calibration is most desirable. Figure 8.24 on page 319 shows C_V for venturi meters of different sizes but having a 2:1 diameter ratio. In most instances of practical interest C_V can be taken to be 0.9 or greater.

The venturi meter is a low-head-loss device, since its diverging downstream section past the throat of the venturi acts to recover kinetic energy. The head loss from inlet to throat can be derived as

$$losses_{1 \to 2} = \left\{ \left(\frac{1}{C_V^2} - 1\right)\left[1 - \left(\frac{A_2}{A_1}\right)^2\right] \right\} \frac{V_2^2}{2g} \qquad (8.33)$$

For most installations the term in braces in equation (8.33) will be from 0.1 to 0.2.

ILLUSTRATION 8.3   THE VENTURI METER

A venturi meter consists of a 125-mm diameter inlet and a 50-mm diameter throat. If h in Figure 8.23 is 250 mm Hg (specific gravity $= 13.6$), determine the ideal quantity of water flowing, that is, $C_V = 1.0$.

ILLUSTRATIVE PROBLEM 8.3

Given: Venturi: $D_1 = 125$ mm, $D_2 = 50$ mm, $h = 250$ mm, sg $= 13.6$, water flowing

Find: Q

Assumptions: Ideal, incompressible flow; $C_V = 1$

Basic Equations: Continuity: $Q = A_1 V_1 = A_2 V_2$;

Manometer Equation (8.32): $Q = \dfrac{A_2}{\sqrt{1 - (A_2/A_1)^2}} \sqrt{2g\left(\dfrac{\gamma_M}{\gamma} - 1\right)h}$

Solution: Evaluating the areas,

$$A_1 = \frac{\pi}{4}(0.125 \text{ m})^2 = 1.227 \times 10^{-2} \text{ m}^2$$

$$A_2 = \frac{\pi}{4}(0.05)^2 = 1.963 \times 10^{-3} \text{ m}^2$$

and

$$\frac{A_2}{A_1} = \frac{1.963 \times 10^{-3} \text{ m}^2}{1.227 \times 10^{-2} \text{ m}^2} = 0.16$$

Using equation (8.32) with $C_V = 1$,

$$Q = \frac{A_2}{\sqrt{1 - (A_2/A_1)^2}} \sqrt{2g\left(\frac{\gamma_M}{\gamma} - 1\right)h}$$

$$Q = \frac{1.963 \times 10^{-3} \text{ m}^2}{\sqrt{1 - (0.16)^2}} \sqrt{2 \times 9.81 \text{ m/s}^2 (13.6 - 1)0.25 \text{ m}}$$

$$Q = 1.56 \times 10^{-2} \text{ m}^3/\text{s}$$

8.9.4 The Orifice

In Chapter 4 we derived Torricelli's theorem, which stated that the velocity of water from an ideal jet would equal the velocity of a body in free fall from the surface of the reservoir to the centerline of the jet. Referring to Figure 8.25, we found

$$V = \sqrt{2gh} \tag{8.34}$$

In deriving equation (8.34) the assumption was made that there were no losses in the flow as it issues from the tank. Actually, even for well-rounded openings, the streamlines will start to converge toward the opening while still in the tank and will continue to converge after leaving the tank. In effect, there is a contraction of the stream after leaving the plane of the opening that gives rise to a section of the stream having an area less than the geometric area of the opening in the tank. For a sharp-edged orifice the streamlines are shown as in Figure 8.26, and the area of minimum cross section is denoted as the *vena contracta*. By making the edges of the opening rounded or by using a tubular insert into the tank, it is possible to cause a rearrangement of the streamlines to yield a larger diameter at the vena contracta than for the case of a sharp-edged orifice. It is usual in practice to modify equation (8.34) to account for these effects by defining a coefficient of contraction,

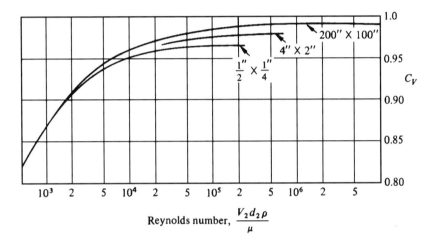

Figure 8.24   Venturi meter coefficients. (Reproduced with permission from *Elementary Fluid Mechanics* by J. K. Vennard, 4th ed., John Wiley & Sons, Inc., New York, 1961, p. 418.)

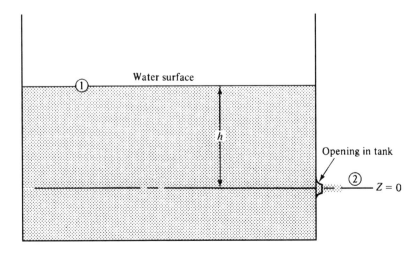

Figure 8.25   Torricelli's theorem.

C_C, which is the ratio of the area of the vena contracta to the area of the orifice (or opening). Thus,

$$C_C = \frac{A_{vc}}{A_0}$$ (8.35)

and we also define the ratio of the actual velocity to the velocity given by equation (8.34) as the coefficient of velocity. This gives

$$C_V = \frac{\text{actual velocity}}{\text{ideal velocity}}$$

and

$$V_2 = C_V \sqrt{2gh}$$ (8.36)

where C_V is the coefficient of velocity. Since the volume discharging through the orifice or opening is the product of the actual velocity in the jet multiplied by the jet area, we arrive at a new coefficient, called the *coefficient of discharge*, which is the ratio of the volume actually flowing in the jet to the volume calculated using the area of the opening and the ideal velocity given by equation (8.34). This also gives us the following relation between these three coefficients:

$$C_D = \frac{\text{actual volume flow}}{\text{ideal volume flow}}$$

and

$$C_D = C_C C_V$$ (8.37)

For sharp-edged orifices some typical values of these coefficients are $C_C = 0.61$, $C_V = 0.98$, and $C_D = 0.60$. For well-rounded openings, C_C can go to 1, which makes C_D go toward unity, since C_V usually lies in the range of 0.95 to 0.98.

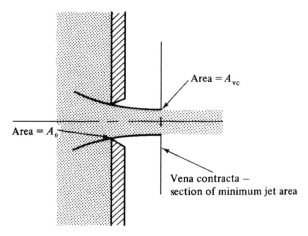

Figure 8.26   Sharp-edged orifice.

ILLUSTRATION 8.4   THE ORIFICE

A large tank has a well-rounded circular opening in its side located 10 ft below the water level in the tank. If the opening is 1 in. in diameter, determine the velocity of the efflux, the area of the vena contracta, and the volume rate of efflux if $C_C = 0.85$ and $C_V = 0.98$.

ILLUSTRATIVE PROBLEM 8.4

Given: $h = 10$ ft, $D = 1$ in., $C_C = 0.85$, $C_V = 0.98$, water flowing

Find: V_{actual}, A_{vc}, Q

Assumptions: Ideal, incompressible flow; Bernoulli equation applicable; C_C and C_V as given

Basic Equations: Equation (8.34): $V = \sqrt{2gh}$

$$V_{actual} = C_V V_{ideal}$$
$$A_{vc} = C_C A_0$$
$$Q = C_D Q_{ideal}$$

Solution: From equation (8.34),

$$V_{2\,ideal} = \sqrt{2gh} = \sqrt{2 \times 32.17 \text{ ft/s}^2 \times 10 \text{ ft}} = 25.4 \text{ ft/s}$$

The actual velocity is $V_{2\,actual} = C_V V_{2\,ideal} = 0.98\,(25.4 \text{ ft/s}) = 24.9$ ft/s. The geometric area of the opening of the orifice is $\dfrac{\pi}{4}\left(\dfrac{1 \text{ in.}}{12 \text{ in./ft}}\right)^2 = 0.00546$ ft^2. The area of

the vena contracta is 0.85×0.00546 ft$^2 = 0.00464$ ft^2.

We can obtain the volume rate of efflux as the product of the actual velocity multiplied by the area of the vena contracta or as C_D multiplied by the product of the area of the opening and the ideal jet velocity. Thus,

$$Q = 24.9 \,\frac{\text{ft}}{\text{s}} \times 0.00464 \text{ ft}^2 = 0.115 \text{ ft}^3/\text{s}$$

and as a check,

$$Q = (A_0 V_2)C_C C_V = (25.4 \text{ ft/s})(0.00546 \text{ ft}^2)(0.85)(0.98) = 0.116 \text{ ft}^3/\text{s}$$

In terms of gallons per minute, we have

$$Q = 0.116 \,\frac{\text{ft}^3}{\text{s}} \times 1728 \,\frac{\text{in.}^3}{\text{ft}^3} \times \frac{1}{231 \text{ in.}^3/\text{gal}} \times 60 \,\frac{\text{s}}{\text{min}} = 52 \text{ gal/min}$$

ILLUSTRATION 8.5   THE ORIFICE

Solve Illustrative Problem 8.4 if the opening is located 3 m below the surface of the water level and if the opening has a 25 mm diameter.

ILLUSTRATIVE PROBLEM 8.5

Given: h = 3 m, *D* = 25 mm, *C_C* = 0.85, *C_V* = 0.98, water flowing

Find: V_{actual}, A_{vc}, Q

Assumptions: Same as Illustrative Problem 8.4

Basic Equations: Same as Illustrative Problem 8.4

Solution: Proceeding as in Illustrative Problem 8.4, we obtain

$$V_{2\,ideal} = \sqrt{2gh} = \sqrt{2 \times 9.81 \text{ m/s}^2 \times 3 \text{ m}} = 7.67 \text{ m/s}$$

The actual velocity is $C_V V_2$ = 0.98 × 7.67 m/s = 7.52 m/s. The area at the vena contracta is

$$C_C A_0 = \frac{\pi}{4} (0.025 \text{ m})^2 \times 0.85 = 4.172 \times 10^{-4} \text{ m}^2$$

the actual discharge will equal the product of the actual velocity and the area of the vena contracta,

$$Q = AV = 4.172 \times 10^{-4} \text{ m}^2 \times 7.52 \frac{\text{m}}{\text{s}} = 3.14 \times 10^{-3} \text{ m}^3/\text{s}$$

Orifices are frequently used in pipelines to meter the quantity of fluid flowing. When used in this manner the orifice usually consists of a thin plate inserted into a pipe and clamped between flanges. The hole in the orifice plate is concentric with the pipe and static pressure taps are provided upstream and down stream of the orifice, as shown in Figure 8.27. The orifice meter causes a constriction in the flow and creates a jet smaller than itself. Applying the Bernoulli equation to this situation between sections ① and ② yields

$$V_2 = \frac{C_V}{\sqrt{1 - C_C^2 (A/A_1)^2}} \sqrt{\frac{2g(p_1 - p_2)}{\gamma}} \tag{8.38}$$

and

$$Q = AV = \frac{A C_C C_V}{\sqrt{1 - C_C^2 (A/A_1)^2}} \sqrt{\frac{2g(p_1 - p_2)}{\gamma}} \tag{8.39}$$

From a practical standpoint it is almost impossible to determine C_C and C_V separately for the orifice meter, and it is equally difficult to locate the pressure tap at the vena contracta. Therefore, the simplification is frequently made to combine these coefficients and to rewrite equation (8.39) as

$$Q = AC \sqrt{\frac{2g(p_1 - p_2)}{\gamma}} \tag{8.40}$$

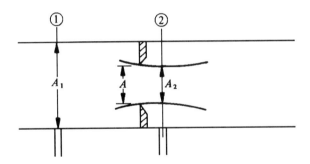

Figure 8.27   Orifice meter.

where C is defined as

$$C = \frac{C_D}{\sqrt{1 - (d_0/d_1)^4}} \qquad (8.40a)$$

and therefore includes the correction for the contraction of the jet. Figure 8.28 shows the co-efficient C for a square-edged orifice as a function of the Reynolds number $d_1 V_1 \gamma/\mu g$. It will be noted that the coefficient becomes constant above certain values of the Reynolds number. The advantages of the orifice meter are its relatively small size, the ease of installation in a pipe, and the fact that "standard" installations can be used without the need for cali-bration. The orifice meter acts as a partially open valve, however, and has a low discharge coefficient and consequently a relatively high head loss. In addition, errors can be caused due to poor placement leading to the orifice being eccentric to the inside diameter of the pipe, inaccurate location of the pressure taps, and a rough edge on the orifice opening.

Some of the problems associated with the orifice meter can be alleviated by using a flow nozzle, as shown in Figure 8.29. The coefficient C is defined as for the orifice meter, and it will be seen that the high value of this coefficient indicates relatively low losses in the nozzle. The manner in which C is defined as well as location of the pressure taps for this flow nozzle lead to values of C that can exceed unity.

The measurement of the three orifice coefficients (either singly or deriving the third from two of the others) presents an interesting problem in experimental fluid mechanics. We can readily visualize weighing the efflux for a given period of time and thus obtaining the actual rate of discharge. Knowing the area of the opening and the height of fluid above the opening we can directly compute C_D from its definition. We can (in principle) also mea-sure the area of the vena contracta by use of a caliper gage if proper care is taken. This mea-surement yields C_C, and C_V is calculated from the values of C_D and C_C. This method of mea-suring C_C is not very accurate, however, and rather than using it, let us attempt to measure C_V and to compute C_C. It is possible to measure C_V in the following manner. Let the jet flow and let us measure the position of a point along the trajectory downstream of the vena con-tracta. We can accomplish this by using a beam and sharp-edged pointers or else by simply noting the point at which the jet strikes a plate placed at some convenient position, as indi-cated in Figure 8.30 on page 325. The time for a particle to fall the distance y (neglecting air friction) is given as

$$t = \sqrt{\frac{2y}{g}}$$

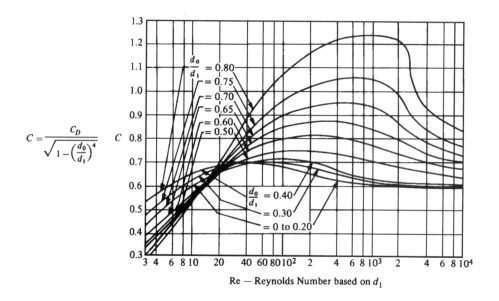

$$C = \frac{C_D}{\sqrt{1 - \left(\frac{d_0}{d_1}\right)^4}}$$

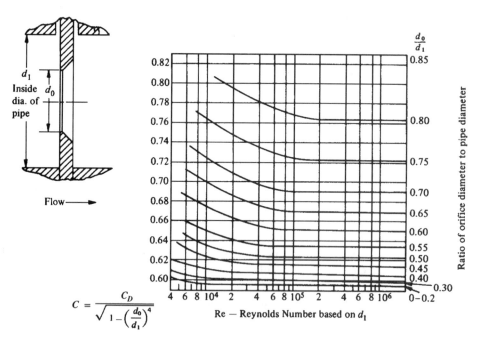

$$C = \frac{C_D}{\sqrt{1 - \left(\frac{d_0}{d_1}\right)^4}}$$

Figure 8.28 Flow coefficient C for square-edged orifices. (Data as modified for *Technical Paper 410,* Crane Co., Chicago, 1957, with permission. Lower-chart data from *Regeln fuer die Durch-flussmessung mit genormtem Duesen und Blenden,* VDI-Verlag G.m.b.H., Berlin, SNW, 7, 1937. Published as *Technical Memorandum 952* by the NACA.

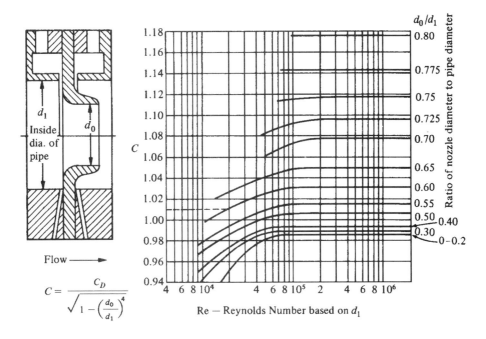

Figure 8.29   Flow coefficient C for nozzles. (Data as modified for *Technical Paper 410*, Crane Co., Chicago, 1957, with permission. Data from *Regeln fuer die Durchflussmessung mit genormtem Duesen und Blenden*, VDI-Verlag G.m.b.H., Berlin, SNW, 7, 1937. Published as *Technical Memorandum 952* by the NACA.)

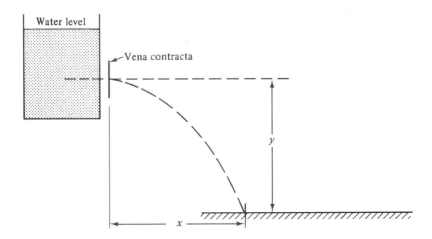

Figure 8.30   Trajectory method of evaluating C_V.

and the time for the particle to travel the distance x with the constant velocity of the vena contracta (V_{vc}) is

$$t = \frac{x}{V_{vc}}$$

Therefore,

$$V_{vc} = \frac{x}{\sqrt{2y/g}} \tag{8.41}$$

and

$$C_V = \frac{V_{vc}}{V_2} \tag{8.41a}$$

by definition. The velocity V_2 is given by equation (8.34).

ILLUSTRATION 8.6   ORIFICE COEFFICIENTS

In a test it is found that the orifice described in Illustrative Problem 8.4 discharges 65 gal/min when subjected to a head of 16 ft of water. The values of x and y were measured as shown in Figure 8.30 and were found to be 17.25 and 5.0 ft, respectively, Determine C_C, C_V, and C_D.

ILLUSTRATIVE PROBLEM 8.6

Given: $Q = 65$ gpm, $h = 16$ ft, $x = 17.25$ ft, $y = 5.0$ ft

Find: C_C, C_V, C_D

Assumptions: Ideal, incompressible flow; flow can be treated as a particle using its trajectory

Basic Equations: Equation (8.32): $V_{2\,ideal} = \sqrt{2gh}$

$$C_D = \frac{\text{actual discharge}}{\text{ideal discharge}}$$

$$C_V = \frac{\text{actual velocity}}{\text{ideal velocity}}$$

$$C_C = \frac{C_D}{C_V}$$

Solution: $V_{2\,ideal} = \sqrt{2gh} = \sqrt{2 \times 32.17 \text{ ft/s}^2 \times 16 \text{ ft}} = 32.1$ ft/s

From Illustrative Problem 8.4, $A_0 = 0.00546$ ft^2. Therefore, the ideal flow is

$$(32.1 \text{ ft/s})(60 \text{ s/min})(0.00546 \text{ ft}^2) \times \frac{1728 \text{ in.}^3}{231 \text{ in.}^3/\text{gal}} = 78.7 \text{ gal/min}$$

From its definition,

$$C_D = \frac{65}{78.7} = 0.826$$

Using trajectory data, we obtain

$$V_{vc} = \frac{x}{\sqrt{2y/g}} = \frac{17.25 \text{ ft}}{\sqrt{(2 \times 5 \text{ ft})/32.17 \text{ ft/s}^2}} = 30.94 \text{ ft/s}$$

and

$$C_V = \frac{30.94}{32.1} = 0.964$$

Having C_D and C_V,

$$C_C = \frac{C_D}{C_V} = \frac{0.826}{0.964} = 0.857$$

8.10 REVIEW

The need to be able to measure such properties as pressure and velocity is almost self-evident. For the serious worker in the field of fluid mechanics, measurements represent both a means to an end as well as the end itself. With this in mind, we started our discussion by considering pressure measurements in static fluids. Certain fundamental items were discussed, and from this we considered the practical equipment that is used, such as pressure gages and pressure transducers. A good part of this chapter was devoted to a qualitative discussion of equipment and a quantitative discussion of the principles of operation of this equipment. Measurements in static fluids, however, are only a part of the need of the technician. We therefore went on to the consideration of making measurements in fluids that are flowing. After noting that the presence of the measuring device that is inserted into a fluid stream will either alter the flow or change the value of the variable being measured, we first considered the piezometer, or an opening in the wall of a pipe that is used to measure the static pressure of the flowing fluid. From this base we then considered the Pitot and the Pitot-static tube. The Pitot tube is used to measure the dynamic or velocity pressure of a stream and the Pitot-static tube is used to measure the velocity or flow. In order to have a permanent device installed in a pipe to monitor the rate of fluid flow continuously with a minimum of losses, we considered both the venturi meter and the orifice meter. Both of these devices were analyzed using the Bernoulli equation, and experimental coefficients were applied to them to bring their accuracy in line with known values. As our final consideration we discussed the trajectory method of obtaining the coefficients of an orifice. This is a simple method which when performed carefully is capable of yielding accurate calibrations of an orifice with reasonable effort.

KEY TERMS

Terms of importance in this chapter:

Barometer: a cistern device used to obtain atmospheric pressure. The working fluid is usually mercury.

Coefficient: (a) contraction: a correction factor used to obtain the actual area from the ideal area of a device; (b) discharge: a correction factor used to obtain the actual discharge quantity from the ideal discharge quantity of a device; (c) velocity: a correction factor used to obtain the actual velocity from the ideal velocity of a device.

Dead-weight piston gage: an accurately machined piston cylinder device used to provide a basic calibration of pressure gages.

Fortin barometer: a particular type of cistern barometer.

Linear variable pressure transducer (LVDT): an electrical device whose output varies linearly with pressure for small displacements of its core. This output can be calibrated and amplified.

McLeod gage: an absolute pressure gage used for very low (vacuum) pressure measurements.

Micromanometer: a manometer used in low-pressure measurements.

Micron: 1000 mm or 10^{-6} m.

Piezometer: an opening in the side of a pipe wall that is used to measure static pressure.

Pitot-static tube: Pitot tube and a piezometer combined in one device.

Pitot tube: an open-ended tube used to determine stream velocities.

Pressure transducer: a device that uses an elastic element to convert fluid energy to mechanical or electrical energy.

Strain gage: an electrical transducer that uses the principle that the electrical resistance of an element varies with its length when it is placed under load.

Trajectory method: a method of calibrating an orifice using the trajectory of the stream that issues from the orifice.

Vena contracta: the section of minimum area of a jet issuing from an orifice.

KEY EQUATIONS

Dead-weight piston gage $$P_{\text{dw}} = \frac{F_e}{A_e} \qquad\qquad (8.1)$$

Manometer $$\Delta p_{\text{mano}} = \gamma_m \Delta h_e \qquad\qquad (8.4)$$

Micromanometer $$\Delta p_{\text{micro}} = Kpn^2 \qquad\qquad (8.11)$$

Barometer $$p_{\text{baro}} = \gamma_{\text{Hg}} h_{t\,0} \qquad\qquad (8.12)$$

McCleod gage $$p_1 = \frac{ah^2}{\bar{V}_1} \qquad\qquad (8.22)$$

| Pitot tube | $V = \sqrt{\dfrac{2g(p_2 - p_1)}{\gamma}}$ | (8.23) |

| Pitot-static tube | $V_1 = \sqrt{2gh_2\left(\dfrac{\gamma_M}{\gamma} - 1\right)}$ | (8.25b) |

| Venturi meter | $AV = \dfrac{A_2}{\sqrt{1 - (A_2/A_1)^2}}\sqrt{\dfrac{2g(p_1 - p_2)}{\gamma}}$ | (8.30) |

| Venturi meter | $(AV)_{actual} = C_V[AV \text{ equation } (8.30)]$ | (8.31) |

| Venturi meter | $(AV)_{actual} = \dfrac{C_V A_2}{\sqrt{1 - (A_2/A_1)^2}}\sqrt{2g\left(\dfrac{\gamma_M}{\gamma} - 1\right)h}$ | (8.32) |

| Venturi meter | $losses_{1 \to 2} = \left\{\left(\dfrac{1}{C_V^2} - 1\right)\left[1 - \left(\dfrac{A_2}{A_1}\right)^2\right]\right\}\dfrac{V_2^2}{2g}$ | (8.33) |

| Orifice | $C_C = \dfrac{A_{vc}}{A_0}$ | (8.35) |

| Orifice | $V_2 = C_V\sqrt{2gh}$ | (8.36) |

| Orifice | $C_D = C_C C_V$ | (8.37) |

| Orifice meter | $Q = AC\sqrt{\dfrac{2g(p_1 - p_2)}{\gamma}}$ | (8.40) |

| Orifice meter | $C = \dfrac{C_D}{\sqrt{1 - [(d_0/d_1)^4}}$ | (8.40a) |

| Trajectory method | $V_{vc} = \dfrac{x}{\sqrt{2y/g}}$ | (8.41) |

| Trajectory method | $C_V = \dfrac{V_{vc}}{V_2}$ | (8.41a) |

QUESTIONS

1. Seven different pressure units are listed in Table 8.1. Look at each unit and try to give a reasonable explanation as to why they originated or how they are used.

2. The micron is an extremely small unit of pressure, 10^{-6} m of mercury. Where would such a small unit be used?

3. Does a Bourdon-type gage yield only gage pressure?

4. If you had a Bourdon-type pressure gage, how would you calibrate it?

5. When compared to the other pressure standards, the McCleod gage has a problem associated with its use. What is this problem?

6. Does a barometer give absolute or gage pressure?

7. Why are strain gages so useful and commonly used?

8. Where would there be an application for an LVDT?

9. Why is a piezometer ring more desirable than a piezometer for measuring the pressure of a fluid in motion?

10. Is a piezometer used to measure the static pressure in a flowing fluid or the dynamic pressure?

11. Most pressure-measuring devices alter the flow or change the pressure being measured in a flowing fluid. What assumption is made that allows the Pitot tube to give a correct measure of the velocity in a stream?

12. What advantages does the Pitot-static tube have over the Pitot tube combined with the piezometer?

13. What are the advantages of the venturi meter over the Pitot-static tube?

14. What advantage does the orifice have over the venturi meter? What disadvantages does it have when compared to the venturi meter?

15. The coefficients of the venturi meter and the orifice depend on which two variables?

16. What are approximate values for both the orifice and venturi discharge coefficients?

17. Can the discharge coefficient of either the venturi meter or the orifice meter exceed unity? Explain your answer.

18. What is the advantage of the trajectory method for calibrating an orifice?

19. How else other than using the trajectory method can you calibrate an orifice?

20. How would you obtain each of the coefficients (contraction, discharge, and velocity) independently of the others?

PROBLEMS

Use $\gamma = 62.4$ lb/ft^3 (9810 N/m^3) for water unless indicated otherwise. Use Tables 2.4 and 2.5 for water properties at stated temperatures.

Nozzles and Orifices

8.1. A tank has a water level of 25 ft above a datum plane. Five feet above the datum plane there exists a 6-in. opening in the tank. If there are no flow losses, determine the velocity of the water leaving the opening and the quantity of water flowing at this instant in gallons per minute.

8.2. Assume that the nozzle shown in Figure P8.2 has no losses. Determine the velocity of the jet.

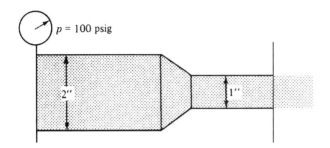

Figure P8.2

8.3. Solve Problem 8.1 if $C_C = 0.61$ and $C_V = 0.95$.

8.4. Solve Problem 8.2 if $C_V = 0.96$.

8.5. A large tank has a 50-mm opening located 4 m below the surface of oil having a specific gravity of 0.85. If $C_C = 0.85$ and $C_V = 0.95$, determine the volume rate of flow through the opening.

*8.6. Derive an equation for the trajectory of a jet issuing from an orifice with a head of h ft and a velocity coefficient of unity. Assume resistance to be negligible.

*8.7. A jet issues from an orifice and strikes horizontally 10 ft and vertically 2 ft from the vena contracta (Figure P8.7). Determine the head h on the jet.

8.8. A 2-in. diameter sharp orifice discharges water under a head of 15 ft. If the discharge rate is 35 lb/s and $x = 15.5$ ft and $y = 4.0$ ft, determine C_V, C_C, and C_D. See also Figure P8.7.

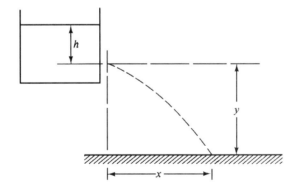

Figure P8.7

8.9. An orifice 5 in. in diameter is placed in a 10-in. diameter pipe in which water at 80°F flows. What difference should be read on a differential mercury–water manometer for 300 gpm?

8.10. Water at 70°F flows in a 6-in. pipeline that has a 4-in. diameter flow nozzle. A mercury–water manometer has a gage deflection of 12 in. of mercury. Calculate the volumetric flow.

8.11. Determine the discharge in a 6-in. diameter pipe with a 4-in. diameter orifice for water at 60°F when a differential mercury–water manometer reads a difference of 3 in.

Pitot and Pitot-Static Tubes

8.12. A Pitot tube is used to determine the velocity of a stream of water. If the specific weight of the water is 60 lb/ft³ and the water rise in a vertical tube attached to the Pitot tube is 5 in., determine the velocity of the stream at this point.

8.13. A body is immersed in a stream of water in such a manner that the water is caused to stagnate against the front face of the body. If the body is 10 ft below the surface of the stream and a Pitot tube in the face of the body has a liquid rise above the stream surface of 1 ft, calculate the stream velocity at this point.

8.14. A Pitot-static tube is directed into a stream of water flowing with a velocity of 10 ft/s. If the manometer fluid is mercury having a specific weight of 850 lb/ft³, determine its coefficient when the manometer differential is 1.8 in.

8.15. A Pitot-static tube is placed in a stream of water that is flowing with a velocity of 9.9 ft/s. If a mercury–water differential manometer reads 1.26 in., determine the coefficient of this Pitot-static tube.

8.16. If a Pitot-static tube has a coefficient of 0.99 when placed in a water stream flowing with a mercury–water manometer differential of 2.2 in., determine the velocity of the stream.

8.17. If the Pitot-static tube in Problem 8.16 has a coefficient of 1.15, determine the velocity of the stream. Is this coefficient, which is greater than unity, possible?

8.18. A Pitot-static tube is inserted into a pipe that carries water. A differential manometer using mercury (sg = 31.6) shows a deflection of 4 in. Calculate the ideal velocity of the water in the pipe.

8.19. If the total pressure on a submerged research vessel is 220 psig at a level of 500 ft, determine its velocity. Assume the specific weight of water to be 62.4 lb/ft³ and that it remains constant.

8.20. An airplane is designed to have a cruising speed of 500 mi/h. If the specific weight of air is 0.075 lb/ft³, determine the pressure between the openings of a Pitot-static tube.

8.21. A Pitot-static tube is used to measure the flow of air in a duct. If the density of the air is 0.075 lb/ft³ and a differential manometer reads 1 in. of water ($\gamma = 62.4$), determine the air velocity.

8.22. An airplane travels at 400 mi/h in air whose density is 0.07 lb/ft³. A Pitot-static tube is installed with the static connection on a wing and the Pitot connection facing forward into the airstream. If the air velocity relative to the wing is 480 mi/h, determine the reading of a differential pressure gage to which this instrument is connected.

Venturi meter

8.23. A venturi meter consists of a 4-in. diameter inlet and a 2-in. diameter throat (Figure P8.23). If 1 ft³/s of water is flowing in the meter, determine h if the specific weight of mercury is 850 lb/ft³ and losses can be neglected.

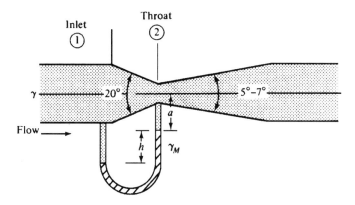

Figure P8.23

8.24. Solve Problem 8.23 if C_V for this meter is 0.9.

8.25. A venturi meter has a pipe diameter of 4 in. and a throat diameter of 2 in. It carries water at 140°F. If a pressure difference of 8 psi is observed between the two metering points, determine the volume discharge rate.

8.26. Calculate the ideal gage difference on a differential mercury–water manometer for 1.4 cfs discharge of water through a 6-in. by 4-in. venturi meter.

8.27. A venturi meter consists of a 100-mm diameter inlet and a 40-mm diameter throat. If h is 200 mm of mercury (sg = 13.6), what is the ideal quantity of water flowing?

8.28. Solve Problem 8.27 if C_V is 0.91.

*8.29. A venturi meter is installed in a pipe so that its axis is vertical. An identical venturi meter is installed in a pipe the same size with its axis horizontal. If each meter gives the same reading on a differential manometer, is the flow the same in both pipes, assuming the density of the fluids to be the same for each installation?

*8.30. Show that independent of the angle of the axis of a venturi meter, equation (8.30) or (8.31) holds.

PUMPS

LEARNING GOALS

After reading and studying this chapter you should be able to:

1. Classify pumps by their displacement, delivery, and motion.
2. Explain why pumps cavitate.
3. Determine the vapor pressure of water as a function of temperature.
4. Calculate the net positive suction head on a system.
5. Calculate pump performance using the terms *required head* and *available head*.
6. Use the affinity laws to evaluate the performance of similar pumps.
7. Use the specific speed to classify pumps.
8. Use the specific speed to select pumps.

9.1   INTRODUCTION

To force a fluid to flow against a resistance, to develop pressure, to transmit power, and so on, some form of machine must be provided. In the pump system mechanical energy is converted to fluid energy. For most modern pumps we find that electric motors provide the mechanical driving force to the device known as the *pump*, which converts the mechanical energy to fluid power. Pumps are made in all sizes, from small pumps used to supply control signals in fluidic systems to huge pumps used in conjunction with earth-moving equipment. In the selection of a pump for a given application, it is necessary to know many factors, such as the capacity required, the properties of fluid being pumped, conditions at inlet and outlet of the pump, the power source for driving the pump, and so on. The purpose of this chapter is to describe the types of pumps in general use and the factors that determine their performance. The reader should note that a good source of performance data for pumps is the literature published by manufacturers.

9.2   CLASSIFICATION OF PUMPS

Although there are many types of pumps, it is possible to classify pumps into three different categories: *displacement*, *delivery*, and *motion*. Table 9.1 shows the various major categories and subcategories under each major division.

Figure 9.1 illustrates four *positive displacement* types of pumps: the external gear pump, the internal gear pump, the vane pump, and the lobe pump. As the name "positive displacement pumps" implies, these pumps are designed to provide a given amount of fluid to a system for each revolution of the pump. The positive displacement pump is made with very close clearances between the rotating and stationary parts to minimize flow back through the pump, or *slip*. In general, these pumps will pump against high pressure but their volumetric capacity is low. Because of the positive displacement per revolution feature, it is necessary to protect positive displacement pumps with relief valves to prevent damage by overpressurization.

The *external gear pump* shown in Figure 9.1a consists of a drive gear and a driven gear enclosed within a precision housing. As the gears rotate, fluid enters the space at the inlet to the pump. The fluid is then trapped between the gear teeth and the casing and is transported around the periphery of both gears to the outlet. The remeshing of the teeth at the outlet forces the fluid out of the outlet of the pump. Care must be taken in operating external gear pumps to ensure the correct rotation of the pump. If the pump is operated in the wrong direction, it can be severely damaged or destroyed. This type of gear pump can be used in fluid power applications for pressures up to 3000 psi (21 MPa) and capacities up to 150 gpm (0.01 m^3/s, or 0.6 m^3/min). This type of pump has a low initial cost, a long operating life, and relatively high efficiencies.

Figure 9.1b shows an *internal gear pump*, where the basic action is similar to that of the external gear pump. The unit has an outer ring gear and an off-center gear with fewer teeth. There is a crescent-shaped spacer around which the fluid is carried.

Table 9.1   Pump Classification

I.	Displacement
	a. Positive
	b. Nonpositive
II.	Delivery
	a. Constant volume
	b. Variable volume
III.	Motion
	a. Rotary
	b. Reciprocating

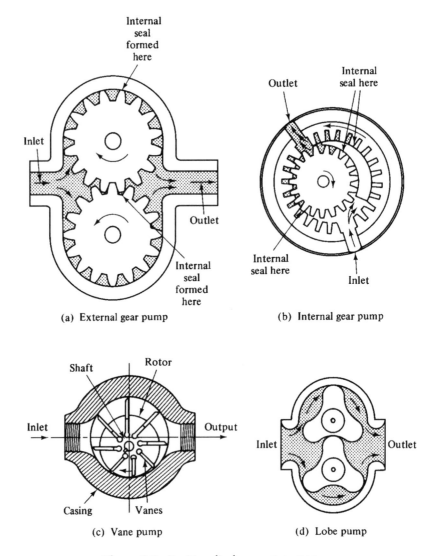

Figure 9.1   Positive displacement pumps.

The *vane pump* is shown in Figure 9.1c. As can be seen, it consists of a series of vanes that are almost radially located in a relatively large diameter rotor. These units use spring force, fluid pressure, and centrifugal force to cause the vanes to stay in contact with the outer casing. This action traps fluid in the space between the vanes and casing and the fluid is transported to the outlet. These pumps show a very good reliability and are not costly. As the vanes wear, the wear is compensated for by the outward motion of the vanes. Vane pumps have been widely used on machine tools and have been used for pressures up to 1500 psi (10.5 MPa).

The pump shown in Figure 9.1d is the *lobe pump*. Although there are similarities between the lobe pump and the external gear pump, there are certain significant differences.

In the lobe pump both rotors are externally driven and neither element contacts the other. The effect of the rotary motion of the lobes is to transfer fluid around their outer peripheries to the outlet port. This type of pump has a higher volume delivery rate per revolution than does the gear pump and is also quieter. Thus, we find it used as a supercharger on engines.

In the *reciprocating pumps* shown in Figure 9.2, the piston is constrained to move back and forth in the cylinder. Figure 9.2a shows schematically a *single acting* (simplex) *reciprocating pump*. In this pump, when the piston moves to the left, the inlet check valve opens to allow fluid into the cylinder and the outlet check valve closes. When the piston moves to the right, the discharge check valve opens and the inlet check valve closes, allowing the fluid to flow out of the cylinder into the discharge pipe. This single acting pump with a crank drive delivers a pulsating flow, since the fluid is delivered only when the piston moves to the right and not during the intake portion of the cycle. Neglecting leakage of the fluid past the piston, the pump discharge volume will equal the volume displaced by the piston, giving us essentially a positive displacement pump.

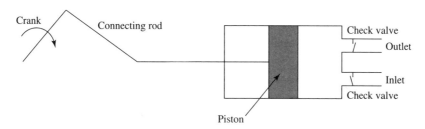

(a) Single acting reciprocating pump

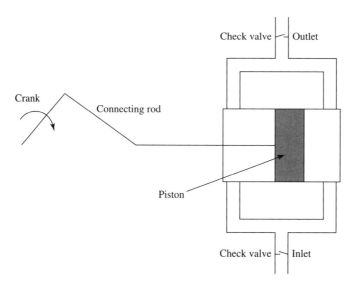

(b) Double acting reciprocating pump

Figure 9.2   Reciprocating pumps

Figure 9.2b shows schematically a *double acting* (duplex) *reciprocating pump.* As the piston moves in this pump we see that one side is discharging fluid while the other side takes fluid in. This yields a continuous discharge of fluid and a smoother delivery. Further smoothing of the flow can be achieved using more pistons.

Reciprocating pumps are used with a wide variety of fluids, but they are not well-suited for handling dirty or very viscous fluids, since these types of fluids tend to clog inlet and outlet valves. Most reciprocating pumps are operated at relatively slow speeds (up to 200 crankshaft revolutions per minute) for small rates of discharge and for high delivery pressures.

Nonpositive displacement pumps are generally used to transfer large volumes of fluids at pressures that are relatively low. Among the many types of nonpositive displacement pumps that can be found, the principal ones are the *centrifugal type* of unit and the *axial flow* type of unit. Figure 9.3 shows these units schematically.

Figure 9.3a shows the type of centrifugal pump known as the *diffuser* type and Figure 9.3b shows the *volute* type. In the diffuser-type pump there is a series of fixed vanes surrounding the impeller. In the diffuser there is a reduction in velocity and an increase in

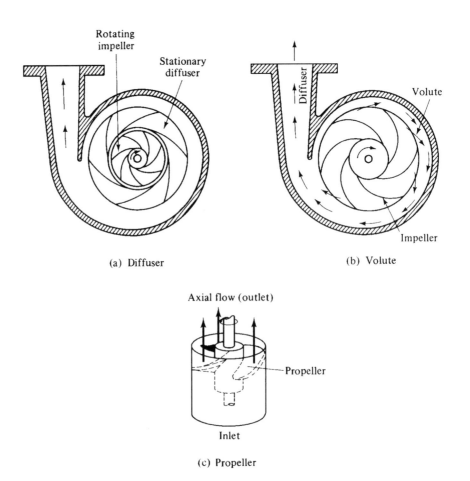

Figure 9.3   Nonpositive displacement pumps.

static pressure, which tends to make the static pressure distribution around the impeller uniform. In Figure 9.3b, which illustrates the volute-type centrifugal pump, the fluid is discharged directly from the impeller into the volute. Notice that in both types of centrifugal pump there is only a single moving part, the impeller. This class of pump (the centrifugal type) takes fluid in at the center of the impeller and ejects it to the discharge. Since there must be large clearances for the centrifugal action to be effective, these pumps have high slip rates. Centrifugal pumps are relatively low in cost, are highly reliable, and can handle almost all types of fluids.

The *propeller-type* pump shown in Figure 9.3c is also a nonpositive displacement pump that is sometimes called an *axial flow* pump. These units are built in a large range of sizes where a large discharge rate is required under a relatively low head. Since the flow is straight through, such pumps can handle solids in suspension quite readily. For this reason they are used in such applications as sewage disposal, irrigation, and drainage.

Figures 9.4 and 9.5 show two large commercial types of centrifugal fans having different types of blading. Figure 9.6 shows three general types of centrifugal fans with different blading configurations. Figure 9.7 indicates a simple type of axial flow fan.

The performance of a fan is usually given graphically. Figure 9.8 shows a typical performance curve for a centrifugal fan. Efficiency, horsepower, and pressure developed are given as a function of capacity (discharge rate).

Figure 9.4   Fan with backward-curved blades and vane-control inlet, suitable for forced-draft service, handling room air. (Courtesy of Babcock & Wilcox Corporation.)

Figure 9.5   Fan with radial-tip blades and double inlet, arranged for induced-draft service. (Courtesy of Babcock and Wilcox Corporation.)

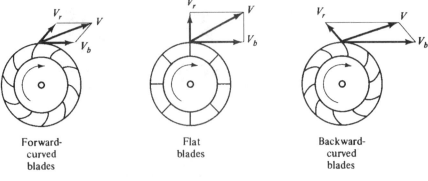

Forward-
curved
blades

Flat
blades

Backward-
curved
blades

V = Absolute velocity of air leaving blade
(shown equal for all three blade types)
V_r = Velocity of air leaving blade relative to blade
V_b = Velocity of blade tip

Figure 9.6   Three general types of centrifugal fans with vectors showing relative tip speed required for equal velocity leaving blades.

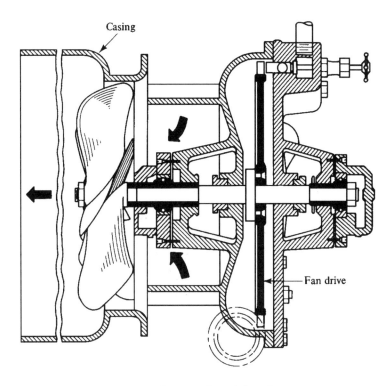

Figure 9.7   Simple type of axial flow fan.

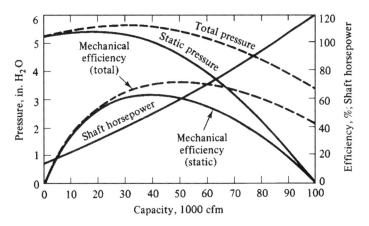

Figure 9.8   Typical characteristic performance curves for a centrifugal pump.

One advantage of the centrifugal pump is that the discharge valve can be closed without damaging the pump. Thus, if for some reason the discharge becomes blocked, the unit will not be damaged. It is unnecessary to protect this pump with a pressure relief valve. The only effect of a blockage is a churning of the fluid with a subsequent rise in the fluid temperature.

9.3   PUMP PERFORMANCE—CAVITATION

Pumps are sometimes operated with the inlet pressure reduced to the point that bubbles of vapor may be formed due to the generation of vapor in the fluid. For any fluid there is a pressure caused by molecules of the fluid that escape from the fluid and exist just above the fluid surface. The pressure of these molecules has been termed the *vapor pressure* of the fluid. For any given fluid, there is a definite relation between the vapor pressure and the temperature of the fluid. Figure 9.9 shows the vapor pressure for water as a function of temperature.

Vapor bubbles generated at the inlet to the pump are carried along until a region of higher pressure is reached and they suddenly collapse. If the vapor bubbles come in contact with the walls at the time they collapse, pitting of the surface can occur due to the high local pressures generated. The entire process of the formation, growth, and collapse of a bubble can occur in milliseconds in a turbomachine. Experimental data indicate pressures on the order of 10^9 Pa (143,000 psi) occurring from the collapse of vapor bubbles, which is consistent with the damage observed due to cavitation. The ultimate failure of the surface of the material is usually a fatigue failure.

If cavitation does occur, there is a very rapid decrease in the performance of a pump. In addition, to this decrease in efficiency of the unit, the flow tends to become unstable.

The basic measure to protect the pump against cavitation is the proper design of the system to avoid pressures that approach the vapor pressure of the fluid. In order to have an

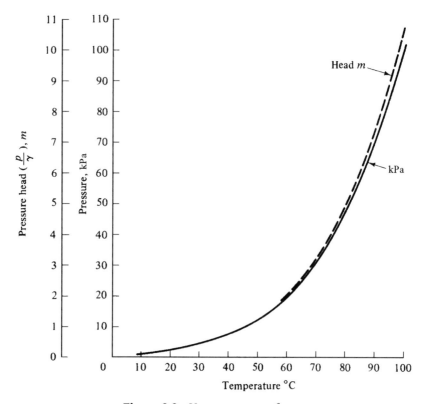

Figure 9.9   Vapor pressure of water.

index of performance to use to specify the minimum suction conditions for a turbomachine, we will have to define certain terms. This can best be done by referring to Figure 9.10, which shows the sketch of a pump system connected between two reservoirs. A term is now defined, called the *net positive suction head* (NPSH), to aid in evaluation of the possibility of cavitation in the pump. Using the pump centerline as a reference,

$$\text{NPSH} = \frac{p}{\gamma} + \frac{V^2}{2g} \mp h - h_L - h_v \qquad (9.1)$$

where

p = absolute pressure on liquid surface (Pa or lb/ft²)

$\dfrac{p}{\gamma}$ = static pressure on the fluid surface (m or ft)

γ = specific weight of the fluid (N/m³ or lb/ft³)

h = distance from centerline of the pump to the fluid level (m or ft); note that if the reservoir is *above* the centerline of the pump, this term will be positive, while a reservoir below the pump causes this term to be negative

h_L = head lost due to friction and other factors in the suction line (m or ft).

h_v = vapor pressure of the liquid corresponding to the liquid temperature (m or ft)

Since water is such a widely used fluid, its vapor pressure is given either from Figure 9.9 or from Table 9.2. Also note that the velocity head term in equation (9.1) is sometimes included by the manufacturer in the NPSH required by the pump. If this is the case it can be omitted from Equation (9.1).

The NPSH can be interpreted as the total suction head of liquid above the vapor-pressure head. It is usual for a manufacturer to run a test to determine the NPSH for the machine that will permit it to operate efficiently and without noise or damage. As long as the machine is operated at a NPSH *above* this value, the operation will be satisfactory. The

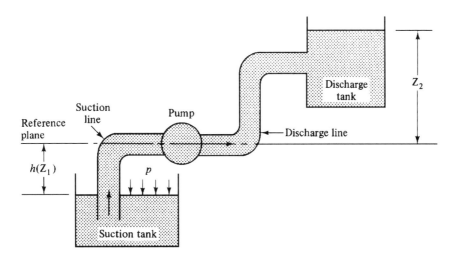

Figure 9.10   Typical pump terminology.

NPSH determined from a test at a given speed (N_1) can be used to obtain the NPSH at a second speed (N_2) by

$$(\text{NPSH})_2 = (\text{NPSH})_1 \left(\frac{N_2}{N_1}\right)^2 \tag{9.1a}$$

Table 9.2   Selected Properties of Water

Temperature (°C)	γ (N/m³)	p(Pa)	h_v(m)
10	9802.08	1227.6	0.125
20	9788.38	2339	0.239
30	9764.01	4246	0.435
40	9730.11	7384	0.759
50	9688.77	12 349	1.275
60	9640.19	19 940	2.068
70	9587.41	31 190	3.253
80	9528.71	47 390	4.973
90	9465.25	70 140	7.410
100	9397.22	100 350	10.785

Source: Data from *Steam Tables,* International Edition, by J. H. Keenan, G. F. Keyes, P. G. Hill, and J. G. Moore, John Wiley & Sons, Inc., New York, 1969.

ILLUSTRATION 9.1   NET POSITIVE SUCTION HEAD

A pump is placed so that its centerline is 1.5 m above the level in a reservoir that is open to the atmosphere. If water is being pumped at the rate of 1 m/s in the suction line and friction losses are estimated at 1.2 m, determine the NPSH if the water is at 40°C.

ILLUSTRATIVE PROBLEM 9.1

Given: $Z_1 = 1.5$ m, $t = 40°C$, $h_L = 1.2$ m, $V = 1$ m/s
Find: NPSH
Assumptions: Steady, incompressible flow

Basic Equations: Equation (9.1): $\text{NPSH} = \dfrac{p}{\gamma} + \dfrac{V^2}{2g} \mp h - h_L - h_v$

Solution: Refer to Figure 9.10. For 40°C (see Table 9.2),

$$\text{NPSH} = \frac{p}{\gamma} + \frac{V^2}{2g} - h - h_L - h_v, \text{ since the pump is above the reservoir}$$

$$\text{NPSH} = \frac{101\ 325 \text{ N/m}^2}{9730.11 \text{ N/m}^3} + \frac{(1 \text{ m})^2}{2 \times 9.81 \text{ m/s}^2} - 1.5 \text{ m} - 1.2 \text{ m} - 0.759 \text{ m}$$

NPSH = 10.41 m + 0.05 m − 1.5 m − 1.2 m − 0.76 m

NPSH = 7.00 m

ILLUSTRATION 9.2   NET POSITIVE SUCTION HEAD

Solve Illustrative Problem 9.1 if the pump is placed so that its centerline is 1.5 m below the level in the reservoir. Assume all other data to be the same.

ILLUSTRATIVE PROBLEM 9.2

Given: Same inputs as Illustrative Problem 9.1 with $Z_1 = -1.5$ m

Find: NPSH

Assumptions: Steady, incompressible flow

Basic Equations: Equation (9.1): $\text{NPSH} = \dfrac{p}{\gamma} + \dfrac{V^2}{2g} \mp h - h_L - h_v$

Solution: Since the pump is below the surface of the reservoir, h will be positive. Thus

$$\text{NPSH} = \frac{p}{\gamma} + \frac{V^2}{2g} - h - h_L - h_v$$

Using the given data,

NPSH = 10.41 m + 0.05 m + 1.5 m − 1.2 m − 0.76 m

∴ NPSH = 10.0 m

Note that, in general, the higher NPSH is desirable.

ILLUSTRATION 9.3   NET POSITIVE SUCTION HEAD

A manufacturer states that the required NPSH for a centrifugal pump is 7 m. Water is at 20°C. If the head loss up to the pump inlet is 2 m, what should the placement of the pump be to avoid cavitation?

ILLUSTRATIVE PROBLEM 9.3

Given: NPSH required = 7 m, $t = 20°C$, $h_L = 2$ m

Find: $h(Z_1)$ in Figure 9.10

Assumptions: Velocity head term is included in NPSH requirement; steady, incompressible flow

Basic Equations: Equation (9.1): $\text{NPSH} = \dfrac{p}{\gamma} + \dfrac{V^2}{2g} \mp h - h_L - h_v$

Solution: Using Table 9.2,

$$\text{NPSH} = \frac{101\ 325\ \text{N/m}^2}{9788.38\ \text{N/m}^3} - h - 2\ \text{m} - 0.239\ \text{m}$$

$$\text{NPSH} = 10.35\ \text{m} - h - 2\ \text{m} - 0.239\ \text{m}$$

$$\text{NPSH} = 8.11 - h$$

For proper operation of the pump, the available NPSH should be greater than the required NPSH. Using this,

$$(\text{NPSH})_{\text{required}} \geq 8.11 - h$$

$$\text{or } h \leq 8.11 - (\text{NPSH})_{\text{required}}.$$

Using the given data, $h = 8.11 - 7 = 1.11$ m

We conclude that proper placement of the pump requires the pump centerline to be equal to or less than 1.11 m above the surface of the suction tank.

9.4   PUMP PERFORMANCE CALCULATIONS

As in any technology, certain terms are used that are by common usage generally understood. The tank from which the fluid is being pumped is called the *suction tank,* and the tank that the fluid is pumped to is called the *discharge tank.* The distance from the pump centerline to the level in the suction tank (Z_1 in Figure 9.10) is known as the *static discharge head.* If we now apply a Bernoulli equation across the pump,

$$\frac{p_1}{\gamma} + \frac{V_1^2}{2g} + Z_1 + E_p = \frac{p_2}{\gamma} + \frac{V_2^2}{2g} + Z_2$$

(9.2)

or

$$E_p = \frac{p_2 - p_1}{\gamma} + \frac{V_2^2 - V_1^2}{2g} + (Z_2 - Z_1)$$

(9.3)

Notice that we have written this equation across the pump and that sections ① and ② refer to the pump inlet and outlet, respectively. The term *total head* or *dynamic head* is used to designate the energy added to the fluid per unit weight and is given by E_p in equation (9.3). If we multiply E_p by the weight flow rate ($\dot{w}$), we obtain the power added to the fluid by the pump as

$$\text{hp} = \frac{\dot{w} E_p}{550} = \frac{\gamma Q E_p}{550}$$

(9.4)

or

$$\text{hp} = \frac{\dot{w}E_p}{746} = \frac{\gamma Q E_p}{746}$$ (9.5)

in English and SI units, respectively.

ILLUSTRATION 9.4   PUMP POWER

A pump steadily pumps water from one reservoir to another. If the pump delivers 0.1 m³/s and has a 150-mm inlet diameter and a 75-mm outlet diameter, determine the power added to the water. Assume the pump outlet to be 1 m above the inlet and that static taps indicate an outlet pressure 70 kPa greater than the inlet pressure. Use $\gamma = 9810$ N/m³.

ILLUSTRATIVE PROBLEM 9.4

Given: $Q = 0.1$ m³/s, $D_1 = 150$ mm, $D_2 = 75$ mm, $p_2 - p_1 = 70$ kPa, $\gamma = 9810$ N/m³, $Z_2 - Z_1 = 1$ m

Find: Power

Assumptions: Steady, incompressible flow

Basic Equations: Continuity: $Q = A_1 V_1 = A_2 V_2$

$$\text{Equation (9.3): } E_p = \frac{p_2 - p_1}{\gamma} + \frac{V_2^2 - V_1^2}{2g} + Z_2 - Z_1$$

$$\text{Equation (9.5): hp} = \frac{\gamma Q E_p}{746}$$

Solution: Since $Q = AV$, $V = Q/A$. At the inlet,

$$V_1 = \frac{0.1 \text{ m}^3}{(\pi/4)(0.15 \text{ m})^2} = 5.7 \text{ m/s}$$

At the outlet,

$$V_2 = \frac{0.1 \text{ m}^3}{(\pi/4)(0.075 \text{ m})^2} = 22.6 \text{ m/s}$$

Using the equation (9.3), we obtain

$$E_p = \frac{p_2 - p_1}{\gamma} + \frac{V_2^2 - V_1^2}{2g} + Z_2 - Z_1$$

$$E_p = \frac{70 \times 10^3 \text{ N/m}^2}{9810 \text{ N/m}^3} + \frac{(22.6 \text{ m/s})^2 - (5.7 \text{ m/s})^2}{2 \times 9.81 \text{ m/s}^2} + 1 \text{ m}$$

$$E_p = 7.14 \text{ m} + 24.4 \text{ m} + 1 \text{ m} = 32.5 \text{ m}$$

The power delivered to the fluid is

$$hp = \frac{\gamma Q E_p}{746}$$

$$hp = \frac{9810 \text{ N/m}^3 \times 0.1 \text{ m}^3/\text{s} \times 32.5 \text{ m}}{746 \text{ (N·m)/(s·hp)}}$$

$$hp = 42.7 \text{ hp}$$

In Illustrative Problem 9.4 we considered the pump only and not the system that it is part of. If it is desired to know the operating point of a pump in a given system, we must consider that the pump will be required to deliver the fluid against a static discharge head and to overcome all the flow losses in the system.

Let us assume that we have a pump that has the characteristics shown by the performance curves of Figure 9.8 connected in the system shown in Figure 9.10. For each assumed value of flow in the system, we will calculate E_p for the system, including all flow losses. If the system is to flow at the assumed rate, the pump is required to provide this head. We will therefore call this the *required head* and the pump head the *available head*. Plotting both curves on the same coordinates gives us Figure 9.11, where we have indicated the head available from the pump and the head required by the system. The head h is the static head that the system must overcome, since it is the value at zero flow. The losses in the system will be the difference between the head required by the system and h. The operating point of the system will be at point ②, where both the head available and the head required curves intersect and both have the same value. By varying the flow resistances in the system (by such methods as opening or closing valves) we can shift the head required curve to give us operating points to the left or right of point ② in Figure 9.11.

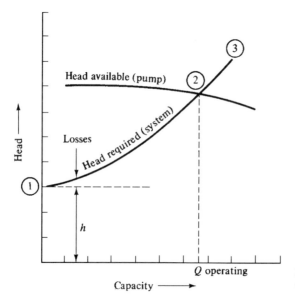

Figure 9.11   Operating characteristics of a system.

9.5   THE AFFINITY LAWS FOR CENTRIFUGAL PUMPS

Data on an original or prototype pump are usually obtained from tests of a model pump. To ensure that the data from the model are applicable to the prototype, it is important that certain similarity conditions be met. In general, we can state that the two pumps must be geometrically similar and that also the flow patterns in the two pumps must be similar. Two units that are geometrically similar and have similar vector diagrams are said to be *homologous*.

The discharge Q equals AV, and, since the area is proportional to the impeller diameter D (or any other characteristic dimension), we say that Q is proportional to VD^2. By noting that the velocity V is proportional to the peripheral speed of the impeller, ND, where N is the rpm of the impeller, we can write the following relation:

$$\frac{Q}{ND^3} = \text{constant} \tag{9.6}$$

which expresses the necessary condition for homologous units (similar geometry and similar flow patterns).

Let us now consider the head developed by a pump. If we start with $Q = AV$ again and note that V is proportional to $\sqrt{H}$, Q will be proportional to $D^2\sqrt{H}$, or

$$\frac{Q}{D^2\sqrt{H}} = \text{constant} \tag{9.7}$$

Eliminating Q in equations (9.6) and (9.7) yields

$$\frac{H}{N^2D^2} = \text{constant} \tag{9.8}$$

The last consideration that we need concerns power, P. Power will be proportional to γQH. If we use equations (9.6) and (9.8) for Q and H, respectively, we have P proportional to N^3D^5, or

$$\frac{P}{N^3D^5} = \text{constant} \tag{9.9}$$

Note that equations (9.6), (9.8), and (9.9) have assumed constant (the same) efficiency.

The utility of equations (9.6), (9.8), and (9.9) lies in the fact that most centrifugal pumps are either run at various speeds to obtain the desired capacity and head or are run at constant speed using different impellers in the same casing to achieve the same results. Let us consider the case of a machine having a given diameter impeller operating at different rotational speeds. From equations (9.6), (9.8), and (9.9), *for constant diameter,*

$$\frac{Q_1}{Q_2} = \frac{N_1}{N_2}$$

$$\frac{H_1}{H_2} = \left(\frac{N_1}{N_2}\right)^2$$

$$\frac{P_1}{P_2} = \left(\frac{N_1}{N_2}\right)^3 \tag{9.10}$$

and *for constant speed,*

$$\frac{Q_1}{Q_2} = \left(\frac{D_1}{D_2}\right)^3$$

$$\frac{H_1}{H_2} = \left(\frac{D_1}{D_2}\right)^2$$

$$\frac{P_1}{P_2} = \left(\frac{D_1}{D_2}\right)^5 \tag{9.11}$$

Equations (9.6), (9.8), and (9.9) or their equivalent forms, equations (9.10) and (9.11), are known as the *affinity laws* for pumps and are extremely useful. They permit tests at a given speed or given impeller diameter to be converted to yield data at another speed or diameter in homologous units.

Equations (9.10) and (9.11) are accurate for fluids having the viscosity of water or light oils and for pumps having reasonably large velocities and dimensions. For small pumps or pumps used with viscous liquids, these relations are inaccurate. When very viscous fluids are pumped there is a noticeable decrease in the pump head as the viscosity increases. In addition, the observed pump power required increases as the viscosity increases.

ILLUSTRATION 9.5   AFFINITY LAWS

A test is conducted on a centrifugal pump, and the pump is found to develop a total head of 40 ft while delivering 100 cfm. The pump was operated at 1100 rpm. What head and capacity would be expected if the pump were operated at 1200 rpm?

ILLUSTRATIVE PROBLEM 9.5

Given: N_1 = 1100 rpm, Q = 100 cfm, H = 40 ft

Find: Q and H if N = 1200 rpm

Assumptions: Fluid is water, $D_1 = D_2$, flow is similar at different speeds

Basic Equations: Equations (9.10): $\dfrac{Q_1}{Q_2} = \dfrac{N_1}{N_2}$, $\dfrac{H_1}{H_2} = \left(\dfrac{N_1}{N_2}\right)^2$

Solution: From Equations (9.10),

$$Q_2 = Q_1\left(\frac{N_2}{N_1}\right) = 100 \text{ cfm}\left(\frac{1200}{1100}\right) = 109.1 \text{ cfm}$$

and
$$H_2 = H_1\left(\frac{N_2}{N_1}\right)^2 = 40 \text{ ft}\left(\frac{1200}{1100}\right)^2 = 47.6 \text{ ft}$$

ILLUSTRATION 9.6   AFFINITY LAWS

A pump has an impeller diameter of 1.5 m and operates at 1200 rpm. If the speed is increased to 1400 rpm, what impeller diameter should be used to maintain a constant power input to the pump?

ILLUSTRATIVE PROBLEM 9.6

Given: $D = 1.5$ m, $N = 1200$ rpm

Find: D for $N = 1400$ rpm

Assumptions: $P =$ constant, units are homologous

Basic Equations: Using Equations (9.10) and (9.11) or (9.9),

$$\frac{P_1}{N_1^3 D_1^5} = \frac{P_2}{N_2^3 D_2^5} \text{ and } N_1^3 D_1^5 = N_2^3 D_2^5$$

Solution: From $N_1^3 D_1^5 = N_2^3 D_2^5$;

$$D_2 = D_1 \left(\frac{N_1}{N_2}\right)^{3/5}$$

Using the data given, $D_2 = 1.5 \text{ m} \left(\dfrac{1200}{1400}\right)^{3/5} = 1.367$ m

---

ILLUSTRATION 9.7   AFFINITY LAWS

A fan operates at 1600 rpm and develops a head of 6 in. of water and delivers 120 cfm. What diameter is required for a geometrically similar fan that will develop 6 in. of water at 1300 rpm? Calculate the volumetric capacity of the new fan. The initial diameter is 15 in.

ILLUSTRATIVE PROBLEM 9.7

Given: $N_1 = 1600$ rpm, $H_1 = H_2 = 6$ in. of water, $Q_1 = 120$ cfm, $N_2 = 1300$ rpm,
         $D_1 = 15$ in.

Find: D_2, Q_2

Assumptions: Homologous units, $H_1 = H_2$

Basic Equations: Equation (9.6): $\dfrac{Q_1}{N_1 D_1^3} = \dfrac{Q_2}{N_2 D_2^3}$; Equation (9.8): $\dfrac{H}{N^2 D^3} =$ constant

Solution: From Equation (9.8),

$$N_1^2 D_1^2 = N_2^2 D_2^2$$

or

$$D_2 = D_1 \times \frac{N_1}{N_2}$$

$$D_2 = 15 \text{ in.} \times \left(\frac{1600}{1300}\right) = 18.46 \text{ in.}$$

From Equation (9.6),

$$\frac{Q_1}{N_1 D_1^{\,3}} = \frac{Q_2}{N_2 D_2^{\,3}}$$

or

$$Q_2 = Q_1 \left(\frac{N_2 D_2^{\,3}}{N_1 D_1^{\,3}}\right)$$

But, from Equation (9.8) (constant head),

$$N_1 D_1 = N_2 D_2$$

Therefore

$$Q_2 = Q_1 \left(\frac{N_1}{N_2}\right)^2 = 120 \text{ cfm} \left(\frac{1600}{1300}\right)^2 = 181.8 \text{ cfm}$$

9.6 SPECIFIC SPEED

We can use equations (9.6) and (9.8) to develop another performance factor that is widely used for both preliminary design and selection of pumps. If we eliminate D from these equations, we obtain

$$\frac{N^{2/3} Q^{1/3}}{H^{1/2}} = \text{constant} \tag{9.12}$$

If we raise both sides of (9.12) to the $\frac{3}{2}$ power,

$$\frac{N\sqrt{Q}}{H^{3/4}} = \text{constant} \tag{9.13}$$

At this point we define the term *specific speed, N_s,* as the value of N in equation (9.13) when $H = 1$ and $Q = 1$; in other words, it is equal to the constant in this equation. Therefore, we can write equation (9.13) as

$$N_s = \frac{N\sqrt{Q}}{H^{3/4}} \tag{9.14}$$

In the conventional English System, N is usually given in rpm, Q is in gal/min, and H is in feet. These values are those at the point of maximum efficiency for the shaft speed used. Unfortunately, the units gal/min and feet do not yield a dimensionless form for N_s. In order to write equation (9.14) in consistent units, such as SI, we can rewrite it as

$$N_s = \frac{N\sqrt{Q}}{gH^{3/4}} \qquad (9.15)$$

Since most published data have been obtained prior to the use of SI units, the published values use equation (9.14) for the definition of specific speed. To convert from equation (9.14) to (9.15), it is necessary to divide the values obtained from equation (9.14) by 17 200 to convert them to SI values consistent with equation (9.15).

The utility of the specific-speed concept is that certain combinations of head, speed, and capacity are typical of a given type of pump. Thus, specific speed can tell us the combinations of these factors that are both possible and desirable. It has been found that the specific speed of centrifugal pumps is low, while axial flow pumps have high specific speeds. Pumps with a combination of these types (mixed flow) have intermediate values of specific speed. For single-stage centrifugal pumps, N_s ranges from 500 to 5000, while single-stage axial flow pumps have specific speeds that lie in the range between 5000 and 10,000 in English conventional units. Figure 9.12 shows how impeller design and optimum efficiency vary as a function of specific speed for centrifugal, mixed flow, and axial flow pumps.

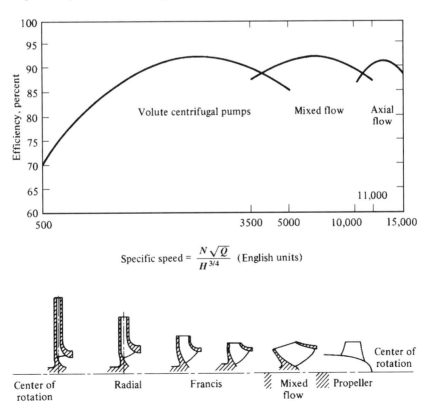

Figure 9.12 Relative shapes and efficiencies versus specific speed.

ILLUSTRATION 9.8   SPECIFIC SPEED

A centrifugal pump operates at 3600 rpm, developing a head of 100 ft while pumping 2000 gal/min. What is its specific speed in both English and SI units? What type of pump do you think that this most likely is?

ILLUSTRATIVE PROBLEM 9.8

Given: $N = 3600$ rpm, $H = 100$ ft, $Q = 2000$ gpm
Find: N_s and type of pump
Assumptions: Figure 9.12 applicable

Basic Equations: Equation (9.14): $N_s = \dfrac{N\sqrt{Q}}{H^{3/4}}$

Solution:

$$N_s = \frac{N\sqrt{Q}}{H^{3/4}}$$

$$N_s = \frac{3600 \text{ rpm } \sqrt{2000 \text{ gpm}}}{(100 \text{ ft})^{3/4}}$$

$$N_s = 5091$$

From Figure 9.12, the most probable unit would be a mixed flow unit operating near its maximum efficiency.
In SI units,

$$N_s = \frac{5091}{17\ 200} = 0.296$$

9.7   REVIEW

This chapter departs slightly from our previous chapters in that it is designed primarily for those persons who will be involved in the selection of pumps rather than in their design. To make such a meaningful selection, we first involved ourselves with the classification of pumps based on their displacement, delivery, and motion. Once we classified the various types of pumps, we concerned ourselves with their performance and the general nomenclature that has become usual in the pump industry. Cavitation in a pump leads to inefficient and noisy performance, and, if continued, can lead to the destruction of the pump.

   After this discussion, we looked at the actual performance of pumps. We were able to use pump performance curves to establish the required head and available head, from which the steady-state operating point of the system can be established. For similar pumps, it was found to be possible to test a pump at a given speed or with a given diameter, and, using the affinity laws, establish the performance at another speed or with a different diameter impeller. Finally, using the concept of specific speed, we found that we could select the pump type for a given application.

KEY TERMS

Terms of importance in this chapter:

Affinity laws: a series of equations permitting one to extrapolate the performance of one size of pump to another. It is assumed that the pumps are geometrically similar and have similar vector diagrams.

Available head: the head that a pump can develop at a given speed.

Cavitation: the formation, growth, and collapse of bubbles in a pump.

Discharge: the outlet of a pump.

Dynamic head: energy added to the fluid being pumped.

Homologous pumps: pumps having geometrically similar construction and similar vector diagrams.

Net positive suction head: the sum of the pressure head plus the velocity head minus the losses in the suction line ∓ the distance from the centerline of the pump to the fluid level minus the vapor pressure of the liquid.

Nonpositive displacement pump: a pump that does not yield a given amount of fluid per revolution.

Positive displacement pump: a pump designed to provide a given amount of fluid per revolution.

Required head: head needed to overcome all losses plus static head plus velocity head at a given rate of flow.

Specific speed: a performance factor used in the preliminary design and selection of pumps.

Static discharge head: the distance from the centerline of the pump to the level in the suction tank.

Suction: the inlet of a pump.

Total head: *see* dynamic head.

Vapor pressure: the pressure exerted by those molecules of fluid that escape from and exist just above the fluid surface.

KEY EQUATIONS

Net positive suction head
$$\text{NPSH} = \frac{p}{\gamma} + \frac{V^2}{2g} \mp h - h_L - h_v \qquad (9.1)$$

Total or dynamic head
$$E_p = \frac{p_2 - p_1}{\gamma} + \frac{V_2^2 - V_1^2}{2g} + (Z_2 - Z_1) \qquad (9.3)$$

Pump power (English) $$hp = \frac{\gamma Q E_p}{550}$$ (9.4)

Pump power (SI) $$hp = \frac{\gamma Q E_p}{746}$$ (9.5)

Affinity laws—constant diameter $$\frac{Q_1}{Q_2} = \frac{N_1}{N_2}$$ (9.10)

$$\frac{H_1}{H_2} = \left(\frac{N_1}{N_2}\right)^2$$

$$\frac{P_1}{P_2} = \left(\frac{N_1}{N_2}\right)^3$$

Affinity laws—constant speed $$\frac{Q_1}{Q_2} = \left(\frac{D_1}{D_2}\right)^3$$ (9.11)

$$\frac{H_1}{H_2} = \left(\frac{D_1}{D_2}\right)^2$$

$$\frac{P_1}{P_2} = \left(\frac{D_1}{D_2}\right)^5$$

Specific speed $$N_s = \frac{N\sqrt{Q}}{H^{3/4}}$$ (9.14)

Specific speed $$N_s = \frac{N\sqrt{Q}}{gH^{3/4}}$$ (9.15)

QUESTIONS

1. Describe the function of a pump.
2. Is a centrifugal pump a positive or a nonpositive displacement pump?
3. Would you expect the pump on a home aquarium to be a positive displacement pump? Why or why not?
4. How does a single acting reciprocating pump differ from a double acting reciprocating pump?
5. Describe the effect of a viscosity increase on the power, capacity, and efficiency of a centrifugal pump.

6. What is the effect of an increase in temperature on the vapor pressure of water?

7. Does a solid such as ice have a vapor pressure?

8. Is it desirable to use a large pipe size for the inlet lines to a pump? Why?

9. How would you define net positive suction head (NPSH)?

10. What can you do to prevent or cure cavitation in a pump?

11. The affinity laws enable one to estimate the performance of one pump from the tests of another. What conditions must be true to enable you to do this?

12. For a given size centrifugal pump (constant diameter), describe the change in (a) capacity, (b) head, and (c) power if the speed is doubled.

13. For a given speed, describe the change in (a) capacity, (b) head, and (c) power if the diameter is halved.

14. Is specific speed a dimensionless quantity like Reynolds number?

15. *True or false:* Axial flow centrifugal pumps have the lowest specific speed.

PROBLEMS

Net Positive Suction Head

9.1. A pump is located with its centerline 2 ft above the level of a reservoir that is open to the atmosphere. Water is being pumped at the rate of 15 ft/s in the suction line, and friction losses are estimated to be 3 ft. Determine the NPSH if the water is at 60°C. ($\gamma = 61.3$ lb/ft^3)

9.2. Solve Problem 9.1 if the pump is located with its centerline below the level in the reservoir.

9.3. A pump delivers water at 68°F from a suction tank to a discharge tank at the rate of 10 liters/s. If the suction line has a 75-mm inside diameter and the pump centerline is located 2 m above the level of the suction reservoir, determine the NPSH. Suction-line losses are estimated to be 1.0 m and the reservoir is open to the atmosphere.

9.4. Solve Problem 9.3 if the pump is located below the level in the reservoir.

9.5. A pump's inlet is 4.5 ft above a tank that is at 50°C. Atmospheric pressure is 14.7 psia. Losses due to flow in the inlet line equal 1.5 ft. Calculate the NPSH for this situation if the water velocity is 4 m/s in the inlet to the pump.

9.6. Solve Problem 9.5 if the pump is located below the level in the reservoir.

9.7. A manufacturer tests a pump and finds that at 1700 rpm it requires a NPSH of 6.6 ft. Calculate the required NPSH at 1400 rpm.

9.8. The required NPSH for a pump is 8.4 ft. At a different speed it is found to be 5.5 ft. Determine the ratio of the two speeds.

Pump Performance

9.9. A pump delivers water from a lower reservoir to an upper reservoir. If the flow rate is 0.05 m^3/s and the pump has a 100-mm inlet diameter and a 50-mm outlet diameter, calculate the power added to the water. The pump outlet is 2 m above the inlet, and static pressure taps indicate an outlet pressure that is 100 kPa greater than the inlet pressure. (γ = 9810 N/m³)

9.10. A pump delivers water from a lower to a higher reservoir. If the pump delivers 100 gal/min and the outlet is 3 ft above the inlet, determine the pump power. The inlet pipe has a 3-in. inside diameter and the outlet pipe has a 1-in. inside diameter. Pressure taps indicate the outlet pressure to be 6 psi higher than the inlet pressure. (γ = 62.4 lb/ft³)

*9.11. A centrifugal turbine develops 200 hp when operated under a head of 10 ft at a speed of 90 rpm. Another homologous turbine having a wheel diameter equal to 75% of the initial turbine's diameter is operated under a head of 20 ft. What speed should the second turbine be operated at and what power do you expect it to develop?

9.12. A centrifugal pump is operated at 1500 rpm against a head of 20 ft. When the pump delivers 3.0 cfs it requires 10 hp. The speed is reduced to 1200 rpm. Calculate Q, H, and P assuming the same efficiency.

9.13. A test is conducted on a pump, and test data show the pump capable of developing a total head of 15 m while delivering 20 liters/s when operated at 1000 rpm. Estimate the head and capacity when the pump is operated at 800 rpm.

9.14. A pump delivers 500 gal/min at 1000 rpm against a total head of 45 ft. Determine its performance at 1100 rpm.

9.15. A pump has an impeller that is 5 ft in diameter and is operated at 1450 rpm. If the speed is changed to 1150 rpm, what impeller diameter should be used to maintain a constant power input to the pump?

9.16. The impeller of a pump is 1.25 m in diameter and the pump is operated at 1050 rpm. If the speed is increased to 1175 rpm, what impeller diameter should be used to maintain a constant power input to the pump?

Specific Speed and Pump Selection

9.17. Determine the initial specific speed of the pump in Problem 9.13.

9.18. Determine the initial specific speed of the pump in Problem 9.14.

9.19. A pump is to deliver 500 gal/min against a head of 40 ft. If it is operated at 1150 rpm, what type of pump should be selected?

9.20. A pump is to deliver 4000 liters/min against a head of 20 m. If the pump is operated at 3600 rpm, select the type of pump to be used.

9.21. A pump operates at 3000 rpm and delivers 400 gal/min against a head of 90 ft. Compute the specific speed and select a pump type to be used.

9.22. If the pump in Problem 9.21 is to be operated at 2500 rpm against a head of 80 ft, determine its specific speed.

9.23. A pump is driven by an electric motor at 1800 rpm. If it can deliver 600 gal/min of water at a total head of 75 ft, determine the specific speed and the type of pump.

10 FLOW ABOUT IMMERSED BODIES

LEARNING GOALS

After reading and studying this chapter you should be able to:

1. Show that *lift* and *drag* are components of the resultant force acting on an immersed body.
2. Use the concept of lift and drag coefficients to calculate the lift and drag of various bodies.
3. Use the empirical data that are found in the literature to calculate the lift and drag on bodies. Realize that just as there are laminar and turbulent flow regimes for flow in pipes, there are laminar and turbulent flow regimes for flow about immersed bodies.
4. Calculate the performance of an airfoil in steady, level flight.
5. Show how induced drag occurs on airfoils.
6. Use the polar curve to determine the performance of airfoils.
7. Calculate the performance of an aircraft during maneuvers.
8. Determine the resultant force that acts on a pilot during a banking maneuver.

10.1 INTRODUCTION

It is a fact that we on the earth live at the bottom of an ocean of air. This fact must be accounted for in the design of all vehicles that move in this environment (i.e., aircraft, trains, automobiles, etc.). Buildings, bridges, and other structures must also be designed to withstand the dynamic forces that this environment imposes on them. Recent advances in oceanography have opened up an entirely new fields, aptly called hydrospace, a relatively viscous, dense, and hostile environment in which man and his vehicles are totally immersed. The flow about an object may be due to the motion either of the object relative to the fluid or of the fluid relative to the object. In this chapter we shall consider the incompressible steady flow of a fluid relative to an object. It should also be noted that the principles developed in this chapter can also be applied to the study of turbines, fans, pumps, propellers, and many other flow systems.

10.2 GENERAL CONSIDERATIONS

When a body is fully submerged in a viscous, incompressible fluid and moves with a velocity V relative to the fluid, the body will experience a resistance to its motion due to the fluid. In general, we can assume that the resultant force on any such body can be represented by two perpendicular components. One of these components is usually taken in a direction parallel to the direction of flow and is known as *drag*. The other component is taken perpendicular to the direction of flow and is known as *lift*. The body (airfoil) shown in Figure 10.1 illustrates these concepts. Quantitatively, we can express the lift and drag on a body in terms of a lift coefficient (C_L) and a drag coefficient (C_D). Equations (10.1) and (10.2) are the defining equations:

$$D = C_D \frac{\rho V^2}{2} A \tag{10.1}$$

and

$$L = C_L \frac{\rho V^2}{2} A \tag{10.2}$$

where D and L are the lift and drag forces, ρ the density of the fluid, V the relative velocity of the fluid to the body, and A a characteristic area that we will discuss in detail later. The term $\rho V^2/2$ is called the *dynamic* or *impact pressure*. The coefficients C_D and C_L are dimen-

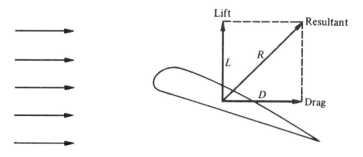

Figure 10.1   Forces on an immersed body.

sionless and do not depend upon the system of units used. Both C_D and C_L can be written as functions of the Reynolds number for incompressible flow,

$$C_D = \phi_1(\text{Re}) \tag{10.3a}$$

$$C_L = \phi_2(\text{Re}) \tag{10.3b}$$

Figures 10.2 and 10.3 show the drag coefficients for spheres, plates, and cylinders. On Figure 10.2 a theoretical equation known as Stokes' law is also plotted. Stokes' law gives $C_D = 24/\text{Re}$ for low Reynolds numbers.

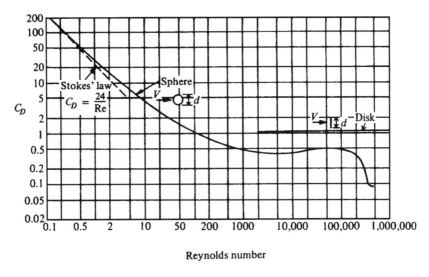

Figure 10.2 Drag coefficients for a sphere and circular disk. (Reproduced with permission from *Fluid Mechanics* by R. C. Binder, 4th ed., Prentice-Hall, Inc., Englewood Cliffs, NJ, 1962, p. 168.)

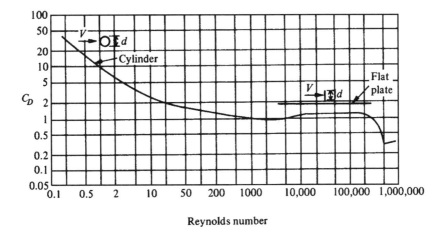

Figure 10.3 Drag coefficients for a cylinder and flat plate. (Reproduced with permission from *Fluid Mechanics* by R. C. Binder, 4th ed., Prentice-Hall, Inc., Englewood Cliffs, NJ, 1962, p. 170.)

When a flat plate is placed parallel to the direction of fluid motion, as shown in Figure 10.4a, it experiences a drag force due to fluid friction only on both faces. As the velocity of the fluid relative to the plate is increased (with a concurrent increase in Reynolds number), the drag force decreases. At some value of Reynolds number (approximately 10^6) a sudden increase in drag occurs, while a still further increase in Reynolds number yields a continual decrease in drag force. If the drag coefficient is plotted as a function of Reynolds number for smooth flat plates where the linear dimension in Reynolds number is taken to be the length of the plate parallel to the flow, the curves in Figure 10.5 are determined. These two curves appear to be similar to the curves for smooth tubes in the Moody diagram, and basically are found to depend on whether the boundary layer is laminar or turbulent.

When the plate is tilted at an angle to the stream, as shown in Figure 10.4b, the resultant force on the plate can be resolved into two component forces, lift (L) and drag (D). The data shown in Figure 10.6 are based upon tests on small rectangular plates whose longer side was six times the shorter side, with air striking the longer side first. Small corrections

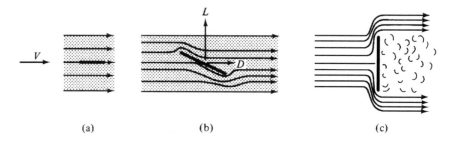

(a)   (b)   (c)

Figure 10.4   Flat plates.

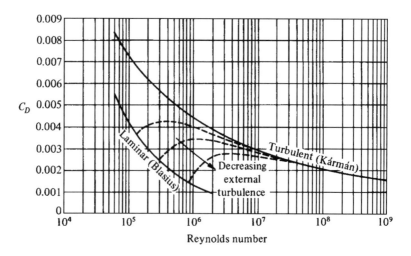

Figure 10.5   Drag on flat plate parallel to flow direction. (Reproduced with permission from *Fluid Mechanics* by R. C. Binder, 4th ed., Prentice-Hall, Inc., Englewood Cliffs, NJ, 1962, p. 173.)

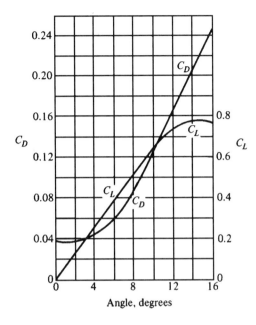

Figure 10.6   Lift and drag coefficients for a flat plate. (Reproduced with permission from *Elements of Practical Aerodynamics* by B. Jones, 3rd ed., John Wiley & Sons, Inc., New York, 1942, p. 21.)

may be needed in applying these data to longer plates or plates of different shapes. It should be noted that these curves show no dependence on Reynolds number and that at an angle of zero degrees, when the plate is parallel to the flow, the drag coefficient is given as a single value. Since the conditions of the test are not specifically given and the Reynolds number is also not given, this curve should be used for design with caution. The qualitative trend, however, shows an increase in both lift and drag coefficients as the angle the plate makes with the stream increases. At some finite angle, approximately 15°, C_L starts to decrease, but the drag coefficient C_D continues to increase.

ILLUSTRATION 10.1   LIFT AND DRAG

Using the data shown in Figure 10.6, determine the lift and drag forces on a plate 6 ft long and 1 ft wide if the plate is set at an angle of 8° to an airstream having a velocity of 35 ft/s. What is the resultant force on the plate? The specific weight of air can be taken as 0.075 lb/ft³.

ILLUSTRATIVE PROBLEM 10.1

Given: Rectangular plate 6 ft × 1 ft, $V = 35$ ft/s, $\gamma = 0.075$ lb/ft³, angle = 8°

Find: Lift, drag, resultant force

Assumptions: Incompressible, steady flow of air; Figure 10.6 applicable

Basic Equations: Lift: $L = C_L \dfrac{\rho V^2}{2} A$

$$\text{Drag: } D = C_D \frac{\rho V^2}{2} A$$

Resultant: $R = \sqrt{L^2 + D^2}$

Solution: Using the data from Figure 10.6 at 8°, $C_L = 0.51$ and $C_D = 0.086$. Therefore,

$$L = C_L \frac{\rho V^2}{2} A = 0.51 \times \frac{0.075 \text{ lb/ft}^3}{2 \times 32.17 \text{ lb/slug}} (35 \text{ ft/s})^2 \times 6 \text{ ft} \times 1 \text{ ft} = 4.37 \text{ lb}$$

$$D = C_D \frac{\rho V^2}{2} A = 0.086 \times \frac{0.075 \text{ lb/ft}^3}{2 \times 32.17 \text{ lb/slug}} (35 \text{ ft/s})^2 \times 6 \text{ ft} \times 1 \text{ ft} = 0.74 \text{ lb}$$

The resultant force is the vector sum of L plus D. Thus,

$$R = \sqrt{L^2 + D^2} = \sqrt{(4.37 \text{ lb})^2 + (0.74 \text{ lb})^2} = 4.43 \text{ lb}$$

ILLUSTRATION 10.2   LIFT AND DRAG

Using SI data, Illustrative Problem 10.1 would have the plate 1.83 m long and 0.305 m wide, the velocity would be 10.67 m/s, and the specific weight of air would be 11.78 N/m³. What is the resultant force on the plate using the data of Figure 10.6?

ILLUSTRATIVE PROBLEM 10.2

Given: Rectangular plate 1.83 m × 0.305 m, $V = 10.67$ m/s, $\gamma = 11.78 \dfrac{\text{N}}{\text{m}^3}$, angle = 8°

Find: Lift, drag, resultant force

Assumptions: Incompressible, steady flow of air; Figure 10.6 applicable

Basic Equations: Lift: $L = C_L \dfrac{\rho V^2}{2} A$

Drag: $D = C_D \dfrac{\rho V^2}{2} A$

Resultant: $R = \sqrt{L^2 + D^2}$

Solution: $C_L = 0.51$ and $C_D = 0.086$ as before

$$L = C_L \frac{\rho V^2}{2} A = 0.51 \times \frac{11.78 \text{ N/m}^3 \times (10.67 \text{ m/s})^2}{2 \times 9.81 \text{ N/kg}} \times 1.83 \text{ m} \times 0.305 \text{ m} = 19.46 \text{ N}$$

$$D = C_D \frac{\rho V^2}{2} A = 0.086 \times \frac{11.78 \text{ N/m}^3 \times (10.67 \text{ m/s})^2}{2 \times 9.81 \text{ N/kg}} \times 1.83 \text{ m} \times 0.305 \text{ m} = 3.28 \text{ N}$$

The resultant force, R, is

$$R = \sqrt{L^2 + D^2} = \sqrt{(19.46 \text{ N})^2 + (3.28 \text{ N})^2} = 19.73 \text{ N}$$

As a check,

$$19.73 \text{ N} \times 2.205 \text{ lb/kg} \times \frac{1}{9.81 \text{ N/kg}} = 4.43 \text{ lb}$$

which agrees with our answer to Illustrative Problem 10.1.

Increasing the plate angle still further, until the plate is perpendicular to the direction of flow, yields the flow pattern shown in Figure 10.4c. For this condition the drag force is due to the pressure difference on both sides of the plate. Figure 10.7 shows the drag coefficient for finite flat plates perpendicular to the direction of flow. Comparison of the data for a square plate ($x = y$) with the data for the disk shown in Figure 10.2 shows good agreement. As the value of x/y increases, the value of C_D approaches the value given for flat plates in Figure 10.3.

We have previously noted that C_D for a sphere was 24/Re, based on Stokes' law. Stokes' law is based upon laminar flow, and it will be seen from Figure 10.2 that for Reynolds numbers greater than 0.5, the discrepancy between values of C_D equal to 24/Re and the experimentally determined value of C_D increases. For small Reynolds numbers, the value of C_D for various bodies is given in Table 10.1. Note that for a flat plate oriented parallel to the direction of flow, the proper area to use in the drag equation is the area of both sides of the plate. It will be noted that C_D in every case is much greater than the values shown for turbulent flow on Figures 10.2 and 10.3. This was also the case for the flow of fluids in pipes, where it will be recalled that the largest values of friction factor occur when the flow is laminar for small values of Reynolds number.

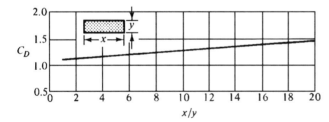

Figure 10.7   Drag coefficients; flat plate perpendicular to the flow. (Reproduced with permission from *Fluid Mechanics* by R. C. Binder, 4th ed., Prentice-Hall, Inc., Englewood Cliffs, N. J., 1962, p. 171.)

Table 10.1

Object	Re	C_D
Sphere	<0.5	24/Re
Disk flow	<0.5	20.4/Re
Disk flow	<0.01	13.6/Re
Circular cylinder	<0.1	$8\pi/\text{Re}[2.0 - \ln \text{Re}]$
Flat plate perpendicular to flow	<0.1	$8\pi/\text{Re}[2.2 - \ln \text{Re}]$
Flat plate parallel to flow	<0.01	4.12/Re

Source: Data for this table are based on *Fluid Mechanics for Engineers* by M. L. Albertson, J. R. Barton, and D. B. Simons, Prentice-Hall, Inc., Englewood Cliffs, N.J., 1960, p. 395.

ILLUSTRATION 10.3   DRAG

A 6-in. by 1-in. flat plate is placed parallel to the flow with the 6-in. edge facing into the flow. If the velocity of the fluid is 0.01 ft/s, evaluate the drag on the plate. Assume that the viscosity of the fluid is 1×10^{-4} lb·s/ft² and that the flow is laminar.

ILLUSTRATIVE PROBLEM 10.3

Given: Plate 6 in. × 1 in. placed parallel to flow; $V = 0.01$ ft/s; $\mu = 1 \times 10^{-4}$ lb·s/ft²,
         flow laminar

Find: Drag

Assumptions: Laminar flow, Table 10.1 applicable

Basic Equations: Drag: $D = C_D \dfrac{\rho V^2}{2} A$

$$C_D = \frac{4.12}{\text{Re}} \text{ (Table 10.1)}$$

Solution: Using $C_D = 4.12/\text{Re}$ from Table 10.1,

$$\text{drag} = D = C_D \frac{\rho V^2}{2}(A)$$

Therefore,

$$D = \frac{4.12}{DV\rho/\mu} \left(\frac{\rho V^2}{2} \right) A$$

and

$$D = \frac{4.12\mu}{(1/12)V\rho} V\rho\frac{V}{2}\left[\frac{1}{12}\left(\frac{6}{12} \right) \right] \times 2$$

Note that twice the area is used for flow parallel to the plate, since the fluid travels over both sides of the plate. Rearranging yields

$$D = 4.12\mu V \left(\frac{6}{12} \right) = 2.06\mu V$$

For this problem,

$$D = 2.06 \times 1 \times 10^{-4} \times 0.01 = 2.06 \times 10^{-6} \text{ lb}$$

It should be noted that whenever C_D can be written as a constant/Re, the resulting equation for the drag will be a function of only the product of viscosity, the velocity, and a characteristic length [i.e., $D = \text{constant}/(L\mu V)$].

The previous discussion has been principally concerned with laminar flow about an immersed object. As the velocity of the fluid relative to the body is increased (with the subsequent increase in Reynolds number), we have already noted that the drag coefficient decreases. At a Reynolds number of approximately 2×10^5, it will be seen from Figures 10.2 and 10.3 that there is a sharp discontinuity, indicating a marked decrease in drag coefficient for both the cylinder and sphere. As a matter of fact, the drag coefficients decrease to approximately one-third of their value just prior to this occurence. It has been found experimentally that the Reynolds number at which this abrupt decrease in drag coefficient occurs is dependent on the turbulence in the undisturbed fluid and the roughness of the body. Thus, if the turbulence in the undisturbed fluid is large and/or the surface of the body is made very rough (relatively), the noted decrease in drag coefficient will occur at Reynolds numbers less than 2×10^5. This behavior is again found to be similar to the effects discussed in Chapter 6 on the incompressible flow of fluids in pipes. In both instances (flow inside pipes and flow around immersed bodies) the underlying behavior is found to reside in the boundary layer adjacent to either the pipe wall or the immersed object. When studying the flow around an immersed body, the abrupt transition in the drag coefficient is found to be caused by the boundary layer changing from a laminar boundary layer to a turbulent boundary layer on the fore part of the body. This same phenomenon has been discussed earlier in this section during our discussion on the flow over flat plates.

Drag coefficients for cylinders and flat plates are given in Table 10.2 on page 368. In the turbulent range the drag coefficient for these bodies decreases with increased Reynolds number.

10.3 LIFT AND DRAG ON AIRFOILS

10.3.1 General

In section 10.2 the discussion was directed to the general problem of forces on an immersed body. At this point the discussion will be directed to the forces (lift and drag) that occur when an airplane moves through the air at speeds much less than the velocity of sound. Prior to our study of specific subtopics under this heading, it will be useful to define and illustrate our terminology.

Airplane.   An airplane is a mechanically driven, fixed-wing aircraft, heavier than air, that is supported by the dynamic reaction of the air against its wings. This definition can be amplified by considering an airplane in steady, level flight. For this airplane to be in equilibrium (it is not accelerating), the lift forces on it must equal its weight, and the drag forces must be countered by an equal but oppositely directed thrust from the airplane's power plant. Figure 10.8 on page 369 illustrates the forces on an airplane in level flight. Figures 10.9 and 10.10 on page 369 show two modern military airplanes.

Airfoil.   An airfoil is any surface, such as the airplane wing, aileron, or rudder, designed to obtain reaction from the air through which it moves. Figure 10.11 on page 369 shows an airfoil and the pressure distribution on this airfoil. Along the upper surface there is a reduced pressure (negative pressure), while underneath the airfoil the pressure is greater than ambient (positive). This pressure distribution occurs from the acceleration of the air over the upper surface of the airfoil and the deceleration of the air over the lower surface of the airfoil. Significantly, the greater portion of the lift is obtained from the upper airfoil surface.

Table 10.2   Drag Coefficients for Cylinders and Flat Plates

Object (flow from L to R)	L/d	Re	C_D
1. Circular cylinder, axis perpendicular to the flow	1	10^5	0.63
	5		0.74
	20		0.90
	∞		1.20
	5	$>5 \times 10^5$	0.35
	∞		0.33
2. Circular cylinder, axis parallel to the flow	0	$>10^3$	1.12
	1		0.91
	2		0.85
	4		0.87
	7		0.99
3. Elliptical cylinder ● (2:1)		4×10^4	0.6
		10^5	0.46
● (4:1)		2.5×10^4 to 10^5	0.32
● (8:1)		2.5×10^4	0.29
		2×10^5	0.20
4. Airfoil (1:3)	∞	$>4 \times 10^4$	0.07
5. Rectangular plate for which L = length and d = width	1	$>10^3$	1.16
	5		1.02
	20		1.50
	∞		1.90
6. Square cylinder ■		3.5×10^4	2.0
◆		$10^4 \times 10^5$	1.6
7. Triangular cylinder 120°		$>10^4$	2.0
			1.72
60°			2.20
			1.39
30°		$>10^5$	1.80
			1.0
8. Hemispherical shell		$>10^3$	1.33
		10^3 to 10^5	0.4
9. Circular disk, normal to the flow		$>10^3$	1.12
10. Tandem disks; spacing is L	0	$>10^3$	1.12
	1		0.93
	2		1.04
	3		1.54

Source: Reproduced with permission from *Fluid Mechanics for Engineers* by M.L. Albertson, J. L. Barton, and D. B Simons, Prentice-Hall Inc., Englewood Cliffs, N.J., 1960, p. 407.

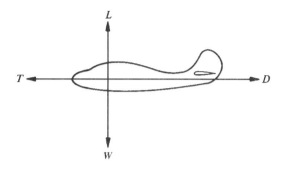

Figure 10.8   Airplane in level flight.

Figure 10.9   E-2C Hawkeye, airborne early warning command and control aircraft with large antenna visible. This airplane uses modern electronics and sensors for data retrieval and automatic control. (Courtesy of Northrup Grumman Corp.)

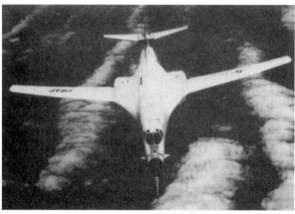

Figure 10.10   B-1B intercontinental bomber. (Courtesy of the United States Air Force.)

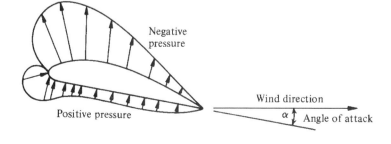

Figure 10.11   Pressure distribution on an airfoil section.

Angle of attack.   The angle of attack is the acute angle between a reference line in a body and the line of the relative wind direction projected on a plane containing the reference line and parallel to the plane of symmetry. The relative wind is the velocity of the air with reference to the body, and is measured at a distance from the body to minimize the disturbing effect of the body. Figure 10.11 also illustrates the angle of attack on an airfoil section.

Angle, zero lift.   The zero-lift angle is that angle of attack of an airfoil when its lift is zero. Figure 10.12 shows the characteristics of a Clark Y airfoil. With an airfoil that is completely symmetrical, the lift coefficient (and consequently the lift) would be zero at a zero angle of attack. This airfoil can be seen to be asymmetric, however, and it is found that C_L is zero at an angle of $-5°$ (i.e., that angle where the sum of the positive lift forces equals the sum of the negative lift forces). For angles of attack greater than the angle of zero lift, it will be seen that the lift coefficient is directly proportional to the angle of attack when this angle is measured relative to the zero-lift angle; that is,

$$C_L = K(\alpha - \alpha_{0L}) \tag{10.4}$$

where K is the slope of the curve of C_L against the angle of attack, α the angle of attack, and α_{0L} the angle of zero lift.

ILLUSTRATION 10.4   LIFT

Using the data for the Clark Y airfoil shown in Figure 10.12, it is found that C_L is 0.5 at an angle of attack of $2°$. Estimate C_L at an angle of attack of $6°$.

ILLUSTRATIVE PROBLEM 10.4

Given: $C_L = 0.5$ at $\alpha = 2°$

Find: C_L at $\alpha = 6°$

Assumptions: Clark Y airfoil, Figure 10.12 applicable

Basic Equations: Equation (10.4): $C_L = K(\alpha - \alpha_{0L})$

Solution: From Figure 10.12, the angle of zero lift is $-5°$. From the stated condition at $2°$,

$$0.5 = K[2 - (-5)]$$

Therefore,

$$K = \frac{0.5}{2 - (-5)} = 0.0714$$

For $6°$,

$$C_L = 0.0714[6 - (-5)] = 0.785$$

From Figure 10.12, C_L for $6°$ is approximately 0.79, which is in good agreement with the calculated value.

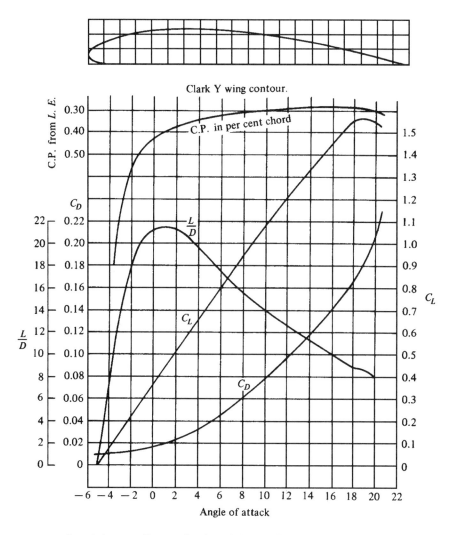

Figure 10.12   Lift and drag coefficients for the Clark Y airfoil. (Reproduced with permission from *Elements of Practical Aerodynamics* by B. Jones, John Wiley & Sons, Inc., New York, 1942, p. 39.)

As the angle of attack is increased the lift coefficient deviates from a linear relationship with respect to the angle of attack. For the Clark Y airfoil the maximum value of C_L occurs at $18\frac{1}{2}°$; above this angle (known as the *burble point* or *stall angle*) the lift decreases with increasing angle of attack.

Chord.   The chord of an airfoil is an arbitrary datum line from which the ordinates and angles of an airfoil are measured. It is usually the straight line tangent to the lower airfoil surface at two points of the straight line joining the leading and trailing edge of the airfoil. This definition is illustrated for the two airfoils shown in Figure 10.13. The chord length (customarily given the symbol c) is the length of the projection of the airfoil profile on its chord.

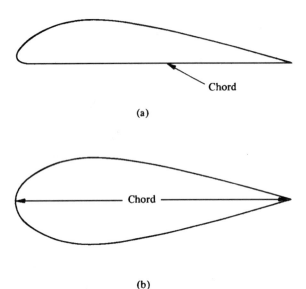

Figure 10.13   Chord for unsymmetric and symmetric airfoils.

Wing area.   To calculate both the lift and drag on an airplane, it is necessary to know the area term in equation (10.2). For purposes of standardization, the wing area is measured from the projection of the actual outline on the plane of the chords without deduction for the area that may be blanketed by the fuselage or nacelles. The wing is thus considered to extend without interruption through the fuselage and nacelles, and the wing area always includes flaps and ailerons.

Aspect ratio (A.R.).   The aspect ratio of a wing is defined as the ratio of the span of the wing to its mean chord. In general, wings are not rectangular when viewed from above, and the aspect ratio in this case is defined as the ratio of the square of the span (the distance from wingtip to wingtip is denoted as b) to the area; that is,

$$\text{A.R.} = \frac{b^2}{A} \tag{10.5}$$

Camber.   The curvature of an airfoil is known as it camber, with upper camber referring to the upper surface, lower camber referring to the lower surface, and mean camber referring to the mean surface. It is usual to express camber as the ratio of the departure of the curved surface from a straight line joining the extremeties of the curve to the length of this straight line.

Center of pressure (C.P.).   The location of the resultant aerodynamic force on an airfoil is not fixed relative to the airfoil. The center of pressure is the point in the chord of the airfoil (prolonged if necessary) that is the intersection of the chord and the line of action of the resultant air force. It is practically expressed in terms of the ratio of the distance of the center of pressure from the leading edge to the chord length and is usually stated as percent of chord. Values of C.P. in percent of chord for the Clark Y airfoil at various angles of attack are shown in Figure 10.12.

   Stall.   When an airfoil is operated at an angle of attack greater than the angle of attack at maximum lift (for the Clark Y airfoil this is $18\frac{1}{2}°$), the airfoil is said to be operating in the stall condition. At stall the air separates from the airfoil or wing and forms an eddying wake with severe turbulence, and as a consequence it is found that as the stall condition is approached there is a pronounced increase in the drag coefficient. Inspection of Figure 10.12 shows the drag coefficient increasing continuously as the angle of attack is increased—the percentage increase becoming larger as the stall condition is approached.

10.3.2 Steady, Level Flight

Let us now return to the basic lift and drag equations and apply them to the airplane in level flight. In the following material the data given in Figure 10.12 for the Clark Y airfoil will be used as illustrative of airfoil data in general. This is not meant to infer that all airfoil data are the same or even close to that shown for the Clark Y airfoil. For other airfoils it is necessary to obtain data for the specific airfoil in question. In many instances such data will be found in the extensive compilations from wind tunnel tests conducted by NASA. From equations (10.2) and Figure 10.8 we can conclude the following:

   1. For an airplane in steady flight the lift will just equal the weight of the airplane. Therefore,

$$W = C_L \frac{\rho}{2} V^2 A \tag{10.6}$$

Rearranging, we obtain

$$V = \sqrt{\left(\frac{W}{A}\right)\frac{2}{\rho C_L}} \tag{10.7a}$$

The term W/A is the weight loading of the wing in lb/ft² or N/m², and therefore the velocity at which a plane must fly in steady, level flight is proportional to the square root of the wing loading.

   Using the specific weight of air to be 0.0765 lb/ft³ at sea level, V in mi/h, W in lb, and A in ft², we can write equation (10.7a) as

$$V = 19.77 \sqrt{\frac{1}{C_L}\left(\frac{W}{A}\right)} \tag{10.7b}$$

In terms of SI units with the specific weight of air taken to be 12.02 N/m³, V in km/h, W in newtons, and A in m², we can write equation (10.7a) as

$$V = 4.6 \sqrt{\frac{1}{C_L}\left(\frac{W}{A}\right)} \tag{10.7c}$$

ILLUSTRATION 10.5   WING LOADING IN STEADY, LEVEL FLIGHT

An airplane is to fly at 300 mi/h at sea level. If the airfoil is a Clark Y airfoil having an angle of 6°, determine the wing loading in pounds per square foot of wing.

ILLUSTRATIVE PROBLEM 10.5

Given: $V = 300$ mi/h at sea level, Clark Y airfoil, $\alpha = 6°$

Find: W/A

Assumptions: Steady, level flight; sea level, Clark Y airfoil

Basic Equations: Equation (10.7b): $V = 19.77\sqrt{\dfrac{1}{C_L}\left(\dfrac{W}{A}\right)}$

Solution: For the units of this problem we will use equation (10.7b),

$$V = 19.77\sqrt{\frac{1}{C_L}\left(\frac{W}{A}\right)} \tag{10.7b}$$

Using the data of this problem and $C_L = 0.79$ for an angle of attack of $6°$, we obtain

$$300 = 19.77\sqrt{\frac{1}{0.79}\left(\frac{W}{A}\right)}$$

Solving for W/A yields

$$\frac{W}{A} = 181.9 \text{ lb/ft}^2$$

ILLUSTRATION 10.6   WING LOADING IN STEADY, LEVEL FLIGHT

An airplane is to fly at 483 km/h at sea level. If the airfoil is a Clark Y airfoil having an angle of $6°$, determine the wing loading in N/m² of wing.

ILLUSTRATIVE PROBLEM 10.6

Given: $V = 483$ km/h at sea level, Clark Y airfoil, $\alpha = 6°$

Find: W/A

Assumptions: Steady, level flight, sea level, Clark Y airfoil

Basic Equations: Equation (10.7c): $V = 4.6\sqrt{\dfrac{1}{C_L}\left(\dfrac{W}{A}\right)}$

Solution: This problem is the same as Illustrative Problem 10.5 with the problem given in SI units. Therefore,

$$V = 4.6\sqrt{\frac{1}{C_L}\left(\frac{W}{A}\right)}$$

and using $C_L = 0.79$ as before, we obtain

$$483 = 4.6\sqrt{\frac{1}{0.79}\left(\frac{W}{A}\right)}$$

Solving yields

$$\frac{W}{A} = 8710 \text{ N/m}^2$$

As a check:

$$181.9 \text{ lb/ft}^2 \times 4.448 \frac{\text{N}}{\text{lb}} \times \left(3.281 \frac{\text{m}}{\text{ft}}\right)^2 = 8710 \text{ N/m}^2$$

The agreement is very good, since 483 km/h is close to 300 mi/h.

2. For an airplane in steady, level flight the velocity of flight is inversely proportional to the square root of the lift coefficient. Since the lift coefficient exhibits a maximum value at the stall point, the minimum velocity (stall speed) that the airplane can have in steady, level flight must occur at an angle of attack coinciding with the maximum value of C_L. For the Clark Y airfoil we have already noted this angle to be $18\frac{1}{2}°$. This minimum velocity will also be (very closely) the lowest landing and/or takeoff speed of the airplane.

ILLUSTRATION 10.7   LANDING AND TAKEOFF SPEED

Using the data given in Illustrative Problem 10.6 and the calculated wing loading determined in this problem, evaluate the landing and takeoff speed of the airplane.

ILLUSTRATIVE PROBLEM 10.7

Given: Inputs as per Illustrative Problem 10.6

Find: Landing and takeoff speed of the airplane

Assumptions: Same as Illustrative Problem 10.6; maximum C_L occurs at $18\frac{1}{2}°$

Basic Equations: Equation (10.7c): $V = 4.6\sqrt{\dfrac{1}{C_L}\left(\dfrac{W}{A}\right)}$

Solution: At $18\frac{1}{2}°$, C_L (max) for the Clark Y airfoil is found to be close to 1.56. Therefore,

$$V_{\text{landing or takeoff}} = 4.6\sqrt{\frac{1}{C_L}\left(\frac{W}{A}\right)}$$

$$= 4.6\sqrt{\frac{1}{1.56}(8710)}$$

$$= 343.7 \text{ km/h}$$

3. When the airplane is in steady, level flight the propulsion system must provide a thrust equal and opposite to the drag force on the plane. An increase in thrust while in level

flight will cause the velocity of the plane to increase until the drag force once again equals the applied thrust. Conversely, a decrease in thrust will cause the airplane to decrease its velocity until the drag equals the thrust. The power input by the propulsion system is the product of the thrust multiplied by the velocity of the airplane. In steady, level flight, however, thrust necessarily equals drag and the power required becomes

$$P = TV = DV \tag{10.8a}$$

In terms of horsepower with D in pounds and V in feet per second,

$$\text{hp} = \frac{DV}{550} = \frac{C_D(\rho/2)V^3A}{550} \tag{10.8b}$$

where 1 hp = 550 ft·lb/s. In terms of miles per hour and a specific weight of 0.0765, we can write

$$\text{hp} = 6.8 \times 10^{-6} C_D V^3 A \tag{10.8c}$$

In SI units, 1 hp = 746 W = 746 N·m/s. In terms of km/h, a specific weight of 12.02 N/m³, and area in m²,

$$\text{hp} = 1.76 \times 10^{-5} C_D V^3 A \tag{10.8d}$$

ILLUSTRATION 10.8   HORSEPOWER REQUIRED

What horsepower is required for the condition given in Illustrative Problem 10.5? Assume that the wing has an area of 550 ft².

ILLUSTRATIVE PROBLEM 10.8

Given: $A = 550$ ft²; other inputs the same as Illustrative Problem 10.5

Find: Horsepower

Assumptions: Same as Illustrative Problem 10.5

Basic Equations: Equation (10.8c): hp = $6.8 \times 10^{-6} C_D V^3 A$

Solution: At an angle of attack of 6°, the Clark Y airfoil has $C_D = 0.045$, and applying equation (10.8c), we have

$$\text{hp} = 6.8 \times 10^{-6} \times 0.045 \times (300)^3 \times 550 = 4540 \text{ hp}$$

ILLUSTRATION 10.9   HORSEPOWER REQUIRED

What horsepower is required for the condition given in Illustrative Problem 10.6 if the wing area is 51 m²?

ILLUSTRATIVE PROBLEM 10.9

Given: $A = 51$ m²; other inputs same as Illustrative Problem 10.6

Find: Horsepower

Assumptions: Same as Illustrative Problem 10.6

Basic Equations: Equation (10.8d); hp $= 1.76 \times 10^{-5} C_D V^3 A$

Solution: The angle of attack is again 6°, and $C_D = 0.045$. Using equation (10.8d), we obtain

$$\text{hp} = 1.76 \times 10^{-5} \times 0.045 \times (483)^3 \times 51 = 4550 \text{ hp}$$

As noted before, this is the SI equivalent of Illustrative Problem 10.8, and the agreement is good.

---

4. The general conclusions reached in the earlier parts of this section are valid even when an airplane is operated at different altitudes. We have already concluded that steady, level flight requires the lift to equal the weight of the airplane, and this must be true at all altitudes. If, however, the angle of attack is constant, then C_L is constant, and the velocity must increase to maintain level flight at altitudes, since the density decreases with increasing altitude. Again, if the angle of attack is constant, then C_D is constant, and for steady, level flight it is necessary to evaluate the drag at increasing altitude, since the velocity has increased but the density has decreased with increasing altitude. Let the subscript 0 denote sea level (zero altitude) and x any altitude; C_L and C_D are constant at all altitudes for a given angle of attack. Consider the lift equation,

$$W = C_L \frac{\rho_0}{2} V_0^2 A = C_L \frac{\rho_x}{2} V_x^2 A \tag{10.9}$$

Therefore,

$$V_x^2 = \frac{\rho_0}{\rho_x} V_0^2 \tag{10.10}$$

From the drag equation,

$$D_0 = C_D \frac{\rho_0}{2} V_0^2 A \tag{10.11}$$

and

$$D_x = C_D \frac{\rho_x}{2} V_x^2 A \tag{10.12}$$

Substituting (10.10) into (10.12), we obtain

$$D_x = C_D \frac{\rho_x}{2} \left(\frac{\rho_0}{\rho_x}\right) V_0^2 A = C_D \frac{\rho_0}{2} V_0^2 A \tag{10.13}$$

or

$$D_x = D_0 \qquad (10.13a)$$

The somewhat surprising conclusion from equation (10.13) or (10.13a) is that for a given angle of attack the drag is the same regardless of altitude. The power required, however, does not stay constant, as can be shown by the following reasoning. From equation (10.8b) we can write

$$(\text{hp})_0 = \frac{D_0 V_0}{550} \qquad (10.14)$$

and

$$(\text{hp})_x = \frac{D_x V_x}{550} = \frac{D_0 V_x}{550} \qquad (10.15)$$

since $D_0 = D_x$.
    From equation (10.10),

$$V_x = V_0 \sqrt{\frac{\rho_0}{\rho_x}}$$

Substitution of (10.16) into (10.15) yields

$$(\text{hp})_x = \frac{D_0 V_0}{550} \sqrt{\frac{\rho_0}{\rho_x}} = (\text{hp})_0 \sqrt{\frac{\rho_0}{\rho_x}} \qquad (10.17)$$

Since ρ_x is always less than ρ_0, the horsepower required at any altitude will be greater than the horsepower required at sea level for a fixed angle of attack.

ILLUSTRATION 10.10 HORSEPOWER AT SEA LEVEL AND ALTITUDE

A Clark Y airfoil having an area of 450 ft² is operated at a 4° angle of attack. If the airplane is designed to fly at 200 mi/h at sea level, determine its velocity at an altitude where the density is eight-tenths of the density at sea level. Also determine the horsepower required at sea level and at altitude. Use $\rho = 0.002378$ slugs/ft³ at sea level.

ILLUSTRATIVE PROBLEM 10.10

Given: Clark Y airfoil, $\alpha = 4°$, $A = 450$ ft², $V = 200$ mi/h at sea level, $\rho = 0.002378$
    slug/ft³ at sea level, $\rho_x / \rho_0 = 0.8$

Find: Horsepower at sea level and altitude

Assumptions: Steady, level flight

Basic Equations: Equation (10.10): $V_x = V_0 \sqrt{\dfrac{\rho_0}{\rho_x}}$

Drag: $D = C_D \dfrac{\rho V^2}{2} A$

$$hp = \dfrac{DV}{550}$$

Solution: At $4°$,

$$C_L = 0.65$$

$$C_D = 0.034$$

From equation (10.10),

$$V_x = V_0 \sqrt{\dfrac{\rho_0}{\rho_x}} = 200 \sqrt{\dfrac{1}{0.8}} = 224 \text{ mi/h}$$

At sea level,

$$D_0 = C_D \dfrac{\rho_0}{2} V_0^2 A \qquad 200 \text{ mi/h} = 294 \text{ ft/s}$$

Therefore,

$$D_0 = 0.034 \dfrac{0.002378 \text{ slugs/ft}^3}{2} (294 \text{ ft/s})^2 \times 450 \text{ ft}^2 = 1560 \text{ lb}$$

$$\dfrac{D_0 V_0}{550} = \dfrac{1560 \text{ lb} \times 294 \text{ ft/s}}{550 \text{ ft·lb/s·hp}} = 835 \text{ hp}$$

At altitude,

$$(hp)_x = 835 \text{ hp} \sqrt{\dfrac{1}{0.8}} = 935 \text{ hp}$$

5. The ability of an airfoil to perform its function of providing lift with the minimum amount of drag is expressed in the ratio L/D. Aerodynamically, the wing having the largest value of this figure of merit would perform in the most desirable manner. The ratio of L/D can also be interpreted by referring to Figure 10.1, where it will be seen that the ratio of lift to drag is the tangent of the angle that the resultant force on the airfoil makes with respect to the line of the relative wind direction. Mathematically, we can write

$$L = C_L \dfrac{\rho}{2} V^2 A$$

$$D = C_D \dfrac{\rho}{2} V^2 A$$

and

$$L/D = \frac{C_L \frac{\rho}{2} V^2 A}{C_D \frac{\rho}{2} V^2 A} = \frac{C_L}{C_D} \qquad (10.18)$$

The ratio of L/D has been plotted in Figure 10.12 for the Clark Y airfoil. For this airfoil section it will be seen that the maximum value of L/D occurs at an angle of attack of approximately $+1°$. For a given angle of attack, an airfoil having the maximum value of L/D will require less thrust from its propulsion system.

ILLUSTRATION 10.11   DRAG

At $18\frac{1}{2}°$ the Clark Y airfoil has its maximum lift and is said to be at the stall point. If an airplane using this airfoil weighs 10,000 lb, determine its drag at the stall point. Also determine the minimum drag for this airplane. Neglect all effects except those due to the airfoil. Discuss the values obtained.

ILLUSTRATIVE PROBLEM 10.11

Given: Clark Y airfoil, $W = 10,000$ lb, $\alpha = 18\frac{1}{2}°$ is angle of maximum lift

Find: D at stall point, $D_{minimum}$

Assumptions: L/D is maximum at $1°$

Basic Equations: Lift and drag

Solution: At $18\frac{1}{2}°$,

$$\frac{L}{D} = 8.8 \text{ (Figure 10.12)}$$

But, $L = W$; therefore,

$$\frac{W}{D} = 8.8$$

$$D = \frac{10,000}{8.8} = 1138 \text{ lb}$$

At $+1°$, L/D is maximum and equal to 21.5.

$$\frac{W}{D} = 21.5$$

$$D = \frac{10,000}{21.5} = 466 \text{ lb}$$

At the higher angle of attack the required thrust is 1135 lb, which is almost $2\frac{1}{2}$ times greater than the value at the lower angle of attack. It must be noted, however, that the velocity of the airplane is not the same for both cases. Since velocity is inversely proportional to the square root of the lift coefficient, we can write

$$\frac{V_{18.5}}{V_1} = \sqrt{\frac{(C_L)_1}{(C_L)_{18.5}}} = \sqrt{\frac{0.45}{1.56}} = 0.54$$

The velocity of the airplane at $18\frac{1}{2}°$ is therefore approximately one-half the velocity that the airplane would have at $1°$. At takeoff, where it is desired that the airplane should become airborne at the minimum possible velocity (using the least length of runway), the high angle of attack is achieved using flaps and other auxiliary devices, but this is achieved only at the expense of requiring more thrust from the airplane's propulsion system.

10.3.3 Polar Diagram

The curves for C_L, C_D, and L/D as a function of angle of attack are satisfactory for evaluating the performance of a given airfoil, and this type of plot can also be useful when comparing airfoils. This requires, however, that three curves must be plotted and evaluated in each case. It is much more convenient to plot these data on a single diagram, called a *polar diagram* (in which C_L is the ordinate and C_D the abscissa), as shown in Figure 10.14 on page 382 for the Clark Y airfoil. If we first consider the plot to have the same scales for C_L and C_D, the ratio of lift to drag (L/D) would be the slope of the straight line drawn from the origin to the curve. The maximum value of L/D would be the slope of the line drawn from the origin tangent to the curve. In every case a line from the origin to the curve would give the direction of the resultant force and be proportional to it in magnitude. The angle of zero lift is readily obtained, and the angle of minimum drag can be found by drawing a vertical tangent to the curve. In addition, the stall point and the maximum value of C_L are readily determined from this curve.

Since values of C_D are approximately one order of magnitude less than the value of C_L at a given angle of attack, it is usual to use scales that are unequal for ease of reading, as done in Figure 10.14. When the scales are unequal, the line drawn from the origin to the curve no longer gives the direction, nor is it proportional to the magnitude of the resultant force on the airfoil. The tangent to the curve drawn from the origin, however, always locates the angle of attack and the values of C_L and C_D for the maximum value of L/D.

ILLUSTRATION 10.12   USE OF POLAR DIAGRAM

From Figure 10.14, determine the stall angle and C_L, C_D, and L/D at this angle.

ILLUSTRATIVE PROBLEM 10.12

Given: Clark Y airfoil

Find: Stall angle, C_L, C_D, $\dfrac{L}{D}$ (at stall angle)

Assumptions: Steady, level flight

Basic Equations: Figure 10.14

Solution: From Figure 10.14, the stall angle is approximately $18\frac{1}{2}°$, $C_L = 1.56$, $C_D = 0.175$, and

$$\frac{L}{D} = \frac{C_L}{C_D} = \frac{1.56}{0.175} = 8.9$$

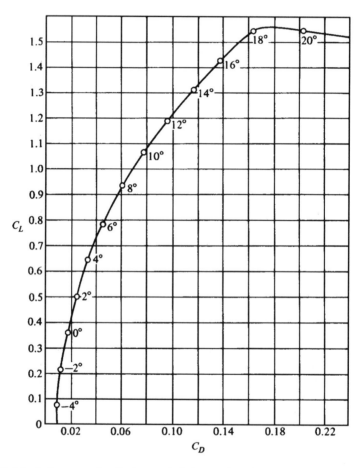

Figure 10.14   Polar curve for Clark Y airfoil. (Reproduced with permission from *Elements of Practical Aerodynamics* by B. Jones, John Wiley & Sons, Inc., New York, 1942, p. 61.)

ILLUSTRATION 10.13   USE OF POLAR DIAGRAM

An airplane operating at sea level utilizes a Clark Y airfoil. If the airplane weighs 10,000 lb and is operated at 120 mi/h, determine the power required to keep it in level flight. The wing area is to be taken as 500 ft². ($\rho = 0.002378$ slug/ft³)

ILLUSTRATIVE PROBLEM 10.13

Given: Clark Y airfoil, $W = 10,000$ lb, $V = 120$ mi/h, $A = 500$ ft², $\rho = 0.002378$ slugs/ft³.

Find: Power for level flight

Assumptions: Steady, level flight

Basic Equations: Lift equation: $L = C_L \dfrac{\rho V^2}{2} A$; Figure 10.14

Solution: It is necessary to determine the angle of attack at which the plane is operating. For this C_L is evaluated, noting that 120 mi/h = 176 ft/s:

$$L = W = C_L \frac{\rho}{2} V^2 A$$

$$10,000 = C_L \left(\frac{0.002378 \text{ slugs/ft}^3}{2} \right)(176 \text{ ft/s})^2 500 \text{ ft}^2$$

and

$$C_L = 0.544$$

From Figure 10.14, the angle of attack is approximately 2.6° and $C_D = 0.027$. Therefore, from equation (10.8c),

$$\text{hp} = 6.8 \times 10^{-6} \times 0.027 \times (120 \text{ mph})^3 \times 500 \text{ ft}^2 = 159 \text{ hp}$$

If the span of an airfoil were infinite and the airfoil had an infinite aspect ratio (A.R.), it would be found to have a uniform lift across the span. For airfoils of finite length the distribution of the lift across the span is not uniform, decreasing in magnitude near the wingtips. As shown in Figure 10.11, the pressure below the airfoil is positive, while the pressure above the airfoil is negative. This pressure difference causes air on the bottom of the wing near the wingtips to flow outward and upward toward the top surface of the wing; on top the incoming air causes the airflow to move inward toward the center of the airfoil. The result of this flow pattern is to produce vortices at the wingtips, which can often be seen as an airplane passes through moist air, due to condensation caused by the reduced pressure and temperature within these vortices.

As a consequence of the foregoing there is a downward flow of air at the wingtips causing changes in both the lift and drag on the wing. The magnitude of this effect is dependent on the span and aspect ratio of the airfoil; the greater the aspect ratio, the smaller is the effect on lift and drag.

Figure 10.15 shows an airfoil of finite span and the velocity of the relative wind in the undisturbed stream. For an infinite span and infinite aspect ratio there is no induced downward motion of the air, so that the relative wind in the undisturbed air forward of the wing is the same as the relative wind at the wing. Due to the downward induced velocity of the finite wing, the relative wind at the airfoil does not equal the relative wind forward of the airfoil in the undisturbed airstream. If we denote the angle between the relative wind V and

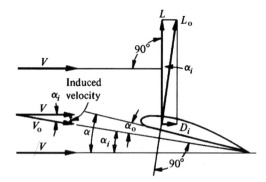

Figure 10.15   Finite airfoil showing induced drag. (Reproduced with permission from *Elementary Fluid Mechanics* by J. K. Vennard, 4th ed., John Wiley & Sons, Inc., New York, 1961, p. 522.)

the true relative wind V_0 to be α_p we obtain the geometric angle of attack α from Figure 10.15 to be

$$\alpha = \alpha_0 + \alpha_i \tag{10.19}$$

where α_0 is the angle of attack for an airfoil of infinite span and infinite aspect ratio. Let us now consider that the airfoil has an infinite span and infinite aspect ratio. If this were the case, α_p, the direction of the relative wind, would coincide with the line of V_0, the effective velocity, and L_0 would be the lift normal to the direction of V_0. Because L_0 is not vertical, we can consider it to be composed of two components; one of these, L, is the lift directed normal to V, and the other, D_i, is called the induced drag and is directed parallel to V. Note that D_i, the induced drag, is in addition to the drag that we obtain for an airfoil of infinite span and infinite aspect ratio. Denoting the drag of an airfoil of infinite span to be the profile drag D_0, the total drag D on a finite wing can be written as the sum of the induced and profile drags.

$$D = D_i + D_0 \tag{10.20}$$

Dividing all the terms in equation (10.20) by $(\rho/2)V^2A$, we obtain

$$C_D = C_{Di} + C_{D0} \tag{10.21}$$

The importance and usefulness of the foregoing lies in the fact that airfoils of finite span that are tested in wind tunnels do not necessarily have the aspect ratio of the airfoil under consideration by the designers. In the past the aspect ratio most commmonly used was 6, and by custom it can be assumed that published airfoil data are for an aspect ratio of 6 unless otherwise stated. It is presently the custom to furnish airfoil data for an infinite aspect ratio and to correct these data to any desired aspect ratio. The correction procedure consists of converting the data from one aspect to an infinite aspect ratio and then reconverting to the desired aspect ratio. By this process it is necessary to report test data only for one aspect ratio (usually infinite), thereby eliminating the necessity for conducting wind tunnel tests at all aspect ratios of interest.

10.4 AIRCRAFT MANEUVERS

The previous discussion has considered only the airplane in steady, level flight. A complete discussion of the mechanics of unsteady flight is beyond the scope of this text, but we have already developed enough information to consider certain steady-state aircraft maneuvers: (1) glide, (2) banked turn, and (3) climb.

When an airplane is said to be in a *glide* we shall assume that the power is off (engine completely throttled or dead) and that the propulsion system does not provide any thrust or drag. The glide angle is defined as the angle below the horizontal that the airplane descends along under these conditions. Denoting this angle as θ, the angle of incidence to be α, and the lift, drag, and weight as L, D, and W, respectively, we obtain the free-body diagram shown in Figure 10.16. In this figure the lift is perpendicular to the flight path, and the drag is parallel to the flight path. For a steady glide, the lift, drag, and weight must be a system of concurrent forces in equilibrium, and from Figure 10.16 and the conditions of equilibrium, we can write

$$L = W \cos \theta \tag{10.22}$$

But

$$L = C_L \frac{\rho}{2} V^2 A \tag{10.23}$$

Therefore,

$$\cos \theta = \frac{L}{W} = \frac{C_L (\rho/2) V^2 A}{W} \tag{10.24}$$

However,

$$D = W \sin \theta \tag{10.25}$$

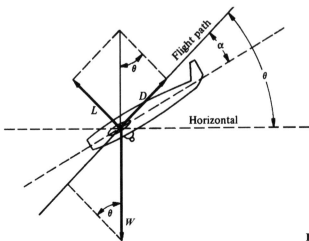

Figure 10.16   Airplane in a glide.

and

$$W = \frac{D}{\sin \theta} = \frac{C_D(\rho/2)V^2 A}{\sin \theta} \tag{10.26}$$

Combining equations (10.24) and (10.26), we obtain

$$\tan \theta = \frac{D}{L} = \frac{C_D}{C_L} \tag{10.27}$$

or

$$\cot \theta = \frac{L}{D} = \frac{C_L}{C_D} \tag{10.28}$$

The lift-to-drag ratio, L/D, can be read directly from curves such as Figure 10.12 or 10.14 for a given airfoil. Note that for each angle of incidence there is a single value of L/D, and therefore there is a single value of the glide angle. Also, since the ratio of lift to drag (and consequently θ) is not a function of the air density, the glide angle is the same for all altitudes.

ILLUSTRATION 10.14   GLIDING

An airplane using a Clark Y airfoil of A.R. 6 has a power plant failure and must glide to a nearby airport. Assuming the airplane to be at an altitude of 1000 ft, can the plane glide to the airport if the airport is 2 mi away?

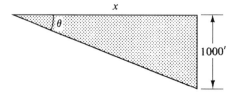

Figure 10.17   Illustrative Problem 10.14.

ILLUSTRATIVE PROBLEM 10.14

Given: Clark Y airfoil, A.R. = 6, altitude = 1000 ft, airport 2 miles away

Find: Can plane glide to airport?

Assumptions: Constant angle of attack

Basic Equations: Equation (10.27): $\tan \theta = \dfrac{C_D}{C_L}$

Solution: The maximum horizontal distance that the plane can glide occurs when the plane is operated at the minimum glide angle, and this corresponds to a maximum value of L/D. For the airplane in question, Figure 10.12 indicates a maximum value of L/D to be close to 21.5 at an angle of attack of approximately 1°. Then $\tan \theta =$

1/21.5 = 0.0465 and $\theta \simeq 2.7°$. Referring to Figure 10.17, the maximum distance the plane can glide is found as follows:

$$\tan \theta = \frac{1000 \text{ ft}}{x} \qquad x = \frac{1000 \text{ ft}}{\tan \theta} = \frac{1000 \text{ ft}}{0.0465} = 21,500 \text{ ft}$$

or 4.08 mi. Therefore, it is possible for the plane to glide safely to the airport.

If the angle of the glide steepens to 90°, a vertical "dive" occurs. Under this condition the lift is zero, the angle of attack corresponds to the angle of zero lift, and the plane will reach a velocity, termed the terminal velocity, V_T, of

$$V_T = \sqrt{\frac{2W}{C_D \rho A}} \tag{10.29}$$

The terminal velocity is seen to decrease as the altitude decreases, since the air density increases with decreasing altitude. If the airplane is placed in a dive (either deliberately or otherwise) with the power on, the thrust of the airplane's power plant must be accounted for. For this condition the drag on the airplane must equal the sum of the weight and thrust; thus,

$$T + W = D = C_D \frac{\rho}{2} V^2 A \tag{10.30}$$

and the terminal velocity with power is given by

$$V_{TP} = \sqrt{\frac{2(W + T)}{C_D \rho A}} \tag{10.31}$$

Depending on the altitude and thrust, it is possible for the plane to continue to accelerate and never reach its terminal velocity. Unless the airplane's controls can be made to bring it out of the dive without structural damage, it will crash and be destroyed.

ILLUSTRATION 10.15   TERMINAL VELOCITY

As a result of a malfunction in the controls of an airplane, it is placed in a dive. If the thrust is 2000 lb, the weight is 3000 lb, and the airfoil is a Clark Y airfoil of A.R. 6, determine the terminal velocity at sea level if the wing area is 600 ft². Use $\rho = 0.02378$ slugs/ft³.

ILLUSTRATIVE PROBLEM 10.15

Given: $T = 2000$ lb, $W = 3000$ lb, Clark Y A.R. = 6, $A = 600$ ft², $\rho = 0.002378$ slugs/ft³

Find: Terminal velocity

Assumptions: Zero lift on the plane; plane can sustain this speed without damage up to impact

Basic Equations: Equation (10.31): $V_{TP} = \sqrt{\dfrac{2(W + T)}{C_D \rho A}}$

Solution: For the Clark Y airfoil of A.R. 6, the angle of zero lift is obtained from Figure 10.12 as approximately −5°. C_D corresponding to this angle of attack is close to 0.01, and the terminal velocity will be

$$V_{TP} = \sqrt{\frac{2(2000 \text{ lb} + 3000 \text{ lb})}{0.01(0.002378 \text{ slugs/ft}^3)600 \text{ ft}^2}} = 836 \text{ ft/s}$$

or

$$V_{TP} = 570 \text{ mi/h}$$

ILLUSTRATION 10.16   TERMINAL VELOCITY

The pilot of the plane described in Illustrative Problem 10.15 is able to cut (turn off) the engine. What will the terminal velocity be?

ILLUSTRATIVE PROBLEM 10.16

Given: Same as Illustrative Problem 10.15 with $T = 0$

Find: Terminal velocity

Assumptions: No lift and engine thrust $= 0$

Basic Equations: Equation (10.29): $V_T = \sqrt{\dfrac{2W}{C_D \rho A}}$

Solution: Since $T = 0$, equation (10.29) applies,

$$V_T = \sqrt{\frac{2W}{C_D \rho A}} = \sqrt{\frac{2(3000 \text{ lb})}{0.01 \times 0.002378 \text{ slugs/ft}^3 \times 600 \text{ ft}^2}}$$

$$V_T = 648 \text{ ft/s}$$

or

$$V_T = 442 \text{ mi/h}$$

While the reduction in terminal velocity is only 22%, it will give the pilot more time, there will be smaller forces on the control surfaces, and the possibility of structural damage will be greatly reduced.

When an airplane in steady, level flight changes its flight direction by turning, it is necessary to consider the centrifugal force acting during the turn. As the airplane is turned in a horizontal plane it is usual to *bank,* that is, to depress the inner wing and elevate the outer wing. Without banking, the airplane tends to move outward; with banking the motion of the airplane depends on the angle of bank, the velocity of the airplane, and the radius of the turn. Consider the free-body diagram of the banked airplane shown in Figure 10.18. For

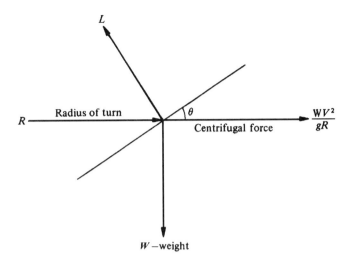

Figure 10.18   Forces in a steady-state banked turn.

equilibrium, the centrifugal force WV^2/gR must be balanced by the horizonal component of the lift, and the weight must be balanced by the vertical component of the lift. Thus,

$$L \sin \theta = \frac{WV^2}{gR} \tag{10.32}$$

and

$$L \cos \theta = W \tag{10.33}$$

From equations (10.32) and (10.33),

$$\tan \theta = \frac{V^2}{gR} \tag{10.34}$$

The angle of the bank is thus seen to be independent of the weight of the airplane, the wing area, and the properties of the airfoil. If the angle of the bank is made less than the value given by equation (10.34), the plane will move outward and is said to skid. If the angle of bank exceeds the value given by equation (10.34), the airplane will move inward and downward and is said to slip.

ILLUSTRATION 10.17   AIRCRAFT IN A STEADY-STATE BANKED CURVE

An airplane in a race over a closed course is required to turn around a pylon at each end of the course. If the plane is traveling at 640 km/h and is restricted to a turn radius that must not exceed 3/4 km, determine the best bank angle.

ILLUSTRATIVE PROBLEM 10.17

Given: $R = \dfrac{3}{4}$ km, $V = 640$ km/h

Find: θ

Assumptions: Steady state with no slipping or skidding

Basic Equations: Equation (10.34): $\tan \theta = \dfrac{V^2}{gR}$

Solution: For proper banking without skidding or slipping,

$$\tan \theta = \frac{V^2}{gR}$$

$$640 \text{ km/h} = \frac{640 \text{ km/h} \times 1000 \text{ m/km}}{3600 \text{ s/h}} = 178 \text{ m/s} \quad \text{and} \quad R = 750 \text{ m}$$

$$\tan \theta = \frac{(178 \text{ m/s})^2}{9.81 \text{ m/s}^2 \times 750 \text{ m}} = 4.31$$

$$\theta = 76.9°$$

The plane must make a very steep bank if it is to perform within the required limitations.

ILLUSTRATION 10.18   TOTAL FORCE IN A TURN

If the pilot in Illustrative Problem 10.17 weighs W, what is his apparent weight in the turn?

ILLUSTRATIVE PROBLEM 10.18

Given: Weight $= W$; other inputs the same as Illustrative Problem 10.17

Find: Total force on W

Assumptions: Same as Illustrative Problem 10.17

Basic Equations: Equation (10.34): $\tan \theta = \dfrac{V^2}{gR}$

Solution: The total force experienced by an object in the airplane is the vector sum of its weight and the centrifugal force acting on the body. Therefore,

$$\text{total force} = \text{apparent weight} = \sqrt{\left(\frac{WV^2}{gR}\right)^2 + W} = W\sqrt{\tan^2 \theta + 1} = W \sec \theta$$

The ratio of the apparent weight to W is the g loading, that is, the force the body would experience in a gravitational field greater than the earth's gravitational field by some multiple of the earth's gravitational field. Therefore, the g load is

$$g \text{ load} = \frac{\text{apparent weight}}{W} = \sec \theta = \frac{1}{\cos \theta}$$

For the problem under consideration, $\theta = 76.9°$, $\cos \theta = 0.2267$, and

$$g \text{ load} = \frac{1}{0.2267} = 4.41$$

Therefore, the apparent weight of the pilot is $4.41\ W$. Notice that the g load is a function only of the bank angle, θ.

10.5 REVIEW

In this chapter we have discussed the flow about immersed bodies from both general and applied aspects. As a starting point we discussed the laminar and turbulent flow regions for flow about immersed bodies. Whenever possible the development was based on reasonable mathematical models, but it was also found necessary to resort to experimentally determined characteristics for each of the topics covered. Spheres, cylinders, and flat plates were discussed in detail.

The airfoil is a form of body that moves through the ocean of air that surrounds the earth. The airfoil is unique, and because of its uniqueness, we defined in detail many of the terms that were used to describe it. The performance of the airfoil during level flight of the aircraft was the first topic that occupied us. After a detailed study of level performance, our attention was directed to the performance of an aircraft during maneuvers. Gliding, diving, and banking of the aircraft were analyzed as the final topics of the chapter.

One word of comment is needed at this time. This subject is vast, and no chapter in a textbook on fluid mechanics can do more than present an introduction to it. Mastery of it, however, will provide a sound basis for further study.

KEY TERMS

Terms of importance in this chapter:

Airfoil: any surface designed to obtain reaction from the air through which it moves.

Airplane: a mechanically driven fixed-wing aircraft, heavier than air, that is supported by the dynamic reaction of the air against its wings.

Angle of attack: the acute angle between a reference line on a body and the line of the relative wind direction projected on a plane containing the reference line and parallel to the plane of symmetry.

Angle of zero lift: the angle of attack when the lift of an airfoil is zero.

Apparent weight: the force on an object during an aircraft maneuver such as a banked turn.

Aspect ratio: the ratio of the span to the mean chord of a wing.

Bank: a turning maneuver with the inner wing depressed and the outer wing elevated.

Camber: the curvature of an airfoil.

Center of pressure: the location of the resultant aerodynamic force on an airfoil.

Chord: an arbitrary datum line from which the ordinates and angles of an airfoil are measured.

Drag: the component of the total force on an airfoil parallel to the direction of flow.

Drag coefficient: a dimensionless coefficient that is used to calculate the drag on an airfoil.

Dynamic pressure: a term that equals $\rho V^2/2$.

Glide: operation of an aircraft with the engine off or throttled to zero thrust.

Impact pressure: see dynamic pressure.

Lift: the component of the total force on an airfoil that is perpendicular to the direction of flow.

Lift coefficient: a dimensionless coefficient that is used to calculate the lift on an airfoil.

Polar diagram: a plot of the lift coefficient versus the drag coefficient with the angle of attack as the third variable.

Stall: operation at an angle of attack greater than the angle of attack at maximum lift.

Stokes' law: for a sphere, $C_D = 24/\text{Re}$ for low values of Re.

Terminal velocity: the equilibrium velocity for an object that is falling; for an airplane occurring when it is in a dive either with or without power.

KEY EQUATIONS

Drag	$$D = C_D \frac{\rho V^2}{2} A$$	(10.1)
Lift	$$L = C_L \frac{\rho V^2}{2} A$$	(10.2)
Lift Coefficient	$$C_L = K(\alpha - \alpha_{0L})$$	(10.4)
Aspect ratio	$$\text{A.R.} = \frac{b^2}{A}$$	(10.5)
Steady, level flight	$$W = C_L \frac{\rho}{2} V^2 A$$	(10.6)
Steady, level flight	$$V = \sqrt{\left(\frac{W}{A}\right)\frac{2}{\rho C_L}}$$	(10.7a)
Steady, level flight (English)	$$V = 19.77\sqrt{\frac{1}{C_L}\left(\frac{W}{A}\right)}$$	(10.7b)

Steady, level flight (SI)	$V = 4.6\sqrt{\dfrac{1}{C_L}\left(\dfrac{W}{A}\right)}$	(10.7c)
Power for level flight	$P = TV = DV$	(10.8a)
Power for level flight	$\text{hp} = \dfrac{C_D(\rho/2)V^3A}{550}$	(10.8b)
Power for level flight (English)	$\text{hp} = 6.8 \times 10^{-6}C_D V^3 A$	(10.8c)
Power for level flight (SI)	$\text{hp} = 1.76 \times 10^{-5}C_D V^3 A$	(10.8d)
Drag at elevation	$D_x = D_0$	(10.13a)
Horsepower at elevation	$(\text{hp})_x = (\text{hp})_0\sqrt{\dfrac{\rho_0}{\rho_x}}$	(10.17)
Lift-to-drag ratio	$\dfrac{L}{D} = \dfrac{C_L}{C_D}$	(10.18)
Induced drag	$\alpha = \alpha_0 + \alpha_i$	(10.19)
Induced drag	$D = D_i + D_0$	(10.20)
Induced drag	$C_D = C_{Di} + C_{D0}$	(10.21)
Glide	$\tan\theta = \dfrac{D}{L} = \dfrac{C_D}{C_L}$	(10.27)
Terminal velocity	$V_T = \sqrt{\dfrac{2W}{C_D\rho A}}$	(10.29)
Terminal velocity	$V_{TP} = \sqrt{\dfrac{2(W + T)}{C_D\rho A}}$	(10.31)
Angle of bank	$\tan\theta = \dfrac{V^2}{gR}$	(10.34)
g load	$g\text{ load} = \sec\theta$	Ill. Prob. 10.18

QUESTIONS

1. *True or false:* The lift and drag coefficients are dimensionless and therefore constant.
2. If for some reason you wanted to have a sensor that would exhibit a constant drag coefficient over a large range of Reynolds number, what shape or shapes would you consider?

3. What area would you consider when calculating the force on a flat plate that is (a) parallel to the direction of flow and (b) perpendicular to the direction of flow?

4. In what way does the drag coefficient behave like the friction factor for pipe flow?

5. When is the lift coefficient zero at zero angle of attack?

6. What condition must be satisfied for an aircraft to operate in steady, level flight?

7. Under what condition does the minimum landing and/or takeoff speed of an aircraft occur?

8. When an aircraft is operated at a given angle of attack the drag is found to be the same regardless of the altitude. Does this mean that the horsepower to operate at any altitude at a given angle of attack is the same?

9. State some reasons why the polar diagram is useful.

10. The total drag on an airfoil consists of the profile drag plus the induced drag. What is the origin of the induced drag?

11. What use can be made of the fact that the total drag on an airfoil consists of the sum of its induced plus its profile drag?

12. The angle of glide of an airplane is a function of what two variables?

13. What is the ultimate limit on the terminal velocity of an airplane?

14. Can you think of another situation that is similar to the banking of an airplane and that will give the same result when analyzed?

15. An airplane will skid and slip under what conditions?

16. Is the operation of an airplane in either a slip or skid condition desirable? Explain your answer.

17. When an airplane is banked at an angle of 30° the total force on the pilot is 15% greater than her weight. At an angle of bank of 60° the total force on the pilot is 100% greater than her weight. Can you give a reason for this?

PROBLEMS

Use

$$\mu = 3.73 \times 10^{-7} \text{ lb·s/ft}^2 \qquad \mu = 1.80 \times 10^{-5} \text{ Pa·s}$$

$$\rho = 0.02378 \frac{\text{lb·s}^2}{\text{ft}^4} \qquad (\gamma = 0.0765 \text{ lb/ft}^3)$$

$$\rho = 1.225 \text{ kg/m}^3 \qquad (\gamma = 12.02 \text{ N/m}^3)$$

for air in the following problems unless otherwise noted.

Lift and Drag

10.1. If a sphere is immersed in a fluid whose relative velocity with respect to the sphere is 0.01 ft/s, determine the drag force on the sphere. Assume the Reynolds number to be less than 0.5, the viscosity of the fluid to be 3×10^{-5} lb·s/ft^2, and the diameter of the sphere to be 0.001 in.

10.2. Using the data for cylinders in Figure 10.3, derive a relationship for the drag coefficient on a cylinder for Re ≤ 0.5 of the form $C_D = $ constant/Re.

10.3. Using the results of Problem 10.2, solve Problem 10.1 for a cylinder placed normal to the flow and having a length of 1 in. This is essentially a short, fine wire having a large ratio of length to diameter.

10.4. A wire, $\frac{1}{2}$ in. in diameter, is stretched taut. In a hurricane, it is exposed to winds with a sustained velocity of 130 ft/s perpendicular to the wire. Calculate the force on the wire per foot of length.

10.5. A smooth sphere, 6 in. in diameter, weighing 10 lb, is kept suspended by an air stream. Calculate the velocity of the stream.

10.6. Air flows at 140 km/h perpendicular to a billboard that is 3 m long and 2 m high. What force is exerted on the billboard?

10.7. What force is exerted on a billboard 10 ft long and 5 ft high by a 100-mi/h wind perpendicular to it?

10.8. A thin plate 3 ft × 3 ft is suspended parallel to a stream of air that is moving relative to the plate at 12 ft/s. Calculate the drag force on the plate.

10.9. A cylinder, 1 ft in diameter and 12 ft long, moves in air at 40 ft/s parallel to its length. Calculate the skin drag on the cylinder.

10.10. Air flows edge-on to a rectangular plate 2 ft × 1 ft. Determine the drag on the plate if the air velocity is 30 mi/h and the stream strikes the 2-ft side first. Assume the flow to be laminar.

10.11. If the plate in Problem 10.10 is to be towed through the air at 30 mi/h, what horsepower is required?

10.12. What is the drag on a spherical ball 3 in. in diameter on top of a flagpole when the relative wind velocity is 60 mi/h?

10.13. A car has a projected area of 30 ft^2 and is moving at 55 mi/h. The manufacturer advertises a C_D of 0.32. Determine the horsepower required to just overcome the air resistance.

10.14. A truck has a projected area of 50 ft^2 and is traveling at 50 mi/h. If $C_D = 0.51$ for the truck, determine the horsepower required to just overcome air resistance.

10.15. A spherical ball, 75 mm in diameter, is placed on top of a flagpole. If the relative wind velocity is 100 km/h, what is the drag force on the sphere?

10.16. Determine the ratio of the drag on a flat plate for angles of the relative wind of 16° and 8°.

10.17. A plate 1 ft long and 8 in. wide is set at an angle of 12° to an air stream with a relative velocity with respect to the plate of 75 ft/s. Determine the lift, drag, and total force on the plate.

10.18. A plate is 0.3 m long and 0.2 m wide. If it is set at an angle of 6° with respect to an airstream with a relative velocity of 20 m/s with respect to the plate, determine the lift, drag, and total force on the plate.

10.19. A hemispherical shell is placed in a fluid so that the relative velocity of the fluid impinges on the circular plane face. Assume the hemisphere to be 4 in. in diameter, the velocity of the fluid to be 10 ft/s relative to the hemisphere, the Reynolds number to be greater than 1000, and the fluid to have a specific weight of 100 lb/ft³. For these conditions determine the drag on the hemisphere.

10.20. If the hemisphere in Problem 10.19 is turned through an angle of 180°, what will the drag be?

In the following problems, set up a free-body diagram. At equilibrium the weight minus the buoyant force will equal the drag force. The drag force can be obtained from Stokes' law, $C_D = 24/Re$. The terminal velocity will be given by $V = \frac{2}{9}(r_0^2/\mu)(\gamma_s - \gamma_F)$, where r_0 is the outside radius of the sphere, μ is the visosity, γ_s is the specific weight of the solid, and γ_F is the specific weight of the fluid. This result is also known as Stokes' law.

*10.21. A sphere whose specific weight is 100 lb/ft³ settles in air at a velocity of 0.3 ft/s. Evaluate the diameter of the sphere. Neglect the buoyant effect of the air. ($\gamma_F = 0$)

*10.22. A sphere having an unkown specific weight settles in still air at a velocity of 0.1 m/s. If the diameter of the sphere is 0.05 mm and the buoyant effect of the air is neglected ($\gamma_F = 0$), determine the specific weight of the material of the sphere.

*10.23. A $\frac{1}{2}$-in. diameter lead sphere having $\gamma = 715$ lb/ft³ falls freely in an oil with a terminal velocity of 0.85 ft/s. If the oil has a specific gravity of 0.88, determine its viscosity.

*10.24. A sphere whose diameter is 0.005 in. settles in air whose viscosity is 4×10^{-7} lb·s/ft² with a terminal velocity of 2 ft/s. Determine the specific weight of the sphere. Include the buoyant effect of the air.

Aircraft Performance

Unless otherwise stated, assume a Clark Y airfoil with an A.R. of 6 for the following problems.

10.25. What are the lift, drag, and ratio of lift to drag at an angle of attack of 9° on a wing having an area of 400 ft² at a relative velocity of 180 mi/h? How much horsepower is required to keep this wing in level flight?

10.26. Determine the lift, drag, and the ratio of lift to drag for a wing having an area of 60 m² when the angle of attack is 9° and the relative wind velocity is 300 km/h.

10.27. Calculate the horsepower required to keep the wing of Problem 10.26 flying on a steady, level course.

10.28. An airplane has a wing loading (W/A) of 100 lb/ft² and is operated at 240 mi/h. What is its angle of attack?

10.29. If the wing loading on an airplane is 4000 N/m² and the plane operates at 360 km/h, what is the angle of attack?

10.30. An airplane weighs 4500 lb and has a wing area of 325 ft². Determine the angle of attack if the plane is in steady horizontal flight at sea level at a speed of 120 mi/h.

10.31. Determine the wing area required for a 6000 lb aircraft. The airplane is flying horizontally at sea level with an angle of attack of 4° at a speed of 200 mi/h.

10.32. An airplane has a wing area of 500 ft² and is flying horizontally at sea level at a speed of 250 mi/h. If the angle of attack of the Clark Y wing is 6°, determine the power required to maintain this speed.

10.33. An airplane has a wing area of 450 ft² and is moving horizontally at sea level at a speed of 264 ft/s. If 500 hp is required to sustain this speed, determine the angle of attack.

10.34. What wing area is required to support an airplane weighing 5000 lb and flying at an angle of attack of 7° with a velocity of 180 mi/h?

10.35. Determine the wing area required to support an airplane weighing 25 kN at an angle of attack of 7° if the velocity is 360 km/h.

10.36. What wing loading will an airplane have if it operates at a angle of attack of 4° at 200 mi/h?

10.37. If the plane in Problem 10.36 operates at an altitude where the density of the air is half that at sea level, determine the permissible wing loading.

10.38. Determine the landing speed of the airplane in Problem 10.28.

*10.39. An airplane takes off from New York (sea level) and is to land at Denver (altitude 5200 ft). The specific weight of air at Denver is approximately 0.8 times the value at New York. Determine the ratio of the takeoff speed from New York to the landing speed at Denver.

*10.40. Using the data given in Problem 10.39, determine the ratio of the horsepower for takeoff from New York to the horsepower required for takeoff from Denver. Assume the same angle of attack for each case.

10.41. It is desired to operate an airplane at the angle of attack for maximum lift. If the plane weighs 6000 lb and has a wing area of 200 ft², determine the operating velocity, the drag, and the horsepower required.

10.42. Determine all of the items in Problem 10.41 if the plane is operated at an angle of attack of 14°. Compare your results with Problem 10.41.

10.43. A plane weighs 30 kN and has a wing area of 25 m². If the plane is operated at the angle of maximum lift, determine the operating velocity, the drag, and the horsepower required.

10.44. The largest engine an airplane can utilize generates 1000 hp. At what angle of attack should the plane be operated with this engine if the design speed is 300 mi/h and the wing area is 200 ft²?

10.45. An airplane operates at a condition such that $C_L = 0.8$ and the drag is 500 lb. What is the lift?

10.46. What is the least glide angle for a wing having an area of 450 ft² and weighing 2500 lb?

10.47. Evaluate the terminal velocity of the airplane in Problem 10.46.

10.48. If the airplane in Problem 10.46 has a thrust of 2000 lb, determine its terminal velocity in a power dive.

10.49. If the airplane in Problem 10.46 is at 750 ft when the glide is started, determine the farthest distance the plane can glide.

10.50. What angle of bank is required for an airplane traveling at 240 mi/h in a horizontal turn of radius equal to 1 mi?

10.51. A plane banks in a turn while traveling at 360 km/h. What is the required angle of bank if the turn radius is 3 km?

10.52. Evalute the *g* force and radius of turn for an airplane weighing 3500 lb that is turning at a bank angle of 45° with a velocity of 180 mi/h.

10.53. A pilot of a racing plane places it in a 75° banked turn. Determine the *g* force on the pilot.

*10.54. A high performance fighter (F-16) has a wing area of 28.0 m² and is configured to achieve a maximum lift coefficient of 1.6. When fully loaded with fuel and armament, its weight is 113.8 kN. During a banking maneuver at sea level the pilot experiences a *g* load of 5. (a) Determine the minimum speed of the plane. (b) Determine the radius of the bank.

OPEN-CHANNEL FLOW

LEARNING GOALS

After reading and studying this chapter you should be able to:

1. Identify the flow regimes that can occur in open channel flow.

2. Show how Manning's formula is arrived at for uniform steady flow.

3. Determine the optimum channel of rectangular cross section.

4. Calculate the flow when a pipe is flowing partially full.

5. Evaluate the flow regime when there is nonuniform steady flow.

6. Compute the height of the hydraulic jump.

7. Calculate the quantity of fluid flowing in an open channel from measurements taken from weirs.

11.1 INTRODUCTION

The study of the fluid mechanics of river, culverts, canals, and conduits flowing partially full comes under the general heading of *open-channel flow*. In this type of flow one surface is free and is not confined or in contact with a wall. Since the flow channel may be irregular and have a varying cross section and depth, the character of the flow differs from one location to another. The general considerations developed for the flow of fluids in pipes are applicable to open-channel flow, but the presence of the free surface and the varying conditions in the channel make the analytical determination of the flow in open channels very difficult.

To approach this subject in a logical manner let us first identify and categorize the many possible flow regimes in a channel. The flow may be classified as being steady (independent of time) or unsteady (time dependent) and uniform or nonuniform; in addition, these flows can be further classified as tranquil or rapid. Steady uniform flow occurs in channels that are very long and whose depth and slope are constant along the length of the channel. In this type of flow, the slope of the free surface is found to be parallel to the slope of the bed of the channel. The flow in this case is also said to be tranquil if the velocity is low enough for a small wave to be able to travel upstream causing upstream conditions to be governed by downstream conditions; the flow is said to be rapid (or shooting) if the stream velocity is so high that a small wave cannot travel upstream. If the flow is such that the stream velocity is just equal to the velocity of a small wave, it is said to be critical.

Steady nonuniform flow occurs where the channel or its depth (or both) change from section to section. In this case the velocity must change from section to section, and the flow can change from tranquil to rapid or from rapid to tranquil at different cross sections along the length of the channel.

11.2 UNIFORM STEADY FLOW

Where fluid is flowing uniformly and steadily in an open channel, the rate of flow of fluid past any cross section as well as the cross section itself will be constant at all times and locations. Thus, every section of the channel will appear to be the same as every other section. The energy that provides the driving force for the flow is the change in potential energy of the channel as it slopes down toward the discharge end. As is shown in Figure 11.1, the channel bed has a constant slope angle of α. Since the flow is uniform and the channel has a constant depth, $V_1 = V_2$ and $V_1^2/2g = V_2^2/2g$. Therefore, the water surface must be parallel to the bed of the channel and have the slope α. If the water has the specific energy $V_1^2/2g + p_1/\gamma + Z_1$ at the inlet, it will have as its specific energy $V_2^2/2g + p_2/\gamma + Z_2 + (FL)$ at the outlet, where F is the loss in energy due to friction per unit length of channel and L the length of the channel. The slope F of the energy line is numerically equal to the slope of the free surface, which in turn is numerically equal to the slope of the channel bed. Thus, these three lines (planes) must be parallel, and are shown to be parallel in Figure 11.1.

With the foregoing in mind, let us consider the case of a rectangular channel with uniform steady flow. A frictional force will exist on the fluid due to the shear in the fluid layers adjacent to the channel wall. Denoting the shear stress by τ and the wetted perimeter by P, the shearing force equals τPL. If the flow is uniform, this shearing force must equal the component of the weight of the fluid in a direction parallel to the flow. This becomes (see Figure 11.2b) equal to $(\gamma LA) \sin \alpha$, where γLA is simply the weight of the fluid. Equating these forces yields

$$\tau PL = \gamma LA \sin \alpha \tag{11.1}$$

and for the shear stress,

$$\tau = \frac{A}{P}(\gamma \sin \alpha) \tag{11.2}$$

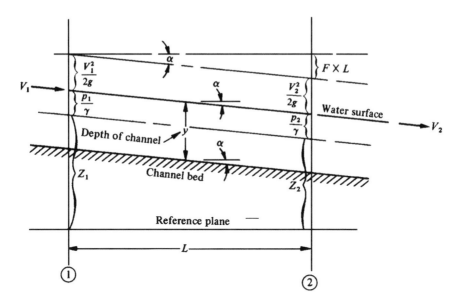

Figure 11.1   Open channel with uniform steady flow.

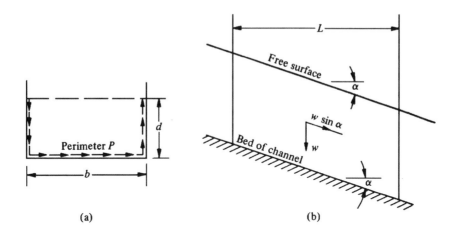

Figure 11.2   Rectangular channel.

The ratio A/P is called the *hydraulic radius* and is denoted by the symbol R. Therefore,

$$R = \frac{\text{flow area}}{\text{wetted perimeter}} = \frac{A}{P} \tag{11.3}$$

From our previous study of flow in pipes, it can be stated that the shear stress will also be some function of $\rho V^2/2$. Using the mathematical symbol Ψ to denote "a function of,"

$$\tau = \Psi \frac{\rho V^2}{2} \tag{11.4}$$

If equations (11.2), (11.3), and (11.4) are combined, we will arrive at a new combined function. Using C to denote the resulting combined function and S the slope of the channel,

$$V = C\sqrt{RS}$$
(11.5)

Equation (11.5) is also known as the *Chezy equation,* and the student will note that the coefficient C is not a constant, and we would expect it to be a function of the Reynolds number, the relative roughness of the channel, and a factor depending on the form of the channel. Manning has empirically determined C as

$$C = \frac{1.486}{n}R^{1/6}$$
(11.6)

where n incorporates the roughness of the channel. Combining equations (11.5) and (11.6) we obtain an equation known as *Manning's formula:*

$$V = \frac{1.486}{n}R^{2/3}S^{1/2}$$
(11.7)

In SI units equation (11.7) can be written as

$$V = \frac{1.00}{n}R^{2/3}S^{1/2}$$
(11.8)

Values for n for various channel surfaces are tabulated in Table 11.1. Also, it should be remembered that

$$Q = AV$$
(11.9)

The student should note that Manning's formula is empirical, and is useful only over certain ranges of operation of channels. Other correlations have been proposed, but these

Table 11.1   Flow of Water in Open Channels, Average Values of Roughness Coefficient (n), for Use in Manning's Formula

Type of open channel	n
Smooth concrete	0.014
Planed timber, asbestos pipe	0.012
Lined cast iron, wrought iron, welded steel	0.015
Vitrified sewer pipe, ordinary concrete, good brickwork, unplaned timber	0.016
Clay sewer pipe, cast iron pipe, cement lining	0.015
Riveted steel, average brickwork	0.018
Rubble masonry, smooth earth	0.025
Firm gravel, corrugated pipe	0.025
Natural earth channels (good condition)	0.025
Natural earth channels (stones and weeds)	0.032
Channels cut in rock, winding river with pools, shoals	0.040
Sluggish river (rather weedy)	0.055

are no more accurate than equation (11.7), and in practice Manning's formula [equation (11.2)] is usually used. It should be further noted that n should have the units of ft$^{1/6}$ or m$^{1/6}$ for dimensional consistency. Manning's equation, however, was not dimensionally consistent, and n is taken to be dimensionless in his empirical equation for conversion to SI. In the English system V will be in ft/s and R will be in feet. In SI units, V will be in m/s and R will be in meters. S, the slope, is in ft/ft or m/m and is dimensionless.

ILLUSTRATION 11.1   UNIFORM STEADY OPEN CHANNEL FLOW

Determine the velocity of flow for a square, smooth, concrete channel flowing full if each side is 2 m and if the slope is 1 m in 1000 m (Figure 11.3).

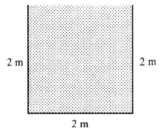

2 m   2 m

2 m   **Figure 11.3**   Illustrative Problem 11.1.

ILLUSTRATIVE PROBLEM 11.1

Given: Smooth concrete channel 2 m × 2 m flowing full; $S = 1$ m in 1000 m

Find: V

Assumptions: n from Table 11.1, Manning's formula applicable, flow uniform and steady

Basic Equations: Equation (11.3): $R = A/P$

$$\text{Equation (11.8): } V = \frac{1.00}{n} R^{2/3} S^{1/2}$$

Solution: From Table 11.1, $n = 0.014$. Using equation (11.8),

$$V = \frac{1.00}{n} R^{2/3} S^{1/2}$$

$$= \frac{1.00}{0.014} \left(\frac{2}{3}\right)^{2/3} (0.001)^{1/2}$$

Therefore,

$$V = 1.72 \text{ m/s}$$

If the volume flow (the discharge) is desired,

$$Q = AV = 4 \text{ m}^2 \times 1.72 \text{ m/s} = 6.88 \text{ m}^3/\text{s}$$

ILLUSTRATION 11.2   UNIFORM STEADY OPEN-CHANNEL FLOW

A concrete trench, square in cross section, is to be built to carry away 2250 gal/min of water. The trench will be 1000 ft long, and the change in elevation between the ends is 10 ft. What must be the dimensions if the trench is designed to operate half-full?

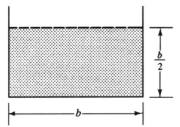

Figure 11.4   Illustrative Problem 11.2.

ILLUSTRATIVE PROBLEM 11.2

Given: Concrete trench—half full, Q = 2250 gpm, S = 10 ft in 1000 ft

Find: Dimensions of trench

Assumptions: n from Table 11.1, flow steady and uniform, Manning's formula applicable

Basic Equations: Equation (11.3): $R = \dfrac{A}{P}$

$$\text{Equation (11.7): } V = \frac{1.486}{n} R^{2/3} S^{1/2}$$

Solution: Denoting the width of the channel as b, the height of the liquid is $b/2$ (Figure 11.4). The area of the channel is $b(b/2) = b^2/2$. The wetted perimeter is $(b/2) + (b/2) + b$, which is $2b$. Therefore,

$$\text{hydraulic radius} = R = \frac{A}{P} = \frac{b^2/2}{2b} = \frac{b}{4}$$

The slope S is given as 10 ft in 1000 ft or $10/1000 = 0.01$. The quantity of water flowing is given as 2250 gal/min. Since 1 gal is 231 in.³, 2250 gal/min is also

$$\text{flow rate} = Q = \frac{\text{gal}}{\text{min}} \times \frac{1}{\text{s/min}} \times \frac{\text{in.}^3/\text{gal}}{\text{in.}^3/\text{ft}^3} = \text{ft}^3/\text{s}$$

$$Q = \frac{2250}{60} = \frac{231}{1728} = 5 \text{ ft}^3/\text{s flowing}$$

Since $Q = AV$,

$$V = \frac{Q}{A} = \frac{5}{b^2/2}$$

where V is in feet per second and A is in square feet. The velocity is also given by equation (11.7). Thus

$$V = \frac{1.486}{n} R^{2/3} S^{1/2} = \frac{10}{b^2}$$

Substituting the data for this problem and using $n = 0.014$ as the average value from Table 11.1 for smooth and ordinary concrete, we obtain

$$\frac{1.486}{0.014} \left(\frac{b}{4}\right)^{2/3} (0.01)^{1/2} = \frac{10}{b^2}$$

Solving yields

$$b = 1.38 \text{ ft}$$

and

$$\frac{b}{2} = 0.69 \text{ ft}$$

The flow area will be equal to $(1.38 \text{ ft}) \times (0.69 \text{ ft}) = 0.952 \text{ ft}^2$ and the velocity is $10/(1.38)^2 = 5.25$ ft/s.

ILLUSTRATION 11.3   UNIFORM STEADY OPEN-CHANNEL FLOW

If the channel in Illustrative Problem 11.2 is designed to flow five-eighths full, determine the dimensions, the velocity of the water, and the area of the channel.

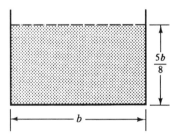

$\frac{5b}{8}$

b

Figure 11.5   Illustrative Problem 11.3.

ILLUSTRATIVE PROBLEM 11.3

Given: Same inputs as Illustrative Problem 11.2 with channel flowing $\frac{5}{8}$ full

Find: Dimensions and area of channel and V

Assumptions: Same as Illustrative Problem 11.2

Basic Equations: Same as Illustrative Problem 11.2

Solution: Proceeding as in Illustrative Problem 11.2 (see Figure 11.5), we obtain

$$\text{hydraulic radius} = R = \frac{A}{P} = \frac{\frac{5}{8}b^2}{\frac{5}{8}b + \frac{5}{8}b + b} = \frac{5}{18}b = 0.278b$$

$$\text{flow rate } = Q = 5 \text{ ft}^3/\text{s} = AV$$

$$\text{velocity } = V = \frac{5}{\frac{5}{8}b^2} = \frac{8}{b^2}$$

Therefore,

$$V = \frac{1.486}{n}R^{2/3}S^{1/2} = \frac{1.486}{0.014} \times (0.278b)^{2/3}(0.01)^{1/2} = \frac{8}{b^2}$$

Solving, we obtain

$$b = 1.238 \text{ ft}$$

$$\tfrac{5}{8}b = 0.772 \text{ ft}$$

$$\text{area} = 0.955 \text{ ft}^2$$

and

$$\text{velocity} = \frac{8}{(1.238)^2} = 5.22 \text{ ft/s}$$

Thus, we see that with the same volume flow and the same slope there is a different set of dimensions that yields the same flow area and the same flow velocity.

ILLUSTRATION 11.4   UNIFORM STEADY OPEN-CHANNEL FLOW

If the channel designed in Illustrative Problem 11.1 is flowing with a depth equal to five-eighths of the width, determine the flow and compare with the results of Illustrative Problem 11.2.

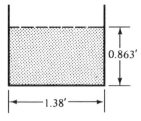

Figure 11.6   Illustrative Problem 11.4.

ILLUSTRATIVE PROBLEM 11.4

Given: Same inputs as Illustrative Problem 11.1 with depth $= \frac{5}{8}$ width

Find: Channel size, Q

Assumptions: Same as Illustrative Problem 11.1

Basic Equations: Same as Illustrative Problem 11.1

Solution: The dimensions of the channel are shown in Figure 11.6. The flow area is
(1.38 ft) × (0.863 ft) = 1.19 ft². Proceeding as before, we obtain

$$\text{hydraulic radius} = R = \frac{A}{P} = \frac{1.19 \text{ ft}^2}{0.863 \text{ ft} \times 0.863 \text{ ft} + 1.38 \text{ ft}} = 0.383 \text{ ft}$$

$$\text{velocity} = V = \frac{1.486}{n} R^{2/3} S^{1/2}$$

All items are the same as in Illustrative Problem 11.2 except that the hydraulic radius
has changed due to the change in dimensions. Therefore, we may form the ratio between the results of Illustrative Problem 11.2 and this problem as follows:

$$\frac{V_2}{V_4} = \left(\frac{R_2}{R_4}\right)^{2/3} = \left(\frac{1.38/4}{0.383}\right)^{2/3} = 0.932$$

where the subscripts 2 and 4 denote, respectively, Illustrative Problems 11.2 and 11.4.
Therefore,

$$V_4 = 1.07 V_2 = 1.07 \times 5.25 = 5.62 \text{ ft/s}$$

$$Q = AV = 1.92 \times 5.62 = 6.7 \text{ ft}^3/\text{s}$$

Thus, increasing the area by 25% leads to a flow increase of 34% when Illustrative
Problem 11.4 is compared to Illustrative Problem 11.2.

11.3 OPTIMUM CHANNEL OF RECTANGULAR CROSS SECTION

In many cases it is desired to design a channel to have the maximum quantity of fluid flow
for a given type of construction and with a specified slope of channel. For a fixed geometry
and flow area, the maximum quantity of flow will occur when the velocity is maximum, and
under these conditions equation (11.9) indicates that this occurs when the hydraulic radius
is a maximum. Let us now consider a rectangular channel having a width b and a depth of
water d, as shown in Figure 11.2a. Since $R = A/P$ and A is given as a fixed quantity, R is
maximum when the wetted perimeter P is a minimum. Thus, we have

$$b + 2d = P \tag{11.10a}$$

$$bd = A \tag{11.10b}$$

Therefore,

$$\frac{A}{d} + 2d = P \tag{11.11a}$$

If for convenience we now let $bd = A = 1$, equation (11.11a) becomes

$$\frac{1}{d} + 2d = P \tag{11.11b}$$

Equation (11.11b) can be solved by the methods of calculus or numerically, as is shown in Table 11.2.

By inspection of Table 11.2 (or by plotting P versus d) it will be seen that the minimum value of P occurs when $d = \frac{1}{2}b$. In other words, the maximum velocity (or flow) for a fixed-area rectangular channel of constant slope occurs when the depth of the water flowing equals half the channel width. In addition, since the perimeter is a minimum with these dimensions, the quantity of material required to construct and line the channel will also be a minimum. These conclusions do not depend on the choice of $A = 1$ in our calculations.

Table 11.2

Assume d is:	b	d	$1/d$	$2d$	P
$0.1b$	3.16	0.316	3.16	0.632	3.792
$0.2b$	2.24	0.448	2.24	0.896	3.136
$0.3b$	1.83	0.549	1.83	1.098	2.928
$0.4b$	1.58	0.632	1.58	1.264	2.844
$0.5b$	1.414	0.707	1.414	1.414	2.828
$0.6b$	1.29	0.774	1.29	1.548	2.838
$0.7b$	1.195	0.837	1.195	1.674	2.869
$0.8b$	1.119	0.895	1.119	1.790	2.909
$0.9b$	1.052	0.947	1.052	0.894	2.946
$1.0b$	1.0	1.0	1.0	2.0	3.0

CALCULUS ENRICHMENT

Since R is maximum when P is minimum, we can solve Equation (11.11b) using calculus.

$$\frac{1}{d} + 2d = P \tag{11.11b}$$

Differentiating both sides of Equation (11.11b) and (11.11b) and setting the result equal to zero,

$$-\frac{1}{d^2} + 2 = 0 \tag{a}$$

or

$$2 = \frac{1}{d^2} \tag{b}$$

and

$$d = \sqrt{\frac{1}{2}} = \frac{\sqrt{2}}{2} \tag{c}$$

But for convenience we assume $bd = 1$. Therefore,

$$b = \frac{1}{d} = \frac{1}{\sqrt{\frac{1}{2}}} = \sqrt{2} \qquad \text{(d)}$$

Comparing Equations (c) and (d) we see that

$$d = \frac{1}{2}b \qquad \text{(e)}$$

We conclude that for the minimum value of P, which gives the maximum value of velocity and the minimum material requirements for a rectangular channel, the depth of water flowing should equal half the width of the channel.

ILLUSTRATION 11.5   OPTIMUM RECTANGULAR CHANNEL FOR STEADY OPEN-CHANNEL FLOW

A rectangular, smooth, concrete channel having a slope of 1 m in 1000 m is designed to carry 6.88 m³/s. Determine the optimum dimensions of the channel.

ILLUSTRATIVE PROBLEM 11.5

Given: Smooth concrete channel, $S = 0.001$, $Q = 6.88$ m³/s

Find: Channel dimensions

Assumptions: Table 11.1 applicable, Manning's equation applicable, depth $= \frac{1}{2}$ width for optimum channel dimensions

Basic Equations: $R = \dfrac{A}{P}$; $b = \dfrac{w}{2}$; $Q = AV$

$$\text{Manning's equation: } V = \frac{1.00}{n} R^{2/3}S^{1/2}$$

Solution: For optimum channel dimensions the depth should be half the width of the channel. This corresponds to the sketch for Illustrative Problem 11.2. Proceeding and using the same notation as in Illustrative Problem 11.2,

$$\text{hydraulic radius} = R = \frac{A}{P} = \frac{b^2/2}{2b} = \frac{b}{4}$$

Since $Q = 6.88$ m³/s,

$$V = \frac{Q}{A}$$

$$V = \frac{6.88}{b^2/2} = \frac{13.76}{b^2}$$

Equating this to equation (11.8) yields

$$\frac{13.76}{b^2} = \frac{1.00}{n} R^{2/3} S^{1/2}$$

For the data of this problem, this becomes

$$\frac{13.76}{b^2} = \frac{1.00}{0.014} \left(\frac{b}{4}\right)^{2/3} (0.001)^{1/2}$$

Solving, we obtain $b = 2.78$ m and $b/2 = 1.39$ m.

   If this problem is compared to Illustrative Problem 11.1, it will be noted that both channels carry the same discharge, 6.88 m³/s. The perimeter in Illustrative Problem 11.1 (and consequently, the amount of material needed) is 6 m; in this problem it is 5.57 m. This agrees with our conclusions that the rectangular channel flowing half-full is the optimum rectangular channel.

ILLUSTRATION 11.6   OPTIMUM RECTANGULAR CHANNEL FOR STEADY OPEN-CHANNEL FLOW

The combined discharge from two 18-in. diameter concrete storm sewer pipes, each flowing one-half full on a 1.25% grade, is to be carried by a cement-lined open rectangular channel on a 1.0% grade (Figure 11.7). Design the rectangular channel so that it will have a depth of flow equal to one-half of its width when carrying this discharge.

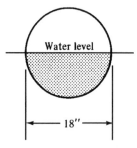

Water level

18"

Figure 11.7   Illustrative Problem 11.6.

ILLUSTRATIVE PROBLEM 11.6

Given: Two 18-in. diameter concrete pipes flowing half full on a 1.25% grade; flow carried by a cement-lined rectangular channel; $d = \dfrac{b}{2}$, $S = 1\%$

Find: d, b

Assumptions: Table 11.1 applicable, Manning's equation applicable

Basic Equations: Continuity: $Q = AV$

$$\text{Manning's equation: } V = \frac{1.486}{n} R^{2/3} S^{1/2}$$

Solution: Consider the 18-in. diameter storm sewer pipes:

$$V = \frac{1.486}{n}R^{2/3}S^{1/2} \qquad \text{from Table 11.1, } n = 0.015$$

$$= \frac{1.486}{0.015}(0.375)^{2/3}(0.0125)^{1/2} = 5.76 \text{ ft/s}$$

Since there are two pipes,

$$\text{total flow} = 2AV = 2\left[\frac{\pi \times (1.5)^2}{8}\right] \times 5.76 = 10.2 \text{ ft}^3/\text{s}$$

The rectangular channel has an area of $\frac{1}{2}b^2$; therefore,

$$\frac{b^2V}{2} = 10.2$$

and

$$b^2V = 20.4$$

or

$$V = \frac{20.4}{b^2}$$

But

$$V = \frac{1.486}{n}R^{2/3}S^{1/2} \qquad n = 0.015 \text{ (Table 11.1)}$$

and

$$R = \frac{b^2/2}{2b} = \frac{b}{4}$$

Therefore,

$$\frac{20.4}{b^2} = \frac{1.486}{0.015}\left(\frac{b}{4}\right)^{2/3}(0.01)^{1/2}$$

$$b \approx 1.85 \text{ ft} \qquad \text{and} \qquad d \approx 0.925 \text{ ft}$$

ILLUSTRATION 11.7   STEADY OPEN-CHANNEL FLOW

How much additional depth of channel in Illustrative Problem 11.6 will be required for the channel to carry the combined discharge when the two pipes are flowing just full?

ILLUSTRATIVE PROBLEM 11.7

Given: Same inputs as Illustrative Problem 11.6—both pipes flow full; channel width same as in Illustrative Problem 1.6

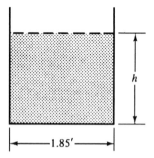

Figure 11.8   Illustrative Problem 11.7.

Find: Height of rectangular channel

Assumptions: Same as Illustrative Problem 11.6

Basic Equations: Same as Illustrative Problem 11.6

Solution: A circular pipe flowing just full or half-full has the same hydraulic radius R and therefore the same velocity V. The total Q is therefore

$$5.77 \times 2 \times \frac{\pi \times (1.5)^2}{4} = 20.4 \text{ ft}^3/\text{s}$$

Denoting the cement-lined channel's depth by h,

$$Q = AV = A\left(\frac{1.486}{0.015}R^{2/3}S^{1/2}\right) = 20.4 \text{ ft}^3/\text{s}$$

Since $R = A/P$,

$$A\left(\frac{1.486}{0.015}\left(\frac{A}{P}\right)^{2/3}(0.01)^{1/2}\right) = 20.4$$

$$\left(\frac{A}{P}\right)^{2/3}A = \frac{20.4 \times 0.015}{1.486 \times (0.01)^{1/2}} = 2.06$$

and

$$\left[\left(\frac{A}{P}\right)^{2/3}A\right]^{3/2} = (2.06)^{3/2} = 2.96 = \frac{A}{P}(A)^{3/2}$$

Therefore,

$$\frac{A}{P} = \frac{2.96}{A^{3/2}}$$

From Figure 11.8, we also have

$$\frac{A}{P} = \frac{1.85h}{2h + 1.85}$$

Therefore,

$$\frac{1.85h}{2h + 1.85} = \frac{2.96}{(1.85h)^{3/2}}$$

To solve for h, let us assume various values of h and evaluate each side of the equation. Try

	Left	Right
$h = 1$	0.48	1.18
$h = 1.5$	0.57	0.64
$h = 2$	0.63	0.42

Plotting these data (Figure 11.9) yields $h \simeq 1.6$. As a check,

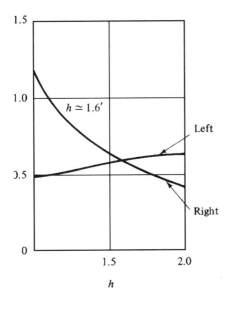

Figure 11.9   Illustrative Problem 11.7.

$$R = \frac{1.6 \times 1.85}{3.2 + 1.85} = 0.59$$

$$V = \frac{1.486}{0.015}(0.59)^{2/3} \times (0.01)^{1/2}$$

$$V = 7.0 \text{ ft/s}$$

$$Q = AV = 1.6 \times 1.85 \times 7.0 = 20.7 \text{ ft}^3/\text{s versus } 20.4 \text{ ft}^3/\text{s}$$

The difference in these solutions is only 1.5%, which is satisfactory.

11.4 CIRCULAR PIPES FLOWING PARTIALLY FULL

Quite often a large circular pipe will be used as a storm drain or sewer. In this case the pipe may not be flowing full and can be treated as an open channel with uniform flow, as was done in Illustrative Problem 11.6. The Chezy formula was applied to this specific situation

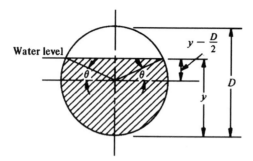

Figure 11.10

to obtain the desired solution, but a more general treatment of this problem will be given in the following paragraphs for pipes of circular cross section.

Consider the circular pipe shown in Figure 11.10, with the liquid flowing above the horizontal centerline. The diameter of the pipe will be denoted as D and the depth of fluid will be denoted as y. To generalize the solution of this problem, it will be convenient to determine the ratio of the quantity of fluid flowing to the quantity when the pipe flows full. From equation (11.7), for a given pipe with a given slope,

$$\frac{Q}{Q_f} = \frac{AV}{A_f V_f} = \frac{A[(1.486/n)R^{2/3}S^{1/2}]}{A_f[(1.486/n)R_f^{2/3}S^{1/2}]} = \frac{A}{A_f}\left(\frac{R}{R_f}\right)^{2/3} \tag{11.12a}$$

where the subscript f denotes conditions where the pipe is flowing full. Since $R = A/P$,

$$\frac{Q}{Q_f} = \left[\left(\frac{A}{A_f}\right)^{5/3}\right]\left(\frac{P_f}{P}\right)^{2/3} \tag{11.12b}$$

From Figure 11.10, the flow area consists of the area of the sector of the circle plus the area of the triangle. In terms of the angle θ (in radians),

$$A = \left(\frac{\pi D^2}{4}\right)\frac{2[(\pi/2) + \theta]}{2\pi} + \left(y - \frac{D}{2}\right)\left(\frac{D}{2}\cos\theta\right) \tag{11.13}$$

The wetted perimeter P is given by

$$P = \frac{\pi D[(\pi/2) + \theta]}{2\pi} \tag{11.14}$$

The angle θ is given in terms of y and D as the angle whose sine is $(y - D/2)/(D/2)$, or

$$\theta = \arcsin\left(\frac{y - D/2}{D/2}\right) \tag{11.15}$$

For the pipe flowing full,

$$A_f = \frac{\pi D^2}{4} \quad \text{and} \quad P_f = \pi D \tag{11.16}$$

Substituting equations (11.13) through (11.16) into equation (11.12b) and simplifying, we obtain

$$\frac{Q}{Q_f} = \left\{ \frac{(\pi/2) + \theta}{\pi} + \frac{\cos\,[(2y/D) - 1]}{\pi} \right\}^{5/3} \left[\frac{\pi}{(\pi/2) + \theta} \right]^{2/3} \qquad (11.17)$$

Similarly,

$$\frac{V}{V_f} = \left\{ \frac{(\pi/2) + \theta}{\pi} + \frac{\cos\,\theta[(2y/D) - 1]}{\pi} \right\}^{2/3} \left[\frac{\pi}{(\pi/2) + \theta} \right]^{2/3} \qquad (11.18)$$

$$= \left\{ \frac{\cos\,\theta[(2y/D) - 1]}{(\pi/2) + \theta} \right\}^{2/3}$$

It will be noted from equations (11.15), (11.17), and (11.18) that Q/Q_f and V/V_f are functions only of the ratio y/D. A similar analysis also shows that Q/Q_f and V/V_f are functions of the ratio y/D when the liquid level is below the horizontal centerline. These solutions are plotted in Figure 11.11, where it is seen that Q/Q_f is maximum at $y = 0.94D$ and V/V_f is maximum at $0.8D$. Figure 11.11 simplifies the work considerably and is sufficiently accurate for most purposes. Since all functions are given as dimensionless ratios, it is applicable to both the English and SI systems of units.

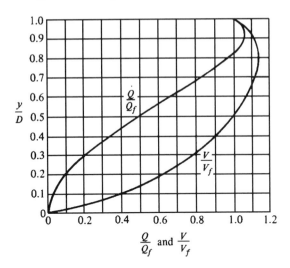

Figure 11.11   Flow in partially full circular pipe.

ILLUSTRATION 11.8   PARTIALLY FULL CIRCULAR PIPE

Water flows at the rate of 0.3 m³/s in a 1-m diameter brick-lined pipe ($n = 0.015$). If the slope is 0.002, calculate the depth of flow.

ILLUSTRATIVE PROBLEM 11.8

Given: $D = 1$ m, brick lined, $n = 0.015$, $S = 0.002$, $Q = 0.3$ m³/s

Find: y, the depth of flow

Assumptions: Uniform steady flow, Figure 11.11 applicable

Basic Equations: $R = \dfrac{A}{P}$; $Q = AV$; Figure 11.4

Solution: If the drain flows full,

$$R = \frac{\pi D^2/4}{\pi D} = \frac{D}{4} = 0.25$$

and

$$Q_f = AV = \left[\frac{\pi(1)^2}{4}\right]\frac{1.00}{0.018}(0.25)^{2/3}\,(0.002)^{1/2}$$

$$= 0.774 \ \text{m}^3/\text{s}$$

Therefore,

$$\frac{Q}{Q_f} = \frac{0.3}{0.774} = 0.387$$

From Figure 11.11, $y/D \approx 0.43$. Thus,

$$y = 0.43 \times 1 = 0.43 \ \text{m}$$

11.5 NONUNIFORM STEADY FLOW

When a fluid flows in an open channel, the velocity is not constant over the cross section of the channel. In general, there are velocity differences from side to side as well as from top to bottom in the channel. If it is desired to express the kinetic energy in terms of the mean (area-weighted) velocity, it is convenient to multiply $V^2/2g$ by a factor α, where V is the average or mean velocity. If the velocity is constant over the area, α equals unity, but depending on the character of the channel, it can equal or exceed 2. Since α cannot be predicted in advance, it is usual to assume its value to be unity; if enough information is available, it should be taken into account. In the following development α will be assumed to be unity.

Let us again consider the case shown in Figure 11.1 and now write an equation for the energy per unit weight, with the reference datum taken as the bottom of the channel. This energy quantity is known as the specific energy and is simply given by

$$E = y + \frac{V^2}{2g} \tag{11.19}$$

In the case of uniform flow, E is a constant, and for nonuniform flow, E may increase or decrease. Since $Q = AV$,

$$V = \frac{Q}{A}$$

and

$$E = y + \frac{Q^2}{2gA^2} \tag{11.20}$$

Per unit width of the channel, $A = y$, and q denotes the volume flow per unit width. Therefore,

$$E = y + \frac{q^2}{2gy^2} \tag{11.21}$$

Figure 11.12 shows a curve of E as a function of y for a fixed value of q. The 45° diagonal line is the potential energy for various values of y, and the horizontal distance between the 45° diagonal and the curve is the kinetic energy. By inspection of Figure 11.12 it can be seen that there is a minimum value of E (at a fixed value of q) that occurs at a single value of the depth, y. This depth is called the *critical depth*, and the flow at the point is called the *critical flow*. For any other value of E greater than this minimum, there can physically exist two values of y. If the depth is less than the critical depth, continuity considerations require the velocity to be greater than the critical velocity, and the flow is called either *supercritical* or *rapid*. If y is greater than the critical depth, the velocity will be less than the critical velocity, and the flow is called *subcritical* or *tranquil*. For a rectangular channel it is not difficult to establish a relation between the minimum value of E and the value of y for the critical depth. Using the methods of calculus, it is found that

$$E_{min} = \tfrac{3}{2} y_c \tag{11.22}$$

where E_{min} is the minimum energy and y_c the critical depth.

The foregoing considerations were for the case of constant q. If we now consider E to be constant and y and q to be variable, q is determined by solving equation (11.21). Thus,

$$q^2 = (E - y)2gy^2 \tag{11.23a}$$

and for constant E, q is a function only of y. Plotting this function yields Figure 11.13. Once again it is possible to solve for the value of maximum q as a function of y, and the critical depth is found to occur at

$$y_c = \tfrac{2}{3}E \tag{11.23b}$$

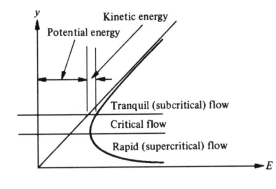

Figure 11.12   Specific energy diagram (fixed q.)

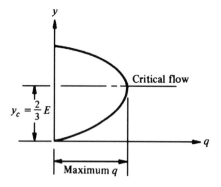

Figure 11.13   Specific energy diagram (fixed *E*.)

which is also the point of maximum *q*. Also, for a given *q*, the critical depth is given by

$$y_c = \left(\frac{q^2}{g}\right)^{1/3} \tag{11.24}$$

In addition to these relations, it is possible to express the character of the flow in terms of the velocity and depth. Thus, from equation (11.24),

$$\frac{q^2}{y_c^3 g} = 1 \tag{11.25}$$

Since $V^2 = q^2/y^2$ or $V^2/y = q^2/y^3$,

$$\frac{q^2}{y_c^3 g} = \frac{V^2}{y_c g} = 1 \tag{11.26}$$

Therefore, if $V^2/gy > 1$, the flow is rapid or supercritical, while $V^2/gy < 1$ is the criterion that the flow is tranquil or subcritical.

ILLUSTRATION 11.9   NONUNIFORM STEADY FLOW

A channel is 10 ft wide and 5 ft deep. The rate of flow is 500 ft³/s. Determine whether the flow is subcritical or supercritical. Also determine the second depth of flow that is possible for the same specific energy.

ILLUSTRATIVE PROBLEM 11.9

Given: Open channel 10 ft × 5 ft, $Q = 500$ ft³/s

Find: Character of flow—subcritical or supercritical; also second depth of flow for the same specific energy

Assumptions: Steady flow

Basic Equations: Continuity: $Q = AV$

$$\text{Equation (11.26):} \quad \frac{V^2}{gy} > 1 \text{ is supercritical,} \quad \frac{V^2}{gy} < 1 \text{ is subcritical}$$

$$\text{Equation (11.21): } y_1 + \frac{q^2}{2gy_1^2} = y_2 + \frac{q}{2gy_2^2}$$

Solution: The quantity of flow is $AV = 500$ ft³/s. Therefore,

$$V = \frac{500}{50} = 10 \text{ ft/s}$$

$$\frac{V^2}{gy} = \frac{(10)^2}{32.17 \times 5} < 1$$

The flow is thus subcritical.

From equation (11.21) (for constant E),

$$y_1 + \frac{q^2}{2gy_1^2} = y_2 + \frac{q^2}{2gy_2^2}$$

Since q is the flow per unit width,

$$q = \frac{500}{10} = 50 \text{ ft}^3/\text{s per foot of width}$$

Therefore,

$$5 + \frac{(50)^2}{2g(5)^2} = y_2 + \frac{(50)^2}{2gy_2^2}$$

$$6.55 = y_2 + \frac{38.9}{y_2^2}$$

Simplifying yields

$$y_2^3 - 6.55y_2^2 + 38.9 = 0$$

Solving numerically, we obtain

$$y_2 \simeq 3.65 \text{ ft}$$

ILLUSTRATION 11.10 NONUNIFORM STEADY FLOW

A rectangular channel is to carry 12 m³/s. Calculate the critical depth and critical velocity if the channel is 6 m wide.

ILLUSTRATIVE PROBLEM 11.10

Given: Rectangular channel, $Q = 12$ m³/s, width = 6 m

Find: Critical depth, critical velocity

Assumption: Steady flow

Basic Equations: Equation (11.24): $y_c = \left(\dfrac{q^2}{g}\right)^{1/3}$

Continuity: $V_c = \dfrac{q}{y_c}$

Solution: $q = \dfrac{12 \text{ m}^3/\text{s}}{6 \text{ m}} = 2 \text{ m}^3/\text{s per meter of width}$

From equation (11.24),

$$y_c = \left(\frac{q^2}{g}\right)^{1/3} = \left(\frac{(2 \text{ m}^3/\text{s per m})^2}{9.81 \text{ m/s}^2}\right)^{1/3} = 0.742 \text{ m}$$

and

$$\text{critical velocity} = V_c = \frac{q}{y_c} = \frac{2 \text{ m}^3/\text{s per m}}{0.742 \text{ m}} = 2.7 \text{ m/s}$$

11.6 HYDRAULIC JUMP

It has already been noted in Section 11.5 that at a fixed value of q there are two equally possible stable depths of flow for a given specific energy. It is possible under certain conditions, however, for the character of this flow to undergo an abrupt change from one of these stable flow states to the other. Thus, a channel in which the depth of flow is less than the critical depth can undergo a sudden change to a depth greater than the critical depth. In the course of this change the velocity is decreased, there is a loss in energy as the depth of flow is increased, and the flow passes through the critical state. These phenomena are known as the *hydraulic jump,* and can be a desirable method of decreasing the velocity in a channel and converting part of the kinetic energy of the flow to potential energy beyond the jump.

To compute the height of this jump (Figure 11.14), certain simplifying assumptions will be made:

1. The frictional forces on the sides and bottom of the channel are negligibly small for the relatively short length of the jump when compared to the other energy terms.

2. The slope of the channel is assumed to be zero (i.e., the channel is horizontal).

3. The hydrostatic pressure forces are assumed to act horizontally.

4. The flow into and out of the jump is steady.

5. For this analysis we shall restrict ourselves to a rectangular channel.

The forces F_1 and F_2 are hydrostatic forces and are simply given as the product of the respective areas multiplied by the hydrostatic pressure exerted by the fluid at the centroid of the area. Mathematically, F_1 and F_2 are given per unit width as

$$F_1 = \frac{\gamma y_1^2}{2} \text{ and } F_2 = \frac{\gamma y_2^2}{2} \tag{11.27}$$

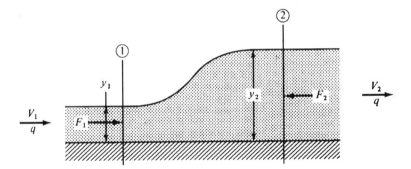

Figure 11.14   Hydraulic jump.

We can now write the equation of impulse and momentum for this problem in the direction of flow as

$$F(\Delta t) = m(\Delta V)$$ (11.28a)

where F is the sum of the forces in the direction of flow, m the mass undergoing acceleration per unit width of channel, and ΔV the change in velocity of the fluid in the direction of flow. Rewriting equation (11.28a), we obtain

$$F_1 - F_2 = \frac{m}{\Delta t}(V_2 - V_1) = \dot{m}(V_2 - V_1)$$ (11.28b)

Substituting for F_1 and F_2 and noting that $\dot{m} = \gamma q/g$ yields

$$\frac{\gamma}{2}(y_1^2 - y_2^2) = \frac{\gamma q}{g}(V_2 - V_1)$$ (11.29)

But V is simply q/y; therefore,

$$\frac{q^2}{gy_1} + \frac{y_1^2}{2} = \frac{q^2}{gy_2} + \frac{y_2^2}{2}$$ (11.30)

Solving equation (11.30), we obtain

$$y_1 = \frac{y_2}{2}\left(-1 + \sqrt{1 + \frac{8q^2}{gy_2^3}}\right)$$ (11.31a)

and

$$y_2 = \frac{y_1}{2}\left(-1 + \sqrt{1 + \frac{8q^2}{gy_1^3}}\right)$$ (11.31b)

From equations (11.31a) and (11.31b), y_2, the depth after the jump for a rectangular channel, is related to y_1, the depth before the jump, and the initial velocity V_1 by

$$y_2 = -\frac{y_1}{2} + \sqrt{\frac{2V_1^2 y_1}{g} + \frac{y_1^2}{4}}$$ (11.32)

The loss of energy, $E_1 - E_2$, is given per unit mass as

$$E_1 - E_2 = \left(y_1 + \frac{V_1^2}{2g}\right) - \left(y_2 + \frac{V_2^2}{2g}\right) \tag{11.33}$$

Expressed per unit width of channel, the loss of energy in the hydraulic jump is

$$E_1 - E_2 = \left(y_1 + \frac{q^2}{2gy_1^2}\right) - \left(y_2 + \frac{q^2}{2gy_2^2}\right) \tag{11.34}$$

ILLUSTRATION 11.11   HYDRAULIC JUMP

A channel is 2 ft wide and the flow is 15 in. above the base of the channel. If 15 million gal/day is flowing, what is the depth of water downstream in the channel? What is the energy loss in this process?

ILLUSTRATIVE PROBLEM 11.11

Given: Channel 2 ft wide, $Q = 15 \times 10^6$ gal/day, $b = 15$ in. above the base of the channel

Find: y_2

Assumptions: Rectangular channel, no friction, horizontal channel

Basic Equations: Equation (11.24): $y_c = \left(\dfrac{q^2}{g}\right)^{1/3}$

$$\text{Equation (11.32):} \quad y_2 = -\frac{y_1}{2} + \sqrt{\frac{2V_1 y_1}{g} + \frac{y_1^2}{4}}$$

$$\text{Equation (11.33):} \quad E_1 - E_2 = \left(y_1 + \frac{V_1^2}{2g}\right) - \left(y_2 + \frac{V_2^2}{2g}\right)$$

Solution: It is first necessary to determine whether the flow is greater or less than the critical flow:

$$y_c = \left(\frac{q^2}{2g}\right)^{1/3}$$

Since there are 231 in.^{3} in 1 gal,

$$q = \frac{\text{gal/day}}{\text{hr/day} \times \text{s/h}} \times \frac{\text{in.}^3/\text{gal}}{\text{in.}^3/\text{ft}^3} \times \frac{1}{\text{channel width (ft)}} = \frac{\text{ft}^3/\text{s}}{\text{channel width (ft)}}$$

$$q = \frac{15{,}000{,}000}{24 \times 3600} \times \frac{231}{1728} \times \frac{1}{2} = 11.6 \text{ ft}^3/\text{s per foot of channel width}$$

Therefore,

$$y_c = \left[\frac{(11.6 \text{ ft}^3/\text{s per foot})^2}{32.17 \text{ ft/s}} \right]^{1/3} = 1.61 \text{ ft}$$

The depth is initially less than the critical depth and will jump to a greater depth. From equation (11.32),

$$y_2 = -\frac{y_1}{2} + \sqrt{\frac{2V_1^2 y_1}{g} + \frac{y_1^2}{4}} \qquad V_1 = \frac{11.6 \text{ ft}^3/\text{s per ft}}{15/12 \text{ ft} \times 1 \text{ ft}} = 9.27 \text{ ft/s}$$

$$= -\frac{1.25 \text{ ft}}{2} + \sqrt{\frac{2 \times (9.27 \text{ ft/s})^2 \times 1.25 \text{ ft}}{32.17 \text{ ft/s}^2} + \frac{(1.25 \text{ ft})^2}{4}}$$

$$= -\frac{1.25}{2} + \sqrt{6.68 + 0.39}$$

$$= -0.625 + 2.66 = 2.03 \text{ ft}$$

The energy lost is

$$\left(y_1 + \frac{V_1^2}{2g} \right) - \left(y_2 + \frac{V_2^2}{2g} \right) \text{ per pound of fluid flowing}$$

But $V_2 = 11.6/2.03 = 5.71$ ft/s:

$$\left[1.25 + \frac{(9.27)^2}{2g} \right] - \left[2.03 + \frac{(5.71)^2}{2g} \right] = 2.59 - 2.54 = 0.05 \text{ ft·lb/lb}$$

The total energy lost is the total mass flowing multiplied by 0.05 ft·lb/lb. Thus,

total energy lost $= (11.6 \text{ ft}^3/\text{s} \times 62.4 \text{ lb/ft}^3 \times 0.05 \text{ ft·lb/lb})(2 \, 24 \text{ h/day} \times 3600 \text{ s/h})$

$$= 3,127,000 \text{ ft·lb/day}$$

horsepower $= 0.066$ hp

11.7 WEIRS

The quantity of fluid flowing in an open channel is usually measured by introducing a mea-suring device into the stream or channel. The simplest and most widely used device is a *weir*, which consists of an obstruction placed in the stream at right angles to the flow. As shown in Figure 11.15, the weir causes the stream to back up and either to flow over it, or if the weir is constructed as a notch, to flow through it. The notch is usually rectangular, tri-angular, or trapezoidal in shape and can be installed in the stream in almost any desirable

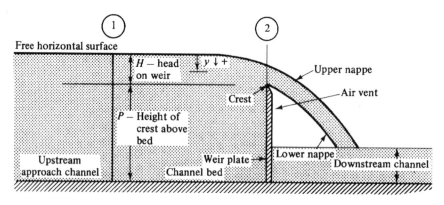

Figure 11.15   Simplified flow over a weir.

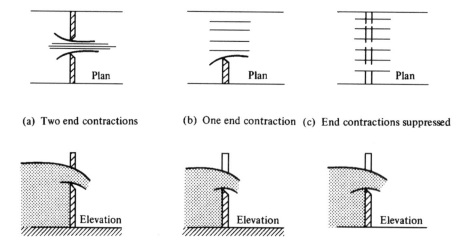

(a) Two end contractions   (b) One end contraction (c) End contractions suppressed

Figure 11.16   Weir types.

manner. Some usual installations are shown in Figure 11.16. In Figure 11.16a the notch does not extend across the entire channel and is said to have two end contractions; Figure 11.16b shows a weir with one end contraction; and Figure 11.16c shows a weir with the notch extending the full width of the channel and having no end contractions. Since the end contractions in Figure 11.16c have been suppressed, it is commonly called a *suppressed weir.*

   To develop a simplified analysis of a weir, the following assumptions will be made:

1. The pressure on the upper nappe and lower nappe is atmospheric. Unless this lower nappe is vented to the atmosphere, the water will tend to adhere to the lower edge of the weir.

2. The weir plate is vertical with a smooth upstream face, and the flow is normal to the plate.

3. The crest is sharp and horizontal, and the flow is normal to the crest.

4. Pressure losses are negligible due to the flow over the weir.

5. The channel is uniform with smooth sides upstream and downstream of the weir.

6. The approach velocity to the weir is uniform and there are no surface waves.

It is obvious that the mathematical model postulated by the foregoing assumptions does not represent the actual flow conditions in weirs. It does, however, permit a rational approach to the problem of computing the flow over a weir, and the results thus obtained can be modified to conform to the experimentally determined flow.

Let us consider a rectangular suppressed weir based upon the assumptions given above, and for the present let us also impose the condition that the velocity of approach to the weir is negligibly small. Writing an energy equation between sections ① and ② shown in Figure 11.15 and recalling the conditions imposed by the assumptions made yields

$$V = \sqrt{2gy} \tag{11.35}$$

where V is the velocity at a depth y in the weir. The velocity is seen to increase as $\sqrt{y}$. Let us now consider the quantity of fluid flowing over the weir. As shown in Figure 11.17, the volume of fluid flowing through the small area $(\Delta y)L$ per unit of time is $(\Delta y)LV$. Since $V = \sqrt{2gy}$, the quantity of fluid flowing through this small area is $\sqrt{y}\Delta y L\sqrt{2g}$. To obtain the total flow through the weir it is necessary to sum up all of these volumes as y goes from zero to H. The complete solution is given by equation (11.36) as

$$Q = \tfrac{2}{3}\sqrt{2g}LH^{3/2} \tag{11.36}$$

Equation (11.36) cannot be expected to yield accurate results when applied to the flow pattern that actually exists in weirs. To account for the assumptions made in the analysis, this equation is usually multiplied by an experimentally determined coefficient C. Thus,

$$Q = C\left[\tfrac{2}{3}\sqrt{2g}LH^{3/2}\right] \tag{11.37}$$

The most common form of equation (11.37) with an empirical value for C incorporated in it is

$$Q = 3.33LH^{3/2} \tag{11.38a}$$

Equation (11.38a) is in the English system of units and with L and H in feet, Q will be given in ft³/s. In SI units, equation (11.38a) becomes

$$Q = 1.84LH^{3/2} \tag{11.38b}$$

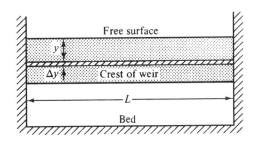

Figure 11.17   Front view of weir.

Equations (11.38a) and (11.38b) are the well-known *Francis formula.* Since many weir equations are empirical, the conversion factor of 1 ft³/s = 0.02832 m³/s is useful in converting from the English system to the SI system.

ILLUSTRATION 11.12   WEIR

Water is flowing through a rectangular sharp-edged weir notch 2 ft wide, 6 in. high, and set 2 ft above the bottom of an approach channel 2 ft wide and 2 ft, 6 in. deep. If the water just fills the approach channel, how much is flowing over the weir notch?

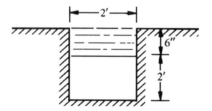

Figure 11.18   Illustrative Problem 11.12.

ILLUSTRATIVE PROBLEM 11.12

Given: Rectangular weir notch 2 ft × 6 in. high, set 2 ft above bottom of approach channel 2 ft wide × 2 ft, 6 in. deep

Find: Q

Assumptions: See beginning of this section; negligible approach velocity

Basic Equations: Equation (11.38a): $Q = 3.33LH^{3/2}$

Solution: Figure 11.18 depicts this suppressed weir. As a first approximation we will neglect the approach velocity. Applying the Francis formula,

$$Q = 3.33LH^{3/2}$$

$$Q = 3.33 \times 2 \text{ ft} \times (\tfrac{1}{2}\text{ft})^{3/2} = 2.35 \text{ ft}^3/\text{s}$$

If the velocity of approach, V_1, is not negligible, the Francis formula is modified to account for this by adding a term incorporating V_1. Thus,

$$Q = 3.33L\left[\left(H + \frac{V_1^2}{2g}\right)^{3/2} - \left(\frac{V_1^2}{2g}\right)^{3/2}\right] \qquad (11.39)$$

ILLUSTRATION 11.13   WEIR WITH APPRECIABLE APPROACH VELOCITY

How does the approach velocity affect the flow in Illustrative Problem 11.12?

ILLUSTRATIVE PROBLEM 11.13

Given: Same inputs as Illustrative Problem 11.12

Find: Q

Assumptions: See beginning of this section; approach velocity appreciable

Basic Equations: Equation (11.39): $Q = 3.33L\left[\left(H + \dfrac{V_1^2}{2g}\right)^{3/2} - \left(\dfrac{V_1^2}{2g}\right)^{3/2}\right]$

Continuity: $Q = AV$

Solution: Using the result of Illustrative Problem 11.12 as a first approximation, we obtain

$$V_1 = \frac{Q}{A} = \frac{2.35 \text{ ft}^3/\text{s}}{2.5 \text{ ft} \times 2 \text{ ft}} = 0.470 \text{ ft/s}$$

$$Q = 3.33 \times 2\left\{\left[\left(\frac{1}{2} + \frac{(0.470)^2}{2g}\right)\right]^{3/2} - \left[\frac{(0.470)^2}{2g}\right]^{3/2}\right\}$$

$$Q = 3.33 \times 2\left[\left(\frac{1}{2} + 0.0034\right)^{3/2} - (0.0034)^{3/2}\right]$$

$$Q = 2.38 \text{ ft}^3/\text{s}$$

It is apparent that the effect of the approach velocity is negligible on the quantity of fluid flowing in this problem.

For small discharge quantities, the V-notch weir or triangular weir is widely used. Referring to Figure 11.19 and denoting the half-angle of the notch by α, the theoretical formula for the discharge through a V-notch is

$$Q = \frac{8}{15} \sqrt{2g} H^{5/2} \tan \alpha \tag{11.40}$$

The actual flow quantity through V-notch weirs has been found to be close to 60% of the value given by equation (11.40). For 90° triangular weirs, the flow can be calculated from

$$Q = 2.5 H^{2.5} \qquad \text{ft}^3/\text{s} \tag{11.41a}$$

In SI Units, equation (11.41a) becomes

$$Q = 1.38 H^{2.5} \qquad \text{m}^3/\text{s} \tag{11.41b}$$

There are several reasons why the actual discharge over a weir differs from the theoretical discharge given by equation (11.36). Among the reasons for this difference are (1) the assumption of a uniform approach velocity is not found in actual weirs; (2) the weir is not perfectly sharp and smooth; (3) viscous effects along the weir face have been neglected;

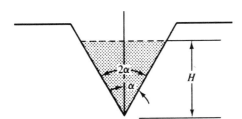

Figure 11.19 V-notch (triangular) weir.

and (4) the channel of approach is nonuniform. Other effects also tend to cause the actual discharge over a weir to deviate from the theoretical discharge based upon the idealized flow model that we have assumed.

Quite often it is found that the use of a weir as a measuring device is either not practical or desirable. The introduction of a weir into a channel with a very small grade may cause backing up upstream of the weir for an undesirable distance. Excessive sedimentation may occur due to the presence of the weir, and the loss of head at the weir may be too great.

To overcome these problems, an instrument called the flume has been developed and is widely used throughout the world. One type of venturi flume, called the *Parshall flume*, has certain outstanding advantages. Among these are the following:

1. Installation is simple.

2. The loss in energy through the flume is very small.

3. Sedimentation in the flume is eliminated.

4. It is easily operated.

As shown in Figure 11.20, the Parshall flume provides a smooth transition from the standard section to a reduced section and another smooth transition back to the standard section. The Parshall flume contracts the flow from the sides and bears the same relation to a sharp-edged weir that a venturi bears to an orifice in a pipe. Calibrations have been made of this flume, yielding the following equations:

$$Q = 4WH_A^{1.522W^{0.26}} \qquad \text{(for } W \text{ from 1 to 8 ft)} \qquad (11.42)$$

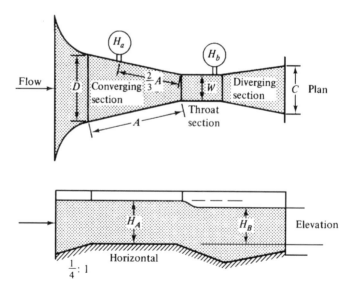

Figure 11.20   Parshall flume.

and

$$Q = (3.688W + 2.5)H_A^{1.6} \qquad \text{(for } W \text{ from 8 to 50 ft)} \qquad (11.43)$$

In equations (11.42) and (11.43), Q is in cubic feet per second, W is in feet, and H_A is measured as shown in Figure 11.20 in feet.

11.8 REVIEW

At the beginning of this chapter we found it necessary to define such new terms as *uniform, nonuniform, steady, unsteady, tranquil,* and *rapid.* Our first topic concerned uniform steady flow. We dealt with this topic in a somewhat simplified manner due to the difficulty of analyzing this problem from a completely theoretical viewpoint. As a manner of fact, we found that this approach was necessary throughout this chapter. Basically, our approach has been to set up a simplified model that could be treated mathematically and then to use the resulting formulation with empirically determined coefficients to make the theoretical equation conform to experimentally determined test data. For uniform steady flow this procedure gave us Manning's formula, an empirical equation that is not dimensionally consistent, yet is probably the most widely used formulation available. We were able, however, to derive the optimum channel shape of rectangular cross section theoretically to be one having a width equal to twice its depth. For pipes flowing partially full we derived a set of curves from theoretical considerations that led to the solution of this otherwise difficult problem.

For nonuniform steady flow we invoked energy considerations to obtain a solution. Several solutions to our equations were found possible, depending on the character of the flow. We then noted that our equations give two equally possible stable depths of flow for a given value of specific energy. It thus is evident that under certain conditions of flow, the character of the flow can undergo an abrupt change from one of these stable flow states to the other. This is known as the hydraulic jump, which is sometimes used to decrease the velocity in a channel and to convert part of the kinetic energy of the flow to potential energy beyond the jump. Using energy and momentum considerations, we determined the conditions in the jump.

In the last part of the chapter we discussed flow-measuring devices known collectively as weirs. For weirs we again invoked the procedure of using some theory and some empirical data to obtain a satisfactory solution. In all of this chapter we have limited ourselves to water as the working fluid and to limited ranges of stream conditions. In any case where greater accuracy is required than these formulas can give, it is necessary to resort to calibration of the device.

KEY TERMS

Terms of importance in this chapter:

Chezy equation: $V = C\sqrt{RS}$ (for rectangular channels).

Critical depth: the depth at which the specific energy is minimum.

Critical flow: the flow at the critical depth.

Francis formula: an empirical formula used to calculate the flow in rectangular weirs.

Hydraulic jump: where the depth in a channel is initially subcritical and suddenly changes to a depth that is greater than the critical depth.

Hydraulic radius: a quantity equal to the ratio of the flow area divided by the wetted perimeter.

Manning's equation: an empirical equation used to calculate the uniform steady flow in a rectangular channel.

Nonuniform flow: flow where the channel or depth or both change from section to section.

Open-channel flow: flow where there is a surface that is free and not in contact with a wall.

Parshall flume: a venturi-like device used to measure the flow in an open channel.

Rapid flow: flow so fast that a small wave cannot travel upstream.

Specific energy: energy per unit weight.

Subcritical flow: *see* tranquil flow.

Supercritical flow: *see* rapid flow.

Tranquil flow: flow slow enough for a small wave to travel upstream.

Uniform flow: in a long channel, where the flow will be parallel to the bed of the channel.

Weir: an obstruction placed into the stream or channel that is used to obtain measurements of the flow.

KEY EQUATIONS

Hydraulic radius	$R = \dfrac{A}{P}$	(11.3)
Chezy equation	$V = C\sqrt{RS}$	(11.5)
Manning's formula	$V = \dfrac{1.486}{n}R^{2/3}S^{1/2}$	(11.7)
Manning's formula (SI)	$V = \dfrac{1.00}{n}R^{2/3}S^{1/2}$	(11.8)
Specific energy	$E = y + \dfrac{V^2}{2g}$	(11.19)
Specific energy	$E = y + \dfrac{q^2}{2gy^2}$	(11.21)
Critical depth	$E_{min} = \tfrac{3}{2}y_c$	(11.22)
Critical depth	$q^2 = (E - y)2gy^2$	(11.23a)

Critical depth	$y_c = \frac{2}{3} E$	(11.23b)
Critical depth	$y_c = \left(\dfrac{q^2}{g}\right)^{1/3}$	(11.24)
Critical depth	$\dfrac{q^2}{y_c^3 g} = \dfrac{V^2}{y_c g}$	(11.26)
Hydraulic jump	$y_1 = \dfrac{y_2}{2}\left(-1 + \sqrt{1 + \dfrac{8q^2}{g y_1^3}}\right)$	(11.31a)
Hydraulic jump	$y_2 = \dfrac{y_1}{2}\left(-1 + \sqrt{1 + \dfrac{8q^2}{g y_2^3}}\right)$	(11.31b)
Hydraulic jump	$E_1 - E_2 = \left(y_1 + \dfrac{V_1^2}{2g}\right) - \left(y_2 + \dfrac{V_2^2}{2g}\right)$	(11.33)
Hydraulic jump	$E_1 - E_2 = \left(y_1 + \dfrac{q^2}{2g y_1^2}\right) - \left(y_2 + \dfrac{q^2}{2g y_2^2}\right)$	(11.34)
Rectangular weir	$Q = \frac{2}{3}\sqrt{2g}LH^{3/2}$	(11.36)
Rectangular weir	$Q = 3.33LH^{3/2}$	(11.38a)
Rectangular weir (SI)	$Q = 1.84LH^{3/2}$	(11.38b)
Triangular weir	$Q = 2.5H^{2.5}$	(11.41a)
Triangular weir (SI)	$Q = 1.38H^{2.5}$	(11.41b)
Parshall flume (1 to 8 ft)	$Q = 4WH_A^{1.522W^{0.26}}$	(11.42)
Parshall flume (8 to 50 ft)	$Q = (3.688W + 2.5)H_A^{1.6}$	(11.43)

QUESTIONS

1. List some of the flow regimes that can be encountered in open-channel flow.
2. Does the term *nonuniform flow* indicate that the flow varies with time?
3. In reference to flow in noncircular pipes, the term *hydraulic diameter* was used. How does hydraulic diameter compare to hydraulic radius?
4. How does Manning's n vary with the roughness of the channel?
5. Can you reason out why the optimum shape of a rectangular channel should have the depth equal to half the width?

6. Would a circular channel also have an optimum shape equal to a flow depth of the radius of the channel?

7. *True or false:* In nonuniform steady flow, the character of the flow depends on the depth of the channel.

8. Can a channel be operated at the critical depth for long periods of time? Explain your answer.

9. *True or false:* Supercritical flow is not desirable, since the water would have large waves on the surface.

10. *True or false:* Tranquil flow is not desirable, since the stream behaves as if it were laminar flow and therefore exhibits a large friction loss.

11. How do you determine the character of the flow in nonuniform steady flow?

12. The hydraulic jump causes a loss in the total energy of the stream. Why is it used?

13. A weir is an obstruction that is placed in a channel. Why would you want to obstruct the channel?

14. In the Francis formula, how does the flow vary as a function of the head on the weir?

15. Why is a weir often an undesirable device?

16. Would you expect that it is easy or difficult to build a Parshall flume? Compare it to a weir.

17. What other instrument does the Parshall flume resemble?

18. Why is the Parshall flume used rather than a weir?

PROBLEMS

Hydraulic Problems

11.1. Determine the hydraulic radius for the shapes in Figure P11.1. Assume each to be flowing full.

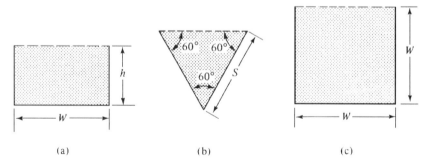

(a)   (b)   (c)

Figure P11.1

11.2. For each of the shapes shown in Figure P11.1, water is flowing half-full. Determine the hydraulic radius in each case.

*11.3. A pipe has a flow that fills it to a level equal to 90% of its diameter. Show that the hydraulic radius is equal to 0.295 × diameter.

Uniform Steady Flow

11.4. A laboratory measures a flow of 14 cfs in a channel that is 4 ft wide and 2 ft deep. If the channel slopes 1 ft in 2000 ft, what is n in Manning's formula?

11.5. A square culvert 6 ft on each side has a slope of 1 ft in 1000 ft and is carrying water to a depth of 4 ft (Figure P11.5). If $n = 0.013$, compute the quantity of water being discharged.

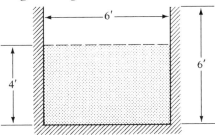

Figure P11.5

11.6. A rectangular channel 2 m wide × 1 m high flows full. If the channel is steel with a slope of 1 m in 1500 m, what is the discharge?

11.7. A square channel is used for irrigation. The original channel is 3 m × 2 m. If there is a lack of water and the channel is flowing half-full, what is the discharge? Assume a brick lining and a slope of 1 m in 5000 m.

11.8. A circular steel culvert is flowing full. If the diameter of the culvert is 6 ft and the slope is 1 ft in 1000 ft, what is the discharge?

11.9. If the culvert of Problem 11.8 is made of brick, determine the discharge.

11.10. A circular steel culvert is flowing half-full. If the diameter of the culvert is 6 ft and the slope is 1 ft in 1000 ft, what is the discharge?

11.11. If the culvert in Problem 11.6 flows with a depth of water of 4 ft, what will the discharge be?

11.12. Solve Problem 11.11 used Figure 11.11. Compare your results.

*11.13. Calculate the uniform flow in a brick-lined trapezoidal canal ($n = 0.018$) whose slope is 1 ft in 10,000 ft (Figure P11.13). Assume the width at the bottom to be 5 ft and the sides to slope at 60° to the horizontal. The depth of flow in the canal is 6 ft.

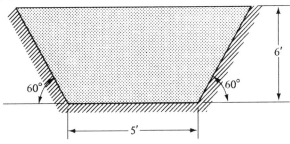

Figure P11.13

*11.14. Solve Problem 11.13 if the canal is lined with ordinary concrete.

*11.15. If the canal of Problem 11.13 is cut in rock, determine the discharge.

11.16. A rectangular irrigation channel is 6 ft wide, has a slope of 1 in 1000, and $n = 0.011$. If Q is 102 ft³/s, what is the depth of water in the channel?

11.17. A wooden flume carries 1.0 m³/s. If the flume is rectangular, with a base of 1 m, what should its slope be if it is made to the best dimensions?

11.18. What are the best dimensions of a rectangular channel whose flow cross-sectional area is 10 m²?

11.19. What will the flow be in Problem 11.18 if the channel has a slope of 1:500 and is made of smooth concrete?

11.20. What are the best dimensions of a rectangular channel whose flow cross-sectional area is 150 ft²?

11.21. What will the flow be in Problem 11.20 if the channel slopes 1 ft in 750 ft and is made of brick?

*11.22. An irrigation channel has been constructed with a bottom 20 ft wide and side slopes at 45°, all lined with concrete (Figure P11.22). If the canal is 10 ft deep and delivers 1050 ft³/s of water, what is the necessary drop of level per mile? Assume that the canal flows full.

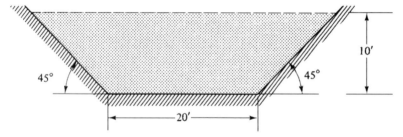

Figure P11.22

*11.23. A steel flume is shaped as shown in Figure P11.23. If it is flowing full and is required to carry 15 ft³/s, what slope is required?

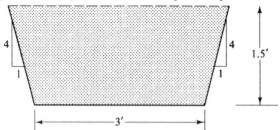

Figure P11.23

*11.24. Solve Problem 11.23 if the flume is constructed of ordinary concrete.

*11.25. If the flume in Problem 11.23 is flowing full and is carrying 10 ft³/s, determine the slope of the flume.

11.26. A rectangular irrigation channel is 4 m wide, has a slope of 1 in 900, and $n = 0.01$. If the flow is 4 m³/s and is uniform, what is the depth of water in the channel?

11.27. A rectangular channel is to carry 80 ft³/s with a slope of 1.5 ft in 1000 ft. If $n = 0.015$, what are the best dimensions of the channel? Assume that the channel is flowing full.

11.28. If $n = 0.020$ in Problem 11.27, determine the best dimensions of the channel. Compare the results of these problems.

Nonuniform Steady Flow

11.29. A rectangular channel is made 10 ft wide. What is the critical depth of flow if 500 ft³/s is being carried?

11.30. A rectangular channel is made 4 m wide. What is the critical depth of flow if 20 m³/s flows?

11.31. A rectangular channel is constructed to be 15 ft wide. At some time 2000 ft³/s is to be carried in the channel with a velocity of 10 ft/s. Determine whether the flow is rapid or tranquil, and also its specific energy.

11.32. A rectangular channel is made 6 m wide. If it is to carry 70 m³/s at a velocity of 4 m/s, will the flow be tranquil or rapid? What is its specific energy?

11.33. The specific energy in a rectangular channel is 5 ft. If the channel is 4 ft wide and the flow is 25 ft³/s, determine the possible flow depths in the channel.

11.34. If a rectangular channel is 10 ft wide, determine the maximum flow for a specific energy of 8 ft.

11.35. A rectangular channel carries a flow of 400 ft³/s. If the channel is 8 ft wide and the water flows at a depth of 2 ft, calculate the height of the flow after a hydraulic jump.

11.36. What is the specific energy loss due to the jump in Problem 11.35?

11.37. A rectangular channel carries 120 cfs. If the channel is 11 ft wide, determine the critical velocity and the critical depth.

11.38. If $n = 0.018$ in Problem 11.37, determine the slope required to produce the critical velocity.

11.39. A rectangular channel is 18 ft wide and carries a flow of 350 cfs. Before the hydraulic jump, the water depth is 2.0 ft. Calculate the height of the flow after the hydraulic jump.

11.40. Calculate the specific energy loss in Problem 11.39 due to the jump.

Weirs

11.41. A sharp-crested weir is constructed at the end of a concrete canal 10 ft wide (Figure P11.41 on page 436). The weir is 8 ft high, and the channel walls extend beyond the top of the weir to a height of 9.44 ft above the bottom of the canal. The nappes are completely ventilated below the weir. How much water is flowing over the weir?

11.42. If the weir described in Problem 11.41 is made 3 m wide and 2.5 m high and if the walls of the weir extend to a height of 3.5 m above the bottom of the canal and the weir is flowing full, how much water is flowing over the weir?

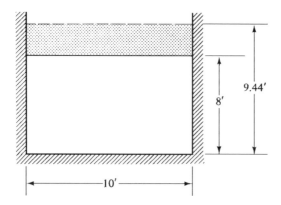

Figure P11.41

11.43. A 90° V-notch weir is to discharge 20 ft³/s. Determine the head on the weir.

11.44. A 90° V-notch weir has a head of 1 m on it. Calculate the discharge through the weir.

11.45. A 90° V-notch weir is to discharge 1 m³/s. Determine the head on the weir.

11.46. A 90° V-notch weir has a head of 2 ft on it. What is the discharge through the weir?

11.47. If the overall notch angle, 2α, is 60° in a V-notch weir, determine the head for a discharge of 20 ft³/s. Compare the result with Problem 11.43.

11.48. If the overall notch angle 2α is 60° in a V-notch weir and the weir has a head of 1 m on it, determine the theoretical discharge through the weir. Compare your result with Problem 11.44.

11.49. If 2α is 60° in Problem 11.45, what will the theoretical head on the weir be?

11.50. Solve Problem 11.46 for α = 60°.

11.51. In connection with a water turbine test, the discharge water goes into a flume 3 ft wide. At the end of this flume the flow is measured by a sharp-crested weir 3 ft high with no end contractions. If the head on the weir is 3.5 ft, what is the flow in cubic feet per second?

SELECTED REFERENCES

1. Fox, Robert W., and Alan T. McDonald. *Introduction to Fluid Mechanics*. 4th ed. New York: John Wiley & Sons, Inc., 1992.

2. Bertin, John J. *Engineering Fluid Mechanics*. 2nd ed. Englewood Cliffs, N.J.: Prentice-Hall, Inc., 1987.

3. Mott, Robert L. *Applied Fluid Mechanics*. 4th ed. New York: Macmillan Publishing Co., 1994.

4. Shames, Irving H. *Mechanics of Fluids*. 3rd ed. New York: McGraw-Hill, Inc., 1992.

5. Streeter, Victor L., and E.B. Wylie. *Fluid Mechanics*. 8th ed. New York: McGraw-Hill, Inc., 1985.

6. Granet, Irving. *Thermodynamics and Heat Power*. 4th ed. Englewood Cliffs, N.J.: Prentice-Hall, Inc., 1990.

7. Sullivan, J.A. *Fundamentals of Fluid Mechanics*. Reston, Va., Reston Publishing Co., 1978.

8. Hanson, A.G. *Fluid Mechanics*. New York: John Wiley & Sons Inc., 1968.

9. Benedict, R.P. *Pressure and Its Measurement*. New York: Electro-Technology, 80, (October 1967), p. 69-90.

10. Kanen, J. D. *Applied Hydraulics for Technology*. New York: Holt, Rinehart and Winston, 1986.

11. Khan, I.A. *Fluid Mechanics*. New York: Holt, Rinehart and Winston, 1987.

12. *Pipe Friction Manual*. 3rd ed. New York: Hydraulic Institute, 1961.

13. Hsu, S.T. *Engineering Heat Transfer*. New York: Litton Educational Publishing, Inc., 1963.

14. Roberson, J.A., and C.T. Crowe. *Engineering Fluid Mechanics.* 3rd ed. Boston: Houghton Mifflin Complany, 1985.

15. "Flow of Fluids Through Valves, Fittings and Pipe," Technical Paper 410, Crane Co., Chicago, 1957.

16. Moody, L.F. "Friction Factors for Pipe Flow,"*Transactions of the ASME,* 66 (1944) p. 671. (also *Mechanical Engineering,* 69 (December 1947, p. 1005.)

17. Jones, B. *Elements of Practical Aerodynamics,* 4th ed. New York: John Wiley & Sons Inc., 1950.

18. Binder, R.C. *Fluid Mechanics,* 4th ed. Englewood Cliffs, N. J., Prentice-Hall, Inc., 1962.

19. *Fluid Meters: Their Theory and Application.* American Society of Mechanical Engineers (ASME). Latest edition.

20. Pease, D.A. *Basic Fluid Power.* Englewood Cliffs, N. J.: Prentice-Hall, Inc., 1967.

21. Keenan, J.H., G.F. Keyes, P.G. Hill, and J.G. Moore. *Steam Tables.* International edition. New York: John Wiley & Sons, Inc., 1969.

22. *Steam—Its Generation and Use.* 38th ed. New York: Babcock & Wilcox Corporation, 1972.

23. Albertson, M.L., J.R. Barton, and D.B. Simons. *Fluid Mechanics for Engineers.* Englewood Cliffs, N.J.: Prentice-Hall, Inc., 1960.

24. Vennard, J.K. *Elementary Fluid Mechanics.* New York: John Wiley & Sons Inc., 1961.

25. Hall, N.A. *Thermodynamics of Fluid Flow.* Englewood Cliffs, N.J.: Prentice-Hall, Inc., 1951.

26. Daugherty, R.L., and J.B. Franzini. *Fluid Mechanics with Engineering Applications.* 8th ed. New York: McGraw-Hill Inc., 1985.

27. White, F.M. *Fluid Mechanics.* 3rd ed. New York: McGraw-Hill Inc., 1995.

APPENDIX A: ANSWERS TO EVEN-NUMBERED PROBLEMS

CHAPTER 1

1.2	1.35 m		1.54	32.81 ft/s
1.4	9.78 cm		1.56	24.59 m/s
1.6	2.894 tons		1.58	28.23 mi/gal
1.8	2.95 ft		1.60	(c)
1.10	3900.9 grams		1.62	(d)
1.12	256.54 cm		1.64	(e)
1.14	943.9 liters/s		1.66	1145 N
1.16	0.2103 liters/s		1.68	4.66 slugs
1.18	0.00333 m^3/s		1.70	(c)
1.20	5.07×10^{-5} m^3/s		1.72	10 kg
1.22	102,600 liters/min		1.74	0.622 slugs
1.24	1579.9 gal/min		1.76	3.73 slugs
1.26	0.00946 m^3/s		1.78	20 m/s^2
1.28	3650 liters/s		1.80	2 kg
1.30	135.92 liters/s		1.82	2.4 N
1.32	3.88 slugs/ft^3		1.84	312.5 kJ
1.34	2576.13 kg/m^3		1.86	400 ft·lb
1.36	0.0774 m^2		1.88	2 m/s
1.38	0.027 m^2		1.90	5.02 ft/s
1.40	0.001571 m^3		1.92	0.69 slugs
1.42	0.0035396 m^3		1.94	7.63 N
1.44	28.32 liters		1.96	23.08 $\dfrac{\text{ft·lb}}{\text{lb}}$
1.46	1.8288 m			
1.48	3.93 liters		1.98	4.33 psi
1.50	2.54×10^{-5} m			
1.52	2.082 m^3			

CHAPTER 2

2.2 $-17.7°C$, $-12.2°C$, $10°C$

2.4 $°C = -106.7$, $°F = -160$

2.6 $°C = \dfrac{20}{11}$ $(°ARB - 20)$

2.8 $40°$

2.10 $\rho = 1000 \ \dfrac{kg}{m^3}$; $v = 0.001 \ \dfrac{m^3}{kg}$

2.12 $\rho = 1200 \ \dfrac{kg}{m^3}$; $\gamma = 11772 \ \dfrac{N}{m^3}$

2.14 88.62 lb

2.16 81.7 lb

2.18 0.692

2.20 48.5 lb/ft³

2.22 10.5 N

2.24 7.17 lb

2.26 $10.79 \ \dfrac{kN}{m^3}$

2.28 $7.65 \ \dfrac{kN}{m^3}$

2.30 derivation

2.32 251.6 kPa

2.34 24.5 kPa

2.36 derivation

2.38 6.70 ft

2.40 67.75 kPa

2.42 15.17 psi

2.44 229.08 kN

2.46 -0.104 ft depression

2.48 0.321 ft

2.50 6.3×10^{-3} m

2.52 57 mm

2.54 derivation

2.56 1×10^6 psi

2.58 0.1136 m³

2.60 $5.5 \times 10^{-6} \ \dfrac{lb \cdot s}{ft^2}$

2.62 $1.2 \times 10^{-4} \ \dfrac{lb \cdot s}{ft^2}$

2.64 $4.1 \times 10^{-7} \ \dfrac{lb \cdot s}{ft^2}$

2.66 2.52×10^{-5} ft²/s; 2.34×10^{-6} m²/s

2.68 $7.31 \times 10^{-4} \ \dfrac{lb \cdot s}{ft^2}$

2.70 5.38×10^{-4} ft²/s

2.72 $6.59 \times 10^{-5} \ \dfrac{lb \cdot s}{ft^2}$

2.74 $2.61 \times 10^{-3} \ \dfrac{lb \cdot s}{ft^2}$

2.76 0.0466 ft/s

2.78 9.81 N

2.80 0.72 lb

2.82 7.394 N

CHAPTER 3

3.2 1571 psf

3.4 3659 psfa

3.6 40.7 psia

3.8 28,224 ft

3.10 infinity

3.12 12.266 psia

3.14 169.98 kPa

3.16 0.68%

3.18 -4.85 lb/ft²

3.20 0.4 in.

3.22 2.06 psig

3.24 10.33 psf

3.26 0.434 psi

3.28 5.07 m

3.30 18.72 psia

3.32 9405 Pa

3.34 103.92 kPa

3.36 17.8 psia

3.38 94.24 kPa

3.40 1.4 psi

3.42 0.641 ft³

3.44 15.85 lb

3.46 gold

3.48 (a) 19 lb (b) 19 lb

3.50 0.953 ft³

3.52 118.1 lb

3.54 44.72 lb/ft³

3.56 46.8 lb/ft³

3.58 0.00068 m³

3.60 643,000 lb

3.62 897.1 in.^{3}
3.64 7.545 kN/m^3
3.66 3.125 ft
3.68 8.8 inches
3.70 6.25 ft
3.72 124.67 lb/ft^3

3.74 $\dfrac{A}{B} = 14.29$

3.76 858.2 lb (No)
3.78 sg = 0.586

3.80 $F = \dfrac{2}{3}\gamma R^3$

3.82 18.76 kN
3.84 2 ft from base
3.86 220.73 kN; 1 m from base
3.88 135.9 kN
3.90 154.81 kN·m
3.92 69.343 kN·m
3.94 70,761 lb; 14.5 ft from top
3.96 34 671 N; 4.5625 m from surface
3.98 0.06 in.
3.100 $\dfrac{3}{4} = \dfrac{\text{wt of sphere}}{\text{wt of cylinder}}$

CHAPTER 4

4.2 166.8 lb/min
4.4 1.89 × 10^{-3} m^3/s
4.6 800 ft/min
4.8 533.3 m/min
4.10 8.33 × 10^{-2} m^3/s
4.12 4.527 m/s; 196.2 N/s
4.14 0.4012 ft^3/s, 180.07 gpm, 25.03 lb/s, 0.778 slugs/s
4.16 15.77 h
4.18 0.849 m/s, 0.377 m/s
4.20 1.283 ft/s, 2.52 ft/s
4.22 0.509 m/s, 3.18 m/s
4.24 13.49 ft/s
4.26 3.342 ft/s, 2.72 ft/s, 5.51 slugs/s
4.28 2.01 m/s
4.30 294.3 J
4.32 7.67 m/s
4.34 250 N·m
4.36 5.1 m
4.38 196.5 ft/s

4.40 1001 gpm
4.42 50 N·m
4.44 −2610 ft·lb/lb
4.46 611.62 N·m/N
4.48 94.1 psig
4.50 2.32 ft
4.52 0.0824 ft^3/s
4.54 43.58 psi
4.56 1.74 × 10^{-2} m^3/s
4.58 17.94 ft/s
4.60 125.8 ft/s
4.62 155.19 kPa
4.64 55.50 ft/s
4.66 33.08 ft
4.68 52.80 m
4.70 253.9 cm (independent of γ)
4.72 200.9 kPa
4.74 110.17 kPa
4.76 0.488 ft^3/s
4.78 0.0834 ft^3/s
4.80 7.13 × 10^{-3} m^3/s
4.82 2.098 ft
4.84 25.8 ft
4.86 1.318 ft^3/s
4.88 2.36 × 10^{-2} m^3/s

CHAPTER 5

5.2 5.28 m/s, 1.42 m, 31.6 m, 36.02 m
5.4 98.2 ft/s
5.6 85.4 psig
5.8 16.54 ft/s
5.10 3.86 ft^3/s
5.12 0.503 ft, from B to A
5.14 30.52 psi
5.16 8.98 ft
5.18 7.11 MPa
5.20 50.1 kPa
5.22 42.4 ft/s
5.24 263 hp
5.26 125.9 ft
5.28 213.9 hp
5.30 1.41 hp
5.32 1.38 hp
5.34 11.3 hp
5.36 28.9 hp
5.38 20.36 hp

5.40   4.12 hp
5.42   16.96 hp
5.44   374.6 ft

CHAPTER 6

6.2   248,262
6.4   0.032
6.6   9078
6.8 V_{max} = 248 ft/s
       V_{avg} = 124 ft/s
6.10   derivation
6.12   turbulent (Re = 76 453)
6.14   1333.3
6.16   laminar (Re = 1578)
6.18   15 240
6.20   28.1 psi
6.22   laminar (Re = 567)
6.24   84 577
6.26   laminar (Re = 1301)
6.28 V_{avg} = 1.68 ft/s
6.30   7911
6.32 V_{max} = 1.2 ft/s
       V_{avg} = 0.6 ft/s
6.34   530
6.36   (c)
6.38 V_{max} = 1.2 ft/s
       V_{avg} = 0.6 ft/s
6.40   28.32 in.
6.42   0.96 ft/s
6.44   laminar (Re = 1865)
6.46   74.2 psi
6.48   36.11 psi
6.50   134 m
6.52   0.074 m³/s
6.54   1.47 ft³/s
6.56   151.7 ft
6.58   7.83 ft
6.60   0.18 psi
6.62   1.59 ft
6.64   4.34 ft³/s
6.66   0.363
6.68   2.06 ft
6.70   9.74 in.
6.72   1.619 ft
6.74   same
6.76   square

6.78   1.656 D
6.80   1.52 in.
6.82 $D_e = D_2 - D_1$
6.84   derivation
6.86   42 ft of air
6.88   1.48 psf
6.90   118.6 ft of water
6.92   46.1 psi
6.94   158 psi
6.96   0.61 ft
6.98   109 ft/s
6.100   13.99 ft of liquid
6.102   4.13 m
6.104   1.4 psi
6.106   0.866 ft³/s
6.108   3.28 ft³/s
6.110   8.85 psi

CHAPTER 7

7.2   impulse x = 2.69 lb·s
       impulse y = 1.01 lb·s
7.4   derivation
7.6   71.9 kN
7.8   90.7 kN
7.10   441.8 N
7.12   342.4 lb
7.14   149.6 lb
7.16   4000 N
7.18   313.9 lb
7.20   61.8 mm
7.22   290.7 lb
7.24   468.8 lb
7.26   112.2 lb
7.28   16.17 psia
7.30   83.3%
7.32   87%; 10.9 kN
7.34   3557 hp
7.36   95.3%

CHAPTER 8

8.2   140.7 ft/s
8.4   135.1 ft/s
8.6   derivation
8.8 C_v = 1.0003, C_C = 0.827,
       C_D = 0.827

8.10   2.59 ft³/s
8.12   5.18 ft/s
8.14   0.9
8.16   12.07 ft/s
8.18   16.44 ft/s
8.20   4.35 psi
8.22   3.74 psi
8.24   35.76 in.
8.26   3.06 in.
8.28 1.82×10^{-2} m³/s
8.30   derivation

CHAPTER 9

9.2   30.24 ft
9.4   11.37 m
9.6   10.91 m
9.8   0.655
9.10   1.08 hp
9.12   5.12 hp
9.14   54.5 ft; 550 gpm
9.16   1.17 m
9.18   1287
9.20   509 (English); 0.296 (SI); mixed flow
9.22   1869

CHAPTER 10

10.2 $C_D = \dfrac{8.5}{Re}$

10.4   1.005 lb/ft
10.6   6002.5 N
10.8   3.57 lb
10.10 0.023 lb
10.12 0.217 lb
10.14 21.71 hp
10.16 2.72
10.18 5.73 N; 0.882 N; 5.80 N
10.20 5.41 lb
10.22 1321 kg/m³
10.24 83.02 lb/ft³
10.26 257.8 kN; 17.3 kN; 14.9
10.28 4.5°
10.30 0°
10.32 2398 hp

10.34 69.4 ft²
10.36 66.8 psf
10.38 157.7 mi/h
10.40 0.894
10.42 138.8 ft/s; 526.7 lb; 132.9 hp
10.44 3°
10.46 2.7°
10.48 917 ft/s
10.50 36.1°
10.52 2166 ft
10.54 144 m/s; 431.4 m

CHAPTER 11

11.2   (a) $\dfrac{Wh}{2(W + h)}$

(b) $\left(\dfrac{\sqrt{3}}{16}\right) S$

(c) $\dfrac{W}{4}$

11.4   0.019
11.6   2.17 m³/s
11.8   116.2 ft³/s
11.10   58 ft³/s
11.12   89.3 ft³/s
11.14   91.4 ft³/s
11.16   approx. 3 ft
11.18   4.47 m × 2.24 m
11.20   17.3 ft × 8.65 ft
11.22 9.53×10^{-5}
11.24 1.188×10^{-3}
11.26   approx. 0.33 m
11.28   6.94 m × 3.47 m
11.30   1.37 m
11.32   3.73 m
11.34   70 ft³/s per ft
11.36   3.22 ft
11.38   0.0057
11.40   0.0094 ft·lb/lb
11.42   5.52 m³/s
11.44   1.38 m³/s
11.46   14.1 ft³/s
11.48   1.36 m³/s
11.50   41.9 ft³/s

APPENDIX B: SELECTED PHYSICAL DATA

Table B.1   Properties of Pipe

Schedules, Wall Thicknesses, and Weights
Conforming to ASA Standard B36.10, 1950, for Wrought Steel and Wrought Iron Pipe

Nominal Pipe Size (in.)	Outside Diameter D (in.)	Wall Thickness t (in.)	Inside Diameter d (in.)	Inside Diameter d^2 (squared)	Inside Diameter d^5 (fifth power)	Area of Metal (in.²)	Internal Cross-Sectional Area in.²	Internal Cross-Sectional Area ft²	External Surface (f^2)	Moment of Inertia (in.⁴)	Weight (Pounds) of Pipe (per ft)	Weight (Pounds) of Water (per ft of pipe)
					Schedule 10							
14 o.d.	14.0	0.250	13.50	182.25	448,403	10.80	143.14	0.994	3.665	255.3	36.71	62.03
16 o.d.	16.0	0.250	15.50	240.25	894,660	12.37	188.69	1.310	4.189	383.7	42.05	81.74
18 o.d.	18.0	0.250	17.50	306.25	1,641,309	13.94	240.53	1.670	4.712	549.1	47.39	104.21
20 o.d.	20.0	0.250	19.50	380.25	2,819,505	15.51	298.65	2.074	5.236	756.4	52.73	129.42
24 o.d.	24.0	0.250	23.50	552.25	7,167,030	18.65	433.74	3.012	6.283	1,315.0	63.41	187.95
30 o.d.	30.0	0.312	29.376	862.95	21,875,768	29.10	677.76	4.707	7.854	3,206.0	98.93	293.72
					Schedule 20							
8	8.625	0.250	8.125	66.02	35,409	6.57	51.85	0.3601	2.258	57.72	22.36	22.47
10	10.75	0.250	10.25	105.06	113,141	8.24	82.52	0.5731	2.814	113.7	28.04	35.76
12	12.75	0.250	12.25	150.06	275,855	9.82	117.86	0.8185	3.338	191.8	33.38	51.07
14 o.d.	14.0	0.312	13.376	178.92	428,185	13.42	140.52	0.975	3.665	314.4	45.68	60.89
16 o.d.	16.0	0.312	15.376	236.42	859,442	15.38	185.69	1.290	4.189	473.2	52.36	80.50
18 o.d.	18.0	0.312	17.376	301.92	1,583,978	17.34	237.14	1.647	4.712	678.2	59.03	102.77
20 o.d.	20.0	0.375	19.25	370.56	2,643,344	23.12	291.04	2.021	5.236	1113	78.60	125.67
24 o.d.	24.0	0.375	23.25	540.56	6,793,832	27.83	424.56	2.948	6.283	1942	94.62	183.95
30 o.d.	30.0	0.500	29.00	841.0	20,511,149	46.34	660.52	4.587	7.854	5,042	157.53	286.23
					Schedule 30							
8	8.625	0.277	8.071	65.14	34,248	7.26	51.16	0.3553	2.258	63.35	24.70	22.17
10	10.75	0.307	10.136	102.74	106,987	10.07	80.69	0.5603	2.814	137.4	34.24	34.96
12	12.75	0.330	12.09	146.17	258,304	12.87	114.80	0.7972	3.338	248.4	43.77	49.74
14 o.d.	14.0	0.375	13.25	175.56	408,394	16.05	137.88	0.9575	3.665	372.8	54.57	59.75
16 o.d.	16.0	0.375	15.25	232.56	824,801	18.41	182.65	1.268	4.189	562.1	62.58	79.12
18 o.d.	18.0	0.437	17.126	293.30	1,473,261	24.11	230.36	1.600	4.712	930.3	82.06	99.84
20 o.d.	20.0	0.500	19.0	361.00	2,476,099	30.63	283.53	1.969	5.236	1,457	104.13	122.87
24 o.d.	24.0	0.562	22.876	523.31	6,264,703	41.39	411.00	2.854	6.283	2,843	140.80	178.09
30 o.d.	30.0	0.625	28.75	826.56	19,642,160	57.68	649.18	4.508	7.854	6,224	196.08	281.30

Table B.1 (continued)

Nominal Pipe Size (in.)	Outside Diameter D (in.)	Wall Thickness t (in.)	Inside Diameter d (in.)	Inside Diameter d^2 (squared)	Inside Diameter d^5 (fifth power)	Area of Metal (in.²)	Internal Cross-Sectional Area (in.²)	(ft²)	External Surface (f²)	Moment of Inertia (in.⁴)	Weight (Pounds) of Pipe (per ft)	Weight (Pounds) of Water (per ft of pipe)
						Schedule 40						
$\frac{1}{8}$	0.405	0.068§	0.269	0.0724	0.00141	0.072	0.0569	0.00040	0.106	0.001064	0.24	0.0250
$\frac{1}{4}$	0.540	0.088§	0.364	0.1325	0.00639	0.125	0.1041	0.00072	0.141	0.003312	0.42	0.0449
$\frac{3}{8}$	0.675	0.091§	0.493	0.2430	0.02912	0.167	0.1909	0.00133	0.177	0.007291	0.57	0.0830
$\frac{1}{2}$	0.840	0.109§	0.622	0.3869	0.09310	0.250	0.3039	0.00211	0.220	0.01709	0.85	0.1317
$\frac{3}{4}$	1.050	0.113§	0.824	0.679	0.3799	0.333	0.5333	0.00371	0.275	0.03704	1.13	0.2315
1	1.315	0.133§	1.049	1.100	1.270	0.494	0.8639	0.00600	0.344	0.08734	1.68	0.3744
$1\frac{1}{4}$	1.660	0.140§	1.380	1.904	5.005	0.669	1.495	0.01040	0.435	0.1947	2.27	0.6490
$1\frac{1}{2}$	1.900	0.145§	1.610	2.592	10.82	0.799	2.036	0.01414	0.497	0.3099	2.72	0.8823
2	2.375	0.154§	2.067	4.272	37.72	1.075	3.356	0.02330	0.622	0.666	3.65	1.454
$2\frac{1}{2}$	2.875	0.203§	2.469	6.096	91.75	1.704	4.788	0.03322	0.753	1.530	5.79	2.073
3	3.5	0.216§	3.068	9.413	271.8	2.228	7.393	0.05130	0.916	3.017	7.58	3.201
$3\frac{1}{2}$	4.0	0.226§	3.548	12.59	562.2	2.680	9.888	0.06870	1.047	4.788	9.11	4.287
4	4.5	0.237§	4.026	16.21	1,058	3.173	12.73	0.08840	1.178	7.233	10.79	5.516
5	5.563	0.258§	5.047	25.47	3,275	4.304	20.01	0.1390	1.456	15.16	14.62	8.674
6	6.625	0.280§	6.065	36.78	8,206	5.584	28.89	0.2006	1.734	28.14	18.97	12.52
8	8.625	0.322§	7.981	63.70	32,380	8.396	50.03	0.3474	2.258	72.49	25.55	21.68
10	10.75	0.365§	10.02	100.4	101,000	11.90	78.85	0.5475	2.814	160.7	40.48	34.16
12	12.75	0.406	11.938	142.5	242,470	15.77	111.93	0.7773	3.338	300.3	53.53	48.50
14 o.d.	14.0	0.437	13.126	172.3	389,638	18.61	135.32	0.9397	3.665	429.1	63.37	58.64
16 o.d.	16.0	0.500	15.000	225.0	759,375	24.35	176.72	1.2272	4.189	731.9	82.77	76.58
18 o.d.	18.0	0.562	16.876	284.8	1,368,820	30.79	223.68	1.5533	4.712	1172	104.75	96.93
20 o.d.	20.0	0.593	18.814	354.0	2,357,244	36.15	278.00	1.9305	5.236	1703	122.91	120.46
24 o.d.	24.0	0.687	22.626	511.9	5,929,784	50.31	402.07	2.7921	6.283	3424	171.17	174.23
						Schedule 60						
8	8.625	0.406	7.813	61.04	29,113	10.48	47.94	0.3329	2.258	88.73	35.64	20.77
10	10.75	0.500¶	9.75	95.06	88,110	16.10	74.66	0.5185	2.814	212.0	54.74	32.35
12	12.75	0.562	11.626	135.16	212,399	21.52	106.16	0.7372	3.338	400.4	73.16	46.00
14 o.d.	14.0	0.593	12.814	164.20	345,480	24.98	128.96	0.8956	3.665	562.3	84.91	55.86

Nominal												
16 o.d.	16.0	0.656	14.688	215.74	683,618	31.62	169.44	1.1766	4.189	932.4	107.50	73.42
18 o.d.	18.0	0.750	16.500	272.25	1,222,981	40.64	213.83	1.4849	4.712	1,515	138.17	92.80
20 o.d.	20.0	0.812	18.376	337.68	2,095,342	48.95	265.21	1.8417	5.236	2,257	166.40	114.92
24 o.d.	24.0	0.968	22.064	486.82	5,229,029	70.04	382.35	2.6552	6.283	4,654	238.11	165.94

Schedule 80

Nominal												
$\frac{1}{8}$	0.405	0.095¶	0.215	0.0462	0.000459	0.093	0.0363	0.00025	0.106	0.001216	0.31	0.0157
$\frac{1}{4}$	0.540	0.119¶	0.302	0.0912	0.002513	0.157	0.0716	0.00050	0.141	0.003766	0.54	0.031
$\frac{3}{8}$	0.675	0.126¶	0.423	0.1789	0.01354	0.217	0.1405	0.00098	0.177	0.008619	0.74	0.0609
$\frac{1}{2}$	0.840	0.147¶	0.546	0.2981	0.04852	0.320	0.2341	0.00163	0.220	0.02008	1.09	0.1013
$\frac{3}{4}$	1.050	0.154¶	0.742	0.5506	0.2249	0.433	0.4324	0.00300	0.275	0.04479	1.47	0.1875
1	1.315	0.179¶	0.957	0.9158	0.8027	0.639	0.7193	0.00499	0.344	0.1056	2.17	0.3112
$1\frac{1}{4}$	1.660	0.191¶	1.278	1.633	3.409	0.881	1.283	0.00891	0.435	0.2418	3.00	0.5553
$1\frac{1}{2}$	1.900	0.200¶	1.500	2.250	7.594	1.068	1.767	0.01225	0.498	0.3912	3.63	0.7648
2	2.375	0.218¶	1.939	3.760	27.41	1.477	2.953	0.02050	0.622	0.8679	5.02	1.279
$2\frac{1}{2}$	2.875	0.276¶	2.323	5.396	67.64	2.254	4.238	0.02942	0.753	1.924	7.66	1.834
3	3.5	0.300¶	2.900	8.410	205.1	3.016	6.605	0.04587	0.917	3.894	10.25	2.859
$3\frac{1}{2}$	4.0	0.318¶	3.364	11.32	430.8	3.678	8.891	0.06170	1.047	6.280	12.51	3.847
4	4.5	0.337¶	3.826	14.64	819.8	4.407	11.50	0.07986	1.178	9.610	14.98	4.976
5	5.563	0.375¶	4.813	23.16	2,583	6.112	18.19	0.1263	1.456	20.67	20.78	7.875
6	6.625	0.432¶	5.761	33.19	6,346	8.405	26.07	0.1810	1.734	40.49	28.57	11.29
8	8.625	0.500¶	7.625	58.14	25,775	12.76	45.66	0.3171	2.257	105.7	43.39	19.79
10	10.75	0.593	9.564	91.47	80,020	18.92	71.84	0.4989	2.817	244.8	64.33	31.13
12	12.75	0.687	11.376	129.41	190,523	26.03	101.64	0.7958	3.338	475.1	88.51	44.04
14 o.d.	14.0	0.750	12.500	156.25	305,176	31.22	122.72	0.8522	3.665	687.3	106.13	53.18
16 o.d.	16.0	0.843	14.314	204.89	600,904	40.14	160.92	1.1175	4.189	1,156	136.46	69.73
18 o.d.	18.0	0.937	16.125	260.05	1,090,518	50.23	204.24	1.4183	4.712	1,833	170.75	88.50
20 o.d.	20.0	1.031	17.938	321.77	1,857,248	61.44	252.72	1.7550	5.236	2,772	208.87	109.51
24 o.d.	24.0	1.218	21.564	465.01	4,662,798	87.17	365.22	2.5362	6.283	5,672	296.36	158.26

Schedule 100

Nominal												
8	8.625	0.593	7.439	55.34	22,781	14.96	43.46	0.3018	2.258	121.3	50.87	18.83
10	10.75	0.718	9.314	86.75	69,357	22.63	68.13	0.4732	2.814	286.1	76.93	29.53
12	12.75	0.843	11.064	122.41	165,791	31.53	96.14	0.6677	3.338	561.6	107.20	41.66
14 o.d.	14.0	0.937	12.126	147.04	262,173	38.45	115.49	0.8020	3.665	824.4	130.73	50.04
16 o.d.	16.0	1.031	13.938	194.27	526,020	48.48	152.58	1.0596	4.189	1,364	164.83	66.12
18 o.d.	18.0	1.156	15.688	246.11	950,250	61.17	193.30	1.3423	4.712	2,180	207.96	83.76
20 o.d.	20.0	1.281	17.438	304.08	1,612,398	75.34	238.82	1.6585	5.236	3,316	256.10	103.65
24 o.d.	24.0	1.531	20.938	438.40	4,024,179	108.07	344.32	2.3911	6.283	6,853	367.40	149.43

Table B.1 (continued)

Nominal Pipe Size (in.)	Outside Diameter D (in.)	Wall Thickness t (in.)	Inside Diameter d (in.)	Inside Diameter d^2 (squared)	Inside Diameter d^5 (fifth power)	Area of Metal (in.²)	Internal Cross-Sectional Area in.²	Internal Cross-Sectional Area ft²	External Surface (ft²)	Moment of Inertia (in.⁴)	Weight (Pounds) of Pipe (per ft)	Weight (Pounds) of Water (per ft of pipe)
						Schedule 120						
4	4.5	0.438	3.625	13.15	626.8	5.578	10.33	0.0717	1.178	11.65	19.01	4.47
5	5.563	0.500	4.563	20.82	1,978	7.953	16.35	0.1136	1.456	25.73	27.04	7.09
6	6.625	0.562	5.501	30.26	5,037	10.705	23.77	0.1650	1.734	49.61	36.39	10.30
8	8.625	0.718	7.189	51.68	19,202	17.84	40.59	0.2819	2.257	140.5	60.63	17.59
10	10.75	0.843	9.064	82.16	61,179	26.24	64.53	0.4481	2.817	324.2	89.20	27.96
12	12.75	1.000	10.750	115.56	143,563	36.91	90.76	0.6303	3.338	641.6	125.49	39.33
14 o.d.	14.0	1.093	11.814	139.57	230,134	44.32	109.62	0.7612	3.665	929.8	150.67	47.57
16 o.d.	16.0	1.218	13.564	183.98	459,133	56.56	144.50	1.0035	4.189	1,555	192.29	62.62
18 o.d.	18.0	1.375	15.250	232.56	824,783	71.82	182.65	1.2684	4.712	2,499	244.14	79.27
20 o.d.	20.0	1.500	17.000	289.00	1,419,857	87.18	226.98	1.5762	5.236	3,754	296.37	98.35
24 o.d.	24.0	1.812	20.376	415.18	3,512,301	126.31	326.08	2.2644	6.283	7,827	429.39	141.52
						Schedule 140						
8	8.625	0.812	7.001	49.01	16,819	19.93	38.50	0.2673	2.257	153.7	67.76	16.68
10	10.75	1.000	8.750	76.56	51,291	30.63	60.13	0.4176	2.817	367.8	104.13	26.06
12	12.75	1.125	10.500	110.25	127,628	41.08	86.59	0.6013	3.338	700.5	139.68	37.52
14 o.d.	14.0	1.250	11.500	132.25	201,136	50.07	103.87	0.7213	3.665	1,027	170.22	45.01
16 o.d.	16.0	1.438	13.125	172.29	389,670	65.74	135.32	0.9397	4.189	1,760	223.50	58.64
18 o.d.	18.0	1.562	14.876	221.30	728,502	80.66	173.80	1.2070	4.712	2,749	274.23	75.32
20 o.d.	20.0	1.750	16.500	272.25	1,222,981	100.33	213.82	1.4849	5.236	4,216	341.10	92.66
24 o.d.	24.0	2.062	19.876	395.09	3,102,022	142.11	310.28	2.1547	6.283	8,625	483.13	134.45
						Schedule 160						
$\frac{1}{2}$	0.840	0.187	0.466	0.2172	0.002197	0.3836	0.1706	0.00118	0.220	0.02212	1.30	0.074
$\frac{3}{4}$	1.050	0.218	0.614	0.3770	0.08726	0.5698	0.2961	0.00206	0.275	0.05269	1.94	0.130
1	1.315	0.250	0.815	0.6642	0.3596	0.8365	0.5217	0.00362	0.344	0.1251	2.84	0.230
$1\frac{1}{4}$	1.660	0.250	1.160	1.346	2.100	1.107	1.057	0.00734	0.435	0.2839	3.76	0.46
$1\frac{1}{2}$	1.900	0.281	1.338	1.790	4.288	1.429	1.406	0.00976	0.498	0.4824	4.86	0.61
2	2.375	0.343	1.689	2.853	13.74	2.190	2.241	0.01556	0.622	1.162	7.44	0.97

$2\frac{1}{2}$	2.875	0.375	2.125	4.516	43.33	2.945	3.546	0.02463	0.753	2.353	10.01	1.54
3	3.5	0.438	2.625	6.896	124.9	4.205	5.416	0.03761	0.917	5.032	14.32	2.35
$3\frac{1}{2}$	4.0	—	—	—	—	—	—	—	—	—	—	—
4	4.5	0.531	3.438	11.82	480.3	6.621	9.283	0.06447	1.178	13.27	22.51	4.02
5	5.563	0.625	4.313	18.60	1,492	9.696	14.61	0.1015	1.456	30.03	32.96	6.33
6	6.625	0.718	5.189	26.93	3,762	13.32	21.15	0.1469	1.734	58.97	45.30	9.16
8	8.625	0.906	6.813	46.42	14,679	21.97	36.46	0.2532	2.257	165.9	74.69	15.80
10	10.75	1.125	8.500	72.25	44,371	34.02	56.75	0.3941	2.817	399.3	115.65	24.59
12	12.75	1.312	10.126	102.54	106,461	47.14	80.53	0.5592	3.338	781.1	160.27	34.89
14 o.d.	14.0	1.406	11.188	125.17	175,292	55.63	98.31	0.6827	3.665	1,117	189.12	42.60
16 o.d.	16.0	1.593	12.814	164.20	345,486	72.10	128.96	0.8955	4.189	1,894	245.11	55.97
18 o.d.	18.0	1.781	14.438	208.46	627,412	90.75	163.72	1.1369	4.712	3,021	308.51	71.05
20 o.d.	20.0	1.968	16.064	258.05	1,069,699	111.49	202.67	1.4074	5.236	4,586	379.01	87.96
24 o.d.	24.0	2.343	19.314	373.03	2,687,570	159.41	292.98	2.0345	6.283	9,458	541.94	127.15

Source: Data taken with permission from *Catalogue 57*, the Walworth Co., New York, 1957.

‡ This column also represents the contents in cubic feet per foot of length.

§ These thicknesses are identical with those listed in ASA B36.10—1950 for standard wall pipe.

¶ These thicknesses are identical with those listed in ASA B36.10—1950 for extra strong wall pipe.

Table B.2   Expansion Coefficient of Liquids at 1 ATM Pressure and 70°F

Liquid	°F⁻¹
Water	1.2×10^{-4}
Ethyl alcohol	6.21×10^{-4}
Freon 12	1.40×10^{-4}
Mercury	1.01×10^{-4}
Silicone oil	4.80×10^{-4}
Petroleum oil	4.0×10^{-4}

Source: Reproduced with permission from *Thermodynamics of Fluid Flow* by N. A. Hall, Prentice-Hall, Inc., Englewood Cliffs, N. J., 1951, p. 11.

Table B.3 Average Properties of Tubes

Diameter		Thickness		External				Internal				Length of Tube Containing One Cubic Foot
						Lineal Feet of Tube per Square Foot of Surface			Volume or Capacity per Lineal Foot			
External (in.)	Internal (in.)	BWG Gage	NOM Wall (in.)	Circumference (in.)	Surface per Lineal Foot (ft²)		Transverse Area (in.²)	In.³	Ft.³	U.S. gal	
$\frac{5}{8}$	0.527	18	0.049	1.9635	0.1636	6.1115	0.218	2.616	0.0015	0.011	661
	0.495	16	0.065	1.9635	0.1636	6.1115	0.193	2.316	0.0013	0.010	746
	0.459	14	0.083	1.9635	0.1636	6.1115	0.166	1.992	0.0011	0.009	867
$\frac{3}{4}$	0.652	18	0.049	2.3562	0.1963	5.0930	0.334	4.008	0.0023	0.017	431
	0.620	16	0.065	2.3562	0.1963	5.0930	0.302	3.624	0.0021	0.016	477
	0.584	14	0.083	2.3562	0.1963	5.0930	0.268	3.216	0.0019	0.014	537
	0.560	13	0.095	2.3562	0.1963	5.0930	0.246	2.952	0.0017	0.013	585
1	0.902	18	0.049	3.1416	0.2618	3.8197	0.639	7.668	0.0044	0.033	225
	0.870	16	0.065	3.1416	0.2618	3.8197	0.595	7.140	0.0041	0.031	242
	0.834	14	0.083	3.1416	0.2618	3.8197	0.546	6.552	0.0038	0.028	264
	0.810	13	0.095	3.1416	0.2618	3.8197	0.515	6.180	0.0036	0.027	280
$1\frac{1}{4}$	1.152	18	0.049	3.9270	0.3272	3.0558	1.075	12.90	0.0075	0.056	134
	1.120	16	0.065	3.9270	0.3272	3.0558	0.985	11.82	0.0068	0.051	146
	1.084	14	0.083	3.9270	0.3272	3.0558	0.923	11.08	0.0064	0.048	156
	1.060	13	0.095	3.9270	0.3272	3.0558	0.882	10.58	0.0061	0.046	163
	1.032	12	0.109	3.9270	0.3272	3.0558	0.836	10.03	0.0058	0.043	172
$1\frac{1}{2}$	1.402	18	0.049	4.7124	0.3927	2.5465	1.544	18.53	0.0107	0.080	93
	1.370	16	0.065	4.7124	0.3927	2.5465	1.474	17.69	0.0102	0.076	98
	1.334	14	0.083	4.7124	0.3927	2.5465	1.398	16.78	0.0097	0.073	103
	1.310	13	0.095	4.7124	0.3927	2.5465	1.343	16.12	0.0093	0.070	107
	1.282	12	0.109	4.7124	0.3927	2.5465	1.292	15.50	0.0090	0.067	111
$1\frac{3}{4}$	1.620	16	0.065	5.4978	0.4581	2.1827	2.061	24.73	0.0143	0.107	70
	1.584	14	0.083	5.4978	0.4581	2.1827	1.971	23.65	0.0137	0.102	73
	1.560	13	0.095	5.4978	0.4581	2.1827	1.911	22.94	0.0133	0.099	75
	1.532	12	0.109	5.4978	0.4581	2.1827	1.843	22.12	0.0128	0.096	78
	1.490	11	0.120	5.4978	0.4581	2.1827	1.744	20.92	0.0121	0.090	83
2	1.870	16	0.065	6.2832	0.5236	1.9099	2.746	32.96	0.0191	0.143	52
	1.834	14	0.083	6.2832	0.5236	1.9099	2.642	31.70	0.0183	0.137	55
	1.810	13	0.095	6.2832	0.5236	1.9099	2.573	30.88	0.0179	0.134	56
	1.782	12	0.109	6.2832	0.5236	1.9099	2.489	29.87	0.0173	0.129	58
	1.760	11	0.120	6.2832	0.5236	1.9099	2.433	29.20	0.0169	0.126	59

Source: Reproduced with permission from *Principles of Heat Transfer* by Frank Kreith, International Textbook Company, Scranton, Pa., 1958, p. 541.

Table B.4 Gas-Constant Values

Substance	Symbol	M	R (ft·lb/lb$_m$ °R)	C_p (Btu/lb °R) at 77°F	C_v (Btu/lb °R) at 77°F	K (C_p/C_v)
Acetylene	C_2H_2	26.038	59.39	0.4030	0.3267	1.234
Air	—	28.967	53.36	0.2404	0.1718	1.399
Ammonia	NH_3	17.032	90.77	0.5006	0.3840	1.304
Argon	A	39.944	38.73	0.1244	0.0746	1.668
Benzene	C_6H_6	78.114	19.78	0.2497	0.2243	1.113
n-Butane	C_4H_{10}	58.124	26.61	0.4004	0.3662	1.093
Isobutane	C_4H_{10}	58.124	26.59	0.3979	0.3637	1.094
1-Butane	C_4H_8	56.108	27.545	0.3646	0.3282	1.111
Carbon dioxide	CO_2	44.011	35.12	0.2015	0.1564	1.288
Carbon monoxide	CO	28.011	55.19	0.2485	0.1776	1.399
Carbon tetrachloride	CCl_4	153.839				
n-Deuterium	D_2	4.029				
Dodecane	$C_{12}H_{26}$	170.340	9.074	0.3931	0.3814	1.031
Ethane	C_2H_6	30.070	51.43	0.4183	0.3522	1.188
Ethyl ether	$C_4H_{10}O$	74.124				
Ethylene	C_2H_4	28.054	55.13	0.3708	0.3000	1.236
Freon, F-12	CCl_2F_2	120.925	12.78	0.1369	0.1204	1.136
Helium	He	4.003	386.33	1.241	0.7446	1.667
n-Heptane	C_7H_{16}	100.205	15.42	0.3956	0.3758	1.053
n-Hexane	C_6H_{14}	86.178	17.93	0.3966	0.3736	1.062
Hydrogen	H_2	2.016	766.53	3.416	2.431	1.405
Hydrogen sulfide	H_2S	34.082				
Mercury	Hg	200.610				
Methane	CH_4	16.043	96.40	0.5318	0.4079	1.304
Methyl fluoride	CH_3F	34.035				
Neon	Ne	20.183	76.58	0.2460	0.1476	1.667
Nitric oxide	NO	30.008	51.49	0.2377	0.1715	1.386
Nitrogen	N_2	28.016	55.15	0.2483	0.1774	1.400
Octane	C_8H_{18}	114.232	13.54	0.3949	0.3775	1.046
Oxygen	O_2	32.000	48.29	0.2191	0.1570	1.396
n-Pentane	C_5H_{12}	72.151	21.42	0.3980	0.3705	1.074
Isopentane	C_5H_{12}	72.151	21.42	0.3972	0.3697	1.074
Propane	C_3H_8	44.097	35.07	0.3982	0.3531	1.128
Propylene	C_3H_6	42.081	36.72	0.3627	0.3055	1.187
Sulfur dioxide	SO_2	64.066	24.12	0.1483	0.1173	1.264
Water vapor	H_2O	18.016	85.80	0.4452	0.3349	1.329
Xenon	Xe	131.300	11.78	0.03781	0.02269	1.667

Source: Reproduced with permission from *Concepts of Thermodynamics* by E. F. Obert, McGraw-Hill Book Company, Inc., New York, 1960, p. 502.

Table B.5 Absolute and Kinematic Viscosities of Water

°F	$\mu \times 10^5$ $(\text{lb}_f \text{ s/ft}^2)$	$\nu \times 10^5$ (ft^2/s)	°F	$\mu \times 10^5$ $(\text{lb}_f \text{ s/ft}^2)$	$\nu \times 10^5$ (ft^2/s)
32	3.75	1.93	120	1.17	0.609
35	3.54	1.82	125	1.12	0.582
40	3.23	1.66	130	1.08	0.562
45	2.97	1.53	135	1.02	0.534
50	2.73	1.41	140	0.981	0.514
55	2.53	1.30	145	0.940	0.493
60	2.35	1.22	150	0.899	0.472
65	2.24	1.13	155	0.868	0.457
70	2.04	1.05	160	0.837	0.440
75	1.92	0.988	165	0.806	0.426
80	1.80	0.929	170	0.776	0.411
85	1.68	0.870	175	0.750	0.397
90	1.60	0.825	180	0.725	0.384
95	1.51	0.782	185	0.701	0.372
100	1.43	0.738	190	0.679	0.362
105	1.35	0.698	195	0.657	0.351
110	1.29	0.668	200	0.637	0.341
115	1.23	0.637	212	0.593	0.318

Source: Reproduced with permission from *Fluid Mechanics* by Arthur G. Hansen, John Wiley & Sons, Inc., New York, 1967, p. 486.

Table B.6 Surface Tension of Water Exposed to Air or Its Own Vapor

°F	Surface Tension σ (lb_f/ft)	N/m
32	5.2×10^{-3}	7.6×10^{-2}
40	5.1	7.4
60	5.0	7.3
80	4.9	7.2
100	4.8	7.0
120	4.7	6.9
140	4.5	6.6
160	4.4	6.4
180	4.3	6.3
200	4.1	6.0
212	4.0	5.8

Source: Reproduced with permission from *Fluid Mechanics* by Arthur G. Hansen, John Wiley & Sons, Inc., New York, 1967, p. 486. SI units added.

Table B.7   Surface Tension of Various Liquids in Contact with Air

Substance (in contact with air)	°F	Surface Tension σ (lb$_f$/ft)	N/m
Mercury	68	0.0324	4.73×10^{-1}
Benzene	68	0.00198	2.89×10^{-2}
Carbon tetrachloride	68	0.00184	2.69×10^{-2}
Glycerine	68	0.00482	7.03×10^{-2}
Ethyl alcohol	68	0.00153	2.23×10^{-2}
Methyl alcohol	68	0.00155	2.26×10^{-2}

Source: Reproduced with permission from *Fluid Mechanics* by Arthur G. Hansen, John Wiley & Sons, Inc., New York, 1967, p. 487. SI units added.

Table B.8   Properties of the Standard Atmosphere

Altitude (ft)	Temperature (°F)	Absolute Pressure (lb$_f$/ft^2)	γ (lb/ft^3)	Speed of Sound (ft/s)
0	59	2116	0.0765	1117
5,000	41	1761	0.0660	1098
10,000	23	1455	0.0566	1078
15,000	6	1194	0.0482	1058
20,000	−12	972	0.0408	1037
25,000	−30	785	0.0343	1017
30,000	−48	628	0.0288	995
35,000	−66	498	0.0238	973
40,000	−68	392	0.0189	971
45,000	−68	308	0.0148	971
50,000	−68	242	0.0117	971
60,000	−68	151	0.0072	971
70,000	−68	94	0.0045	971
80,000	−68	58	0.0028	971
90,000	−68	36	0.0017	971
100,000	−68	22	0.0011	971
150,000	114	3	9.8×10^{-5}	1174
200,000	159	0.7	2.0×10^{-5}	1220
250,000	−8	0.1	4.8×10^{-6}	1042
500,000	450	10^{-4}	3.1×10^{-9}	—

Source: Reproduced with permission from *Heat, Mass and Momentum Transfer* by W. M. Rohsenow and H. Y. Choi, Prentice-Hall, Inc., Englewood Cliffs, N. J., 1961, p. 523.

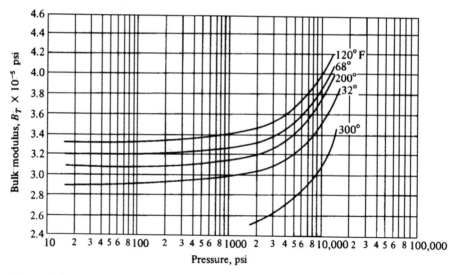

Figure B.1 Variation of the bulk modulus of water with pressure and temperature. (Reproducted with permission from *Fluid Mechanics* by Aruthur G. Hansen, John Wiley & Sons, Inc., New York, 1967, p. 487.)

APPENDIX C: CONVERSION FACTORS

LENGTH

1 m = 39.37 in. = 3.281 ft
1 in. = 2.54 cm
1 km = 0.621 mi
1 mi = 5280 ft = 1.609 km
1 ft = 12 in. = 0.3048 m = 30.48 cm

MASS

1 kg = 10^3 g = 6.85×10^{-2} slug
1 slug = 14.59 kg

VOLUME

1 liter = 1000 cm^3 = 3.531×10^{-2} ft^3
1 ft^3 = 2.832×10^{-2} m^3
1 gallon = 231 in.^{3}
1 gallon = 3.786 liters
1 ft^3 = 7.48 gallons
1 liter = 10^{-3} m^3

FORCE

1 N = 0.2248 lb
1 lb = 4.448 N
1 dyne = 10^{-5} N
1 N = 10^5 dynes

1 dyne = 2.248×10^{-6} lb
1 ton = 2000 lb
1 ton (metric) = 1000 kg

SPEED

1 km/h = 0.621 mi/h
1 mi/h = 1.61 km/h = 0.447 m/s = 1.467 ft/s

VOLUME FLOW RATE

1 gal/min = 6.309×10^{-5} m^3/s
1 liter/min = 16.67×10^{-6} m^3/s
1 gal/min = 3.785 liters/min

PRESSURE

1 Pa = 1 N/m^2
1 atm = 1.013×10^5 Pa
1 atm = 14.70 lb/in.^{2}
1 lb/in.^{2} = 6895 Pa
1 bar = 10^5 Pa
1 bar = 14.5 lb/in.^{2}

POWER

1 hp = 550 ft·lb/s = 0.746 kW

1 W = 1 J/s = 0.738 ft·lb/s
1 ft·lb/s = 1.356 W

WORK AND ENERGY

1 J = 0.738 ft·lb
1 cal = 4.186 J
1 ft·lb = 1.356 J
1 Btu = 252 cal
1 kW·h = 3.60×10^6 J
1 Btu = 1.054×10^3 J

VISCOSITY

Dynamic

lb·s/ft^2 × 47.88 = Pa·s
lb·s/in.^{2} × 6895 = Pa·s
poise × 10 = Pa·s
100 cP = 1 P
cP × 10^{-3} = Pa·s
cP × 2.09×10^{-5} = lb·s/ft^2

Kinematic

ft^2/s × (9.29×10^{-2}) = m^2/s
in.^{2}/s × (6.45×10^{-4}) = m^2/s
stoke × 10^{-4} = m^2/s
100 cSt = 1 stoke
m^2/s × 10.764 = ft^2/s
Saybolt Seconds Universal × (2.33×10^{-6}) = ft^2/s
for Saybolt Seconds Universal > 100, stoke × 1.076×10^{-3} = ft^2/s

INDEX